AN ELEMENTARY TREATISE ON
THEORETICAL MECHANICS

J. H. JEANS

DOVER PUBLICATIONS, INC.
Mineola, New York

DOVER PHOENIX EDITIONS

Bibliographical Note

This Dover edition, first published in 1967 and reissued in 2005, is an unabridged and unaltered republication of the work originally published in 1907 by Ginn and Company, Boston.

International Standard Book Number: 0-486-44179-2

Manufactured in the United States of America
Dover Publications, Inc., 31 East 2nd Street, Mineola, N.Y. 11501

PREFACE

The primary aim of the present book is to supply for students beginning the study of Theoretical Mechanics a course of such a nature as shall emphasize the fundamental physical principles of the subject. Different students will of course approach the study of mechanics with different interests, different aims, and different amounts of mathematical equipment, so that it may not be possible to produce a single book which shall exactly fit the requirements of every class of student. But I believe that *all* students of mechanics, no matter what their aims and intentions may be, will be in the same position in one respect, namely that they will best begin the study of the subject by trying to acquire a firm grasp of the physical principles, leaving aside at first all mathematical developments and all practical applications, except in so far as these contribute to the elucidation of the fundamental physical principles.

I am aware that this belief is not held by all teachers of mechanics, some of whom regard the laws of mechanics simply as working rules to be acquired as rapidly as possible for their utilitarian value, while others appear to regard them in the same light as the rules of a game, the game consisting in the solution of mathematical puzzles, most of which have no conceivable reference to the facts of nature. I find it hard to believe that there can be any considerable class of students for whom either of these points of view is the best. As regards the former, I feel that a student who cannot get, or does not wish to get, a clear understanding of mechanical principles would be well advised not to enter a profession in which his work will consist in the handling of mechanical problems; and as regards the latter, that a student who wishes merely to obtain material for puzzle solving would do better to turn his attention to chess or double acrostics.

PREFACE

If I have taken some space to express my private convictions, it is because the method I have embodied in the present book arises directly out of these convictions. Mathematical analysis is, of course, not excluded from the book, because without mathematics there can be no serious study of mechanics, but I have tried to reduce the amount of mathematics to a minimum, and I have regarded it (in the present book) as the servant and not as the master. Again, practical applications of mechanics have not been excluded, — on the contrary, these have been introduced wherever possible as illustrations of principles or results, — but I have tried to place principles first and applications second. And problems have not been excluded: I have inserted a great number, because the solution of problems seems to me to be the one and indispensable way of emphasizing a group of abstract principles and of fixing them in the mind of the student. But I have regarded the problems as an adjunct to the study of the principles, and not the principles as a framework round which to build problems.

Besides explaining the method and objects of a book, a preface may be expected to explain where the book starts and where it ends. The present book is intended to start from the very beginning of its subject, assuming no previous knowledge of mechanics on the part of the student. The question of how much knowledge of mathematics ought to be assumed has been a more difficult one to settle. I finally decided to rely as little as possible on the student's knowledge of trigonometry, and to employ the calculus as little as possible in the earlier chapters, but felt that the subjects of the later chapters could not be advantageously treated without a very considerable use of the calculus. Until the later chapters the use of the calculus is confined almost exclusively to unimportant branches and extensions of the subject, and to the working of illustrative examples. Thus a student who has no knowledge at all of the calculus will, I hope, be able to omit the sections of the book in which it is used, while at the same time acquiring a considerable and continuous knowledge of the essentials of theoretical mechanics.

PREFACE

The point at which the book ought to close seemed in the present instance to be determined by the method of the book itself. If, as I believe, a study of physical principles ought to be the common preliminary to the study of every branch and every application of mechanical science, then the book might clearly try to cover all this common ground, and ought to stop at the point at which detailed specialization becomes feasible and profitable. It ought, in fact, to cover the range which will be covered by all students, and stop short of subjects which will be of interest or importance only to a few. Judged by this criterion the book will perhaps be thought by some to be open to the criticism of covering too much ground; it may be thought that the final chapter on generalized coördinates can hardly be regarded as essential to the student whose study of mechanics is a preliminary to his entering the profession of, say, engineering. I am nevertheless convinced that, even if the study of generalized coördinates is not absolutely indispensable to such students, it is of extreme value and ought not to be neglected by a student, possessed of the requisite ability, who can possibly find time for it. The student who omits it shuts himself off from a point of view which sums up and illuminates the whole of dynamical theory; at the same time he denies himself the opportunity of studying, or at least of fully understanding, the theory of electricity and magnetism. And as regards the student who intends to continue his studies in the direction of theoretical physics, the theory of generalized coördinates forms so essential a preliminary to the study of most branches of physics that the advantages of including a short treatment of this subject in the preliminary mechanics course will hardly be disputed.

J. H. JEANS

Princeton
November, 1906

CONTENTS

	PAGE
CHAPTER I. REST AND MOTION	1

    Introduction. Motion of a point. Velocity. Acceleration. Vectors.

CHAPTER II. FORCE AND THE LAWS OF MOTION 26

    Newton's laws. Frame of reference. Laws applicable only to the motion of a particle.

CHAPTER III. FORCES ACTING ON A SINGLE PARTICLE 37

    Composition and resolution of forces. Particle in equilibrium. Types of forces, — weight of a particle, tension of a string, reaction between two bodies. Friction.

CHAPTER IV. STATICS OF SYSTEMS OF PARTICLES 59

    Moments. System of particles in equilibrium. Forces in one plane. Strings, — the suspension bridge, the catenary.

CHAPTER V. STATICS OF RIGID BODIES 90

    Rigidity. Conditions of equilibrium for a rigid body. Transmissibility of force. Composition of forces acting in a plane. Parallel forces. Couples. Forces in space.

CHAPTER VI. CENTER OF GRAVITY 117

    Center of gravity of a lamina. Center of gravity obtained by integration. Center of gravity of areas and volumes.

CHAPTER VII. WORK 145

    Measurement and units. Work done against a variable force. Work done in stretching an elastic string. Work represented by an area. The principle of virtual work. Potential energy. Kinetic energy. Conservation of energy. Stable and unstable equilibrium.

CONTENTS

PAGE

CHAPTER VIII. MOTION OF A PARTICLE UNDER CONSTANT FORCES 188

Body falling under gravity. Motion on an inclined plane. Atwood's machine. Motion referred to a moving frame of reference. Frictional reactions between moving bodies. Flight of projectiles.

CHAPTER IX. MOTION OF SYSTEMS OF PARTICLES 220

Equations of motion. Conservation of momentum. Motion of center of gravity. Kinetic energy. Impulsive forces. Elasticity.

CHAPTER X. MOTION OF A PARTICLE UNDER A VARIABLE FORCE 254

Equations of motion. The simple pendulum. Simple harmonic motion. The cycloidal pendulum. Motion of a particle about a center of force, — force proportional to the distance. General theory of motion about a center of force. The law of the inverse square.

CHAPTER XI. MOTION OF RIGID BODIES 286

Angular velocity. Kinetic energy. Radii of gyration. Moment of momentum. General theory of moments of inertia. General equations of motion of a rigid body. Euler's equations. Rotation of a planet. Motion of a top.

CHAPTER XII. GENERALIZED COÖRDINATES 320

Hamilton's principle. Principle of least action. Lagrange's equations. Small oscillations. Stability and instability of equilibrium. Forced oscillations. The canonical equations.

INDEX 361

AN ELEMENTARY TREATISE ON
THEORETICAL MECHANICS

CHAPTER I

REST AND MOTION

Introduction

1. Uniformity of nature. If we place a stone in water, it will sink to the bottom; if we place a cork in water, it will rise to the top. These two statements will be admitted to be true not only of stones and corks which have been seen to sink or rise in water but of all stones and corks. Given a piece of stone which has never been placed in water, we feel confident that if we place it in water it will sink. What justification have we for supposing that this new and untried piece of stone will sink in water? We know that millions of pieces of stone have at different times been placed in water; we know that not a single one of these has ever been known to do anything but sink. From this we infer that nature treats all pieces of stone alike when they are placed in water, and so feel confident that a new and untried piece of stone will be treated by the forces of nature in the same way as the innumerable pieces of stone of which the behavior has been tested, and hence that it will sink in water. This principle is known as that of the *uniformity of nature;* what the forces of nature have been found to do once, they will, under similar conditions, do again.

2. Laws of nature. The principle just stated amounts to saying that the action of the forces of nature is governed by certain laws; these we speak of as *laws of nature*. For instance, if it has been found that every stone which has ever been placed in water has sunk to the bottom, then, as has already been said, the principle of uniformity of nature leads us to suppose that every stone which at any future time is placed in water will sink to the bottom; and we can then announce, as a law of nature, that any stone, placed in water, will sink to the bottom.

That part of science which deals with the laws of nature is called *natural science*. Natural science is divided into two parts, experimental and theoretical. *Experimental science* tries to discover laws of nature by observing the action of the forces of nature time after time. *Theoretical science* takes as its material the laws of nature discovered by experimental science, and aims at reducing them, if possible, to simpler forms, and then discovering how to predict from these laws what the action of the forces of nature will be in cases which have not actually been subjected to the test of experiment. For example, experimental science discovers that a stone sinks, that a cork floats, and a number of similar laws. From these *theoretical physics* arrives at the simple laws of nature which govern all phenomena of sinking or floating, and, going further, shows how these laws enable us to predict, before the experiment has been actually tried, whether a given body will sink or float. For instance, experimental science cannot discover whether a 50,000-ton ship will float or sink, because no 50,000-ton ship exists with which to experiment. The naval architect, relying on the uniformity of nature, on the laws of nature determined by experimental science, and on the method of handling these laws taught by theoretical science, may build a 50,000-ton ship with every confidence that it will behave in the way predicted by theoretical science.

3. The science of mechanics. The branch of science known as *mechanics* deals with the motion of bodies in space, and with the forces of nature which cause or tend to cause this motion. The laws of nature which govern the action of these forces and the motion of bodies have long been known, and were reduced to their simplest form by Newton. Thus we may say that *experimental mechanics* is a completed branch of science.

The present book deals with *theoretical mechanics*. We start from the laws supplied by experimental mechanics, and have to discuss how these laws can be used to predict the motion of bodies, — for instance, the falling of bodies to the ground, the firing of projectiles, the motion of the earth and the planets round the sun. An important class of problems which we shall have to discuss will be those in which no motion takes place, the forces of nature which tend to cause motion being so evenly balanced that no motion occurs. Such problems are known as *statical*.

Motion of a Point

4. State of rest. Before we can reason about the motion of a body we have to determine what is meant by a body being at rest. In ordinary language we say that a train is at rest when the cars are not moving over the rails. We know, however, that the train, in common with the rest of the earth, is not actually at rest, but moving round the sun with a great velocity. Again, a fly crawling on the wall of a railway car might in one sense be said to be at rest, if it remained standing on the same spot of the wall. The fly, however, would not actually be at rest; it would share in the motion of the train over the country, the country would share in the motion of the earth round the sun, and the sun would share in the motion of the whole solar system through space.

These instances will show the necessity of attaching a clear and exact meaning to the conceptions of rest and motion. Obviously our statements would have been exact enough if we had said that in the first case the train was at rest *relatively to the earth*, and that in the second case the fly was at rest *relatively to the car*.

5. Frame of reference. Thus we find it necessary, before discussing rest and motion, to introduce the conception of a *frame of reference*. The earth supplied a frame of reference for the motion of the train, and when a train is not moving over the rails we may say that it is *at rest, the earth being taken as frame of* *reference*. So also we could say that the fly was *at rest, the car being taken as frame of reference*. Obviously any framework, real or imaginary, or any material body, may be taken as a frame of reference, provided that it is rigid, i.e. that it is not itself changing its shape or size.

We may accordingly say that a point is at rest relatively to any frame of reference when the distance of the point from each point of the frame of reference remains unaltered.

6. Motion relative to frame of reference. Having specified a frame of reference, we can discuss not only rest but also motion relative to the frame of reference. When the train has moved a mile over the tracks we say that it has moved a mile relatively to its frame of reference, the earth. When the fly has crawled from floor to ceiling of the car we say that it has moved, say, eight feet relatively to its frame of reference, the car.

In fixing the distance traveled by the fly relatively to the train in an interval between two instants t_1, t_2, we notice that the actual point from which the fly started is, say, a mile behind the present position of the train; but the point from which we measure is the point which occupies the same position in the car at time t_2 as this point did at time t_1. So, in general, to fix the distance moved relatively to a given frame of reference in the interval between times t_1 and t_2, we first find the point A which stands in the same position relative to the frame of reference at time t_2 as did the point from which the moving point started at time t_1. The distance from this point A to the point B, which is occupied by the moving point at instant t_2, is the distance moved relatively to the moving frame of reference.

By the motion of a particle B relative to a particle A, is meant the motion of B relative to a frame of reference moving with A.

7. Composition of motions. Suppose that in a given time the moving point moves a certain distance relatively to its frame of reference, while this frame of reference itself moves some other distance relatively to a second frame of reference, — as will, for instance, occur if a fly climbs up the side of a car while the car moves relatively to the earth.

Let us suppose that there is a frame of reference moving in the plane of the paper on which fig. 1 is drawn, and that the paper itself supplies a second frame of reference. Suppose that the moving point starts at A, and that during the motion that point of the first frame of reference which originally coincided with the moving point has moved from A to B, while the point itself has moved to C. Then the line AB represents the motion of frame 1 relative to frame 2, while BC represents the motion of the moving point relative to frame 1. The whole mo-

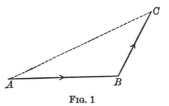

Fig. 1

tion of the point relative to frame 2 is represented by AC. The motion AC is said to be *compounded* of the two motions AB, BC, or is said to be the *resultant* of the two motions. Thus:

If a point moves a distance BC relatively to frame 1, while frame 1 moves a distance AB relatively to frame 2, the resultant motion of the point relative to frame 2 will be the distance AC, obtained by taking the two distances AB, BC and placing them in position in such a way that the point B at which the one ends is also the point at which the other begins.

There is a second way of compounding two motions. Let x, y represent the two motions. The rule already obtained directs us to construct a triangle ABC, to have x, y for the sides AB, BC, and then AC will be the motion required. Having constructed such a triangle ABC, let us complete the parallelogram $ABCD$ by drawing AD, CD parallel to the side of the triangle. Then AD, being

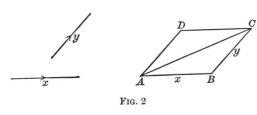

Fig. 2

equal to BC, will also represent the motion y, so that we may say that the two edges of the parallelogram which meet in A represent the two motions to be compounded, while the diagonal AC through A has already been seen to represent the resultant motion. Thus we have the following rule for compounding two motions x, y:

Construct a parallelogram ABCD such that the two sides AB, AD which meet in A represent the two motions x, y to be compounded, as regards both magnitude and direction; then the diagonal AC which passes through A will represent the resultant obtained by compounding these two motions.

Velocity

8. Uniform and variable velocity. Velocity means simply rate of motion. It may be either uniform or variable. If a point moves in such a way that a feet are described in each second of its motion, no matter which second we select, we say that the velocity of the point is a *uniform* velocity of a feet per second. If, however, the point moves a feet in one second, b feet in another, c feet in a third, and so on, we cannot say that any one of the quantities a, b, or c measures the velocity. The velocity is now said to be *variable*: it is different at different stages of the motion. To define the velocity at any instant, we take an infinitesimal interval of time dt and measure the distance ds described in this time. We then define the ratio $\frac{ds}{dt}$ to be the velocity at the instant at which the interval dt is taken. If the velocity is uniform, $\frac{ds}{dt}$ is the space described in unit time, and so the present definition of velocity becomes the same as that already given.

Average velocity. If a point moves with variable velocity, and describes a distance of a feet in t seconds, we speak of $\frac{a}{t}$ as the "average velocity" of the moving point during the time t. This average velocity is the velocity which would have to be possessed by an imaginary point moving with uniform velocity, if it were to cover the same distance in time t as the actual point moving with variable velocity.

Units. In measuring a velocity we need to speak in terms of a unit of length and of a unit of time; for instance, in saying that a point has a velocity of a feet per second we have selected the foot

VELOCITY

as unit of length and the second as unit of time. We can find the amount of this same velocity in other units by a simple proportion.

Thus suppose it is required to express a velocity of a feet per second in terms of miles and hours.

The point moves a feet in one second, and therefore $a \times 60 \times 60$ feet in one hour, and therefore

$$\frac{a \times 60 \times 60}{3 \times 1760} = \frac{15\,a}{22} \text{ miles}$$

in one hour. Thus the velocity is one of $\dfrac{15\,a}{22}$ miles per hour.

EXAMPLES

1. A railway train travels a distance of 918 miles in 18 hours. What is its average velocity in feet per second?

2. Compare the velocities of a train and an automobile which move uniformly, the former covering 100 feet a second and the latter 1500 yards a minute.

3. A man runs 100 yards in $9\frac{4}{5}$ seconds. What is his average speed in miles per hour?

4. The two hands of a town clock are 10 and 7 feet long. Find the velocities of their extremities.

5. Taking the diameter of the earth as 7927 miles, what is the velocity in foot-second units of a man standing at the equator (in consequence of the daily revolution of the earth about its axis)?

6. Two trains 230 and 440 feet long respectively pass each other on parallel tracks, the former moving with twice the speed of the latter. A passenger in the shorter train observes that it takes the longer train three seconds to pass him. Find the velocities of both trains.

9. Composition of velocities. All motion, as we have seen, must be measured relatively to a frame of reference. Thus velocity, or rate of motion, must also be measured relatively to a frame of reference. A point may have a certain velocity relative to a frame of reference, while the frame of reference itself has another velocity relative to a second frame. It may be necessary to find the velocity of the moving point with reference to the second frame, in other words, to *compound* the two velocities.

To do this we consider the motions which take place during an infinitesimal interval of time dt. Let the moving point have a velocity v_1 in a direction AB relative to the first frame, while the frame has a velocity v_2 in a direction AC relative to the second frame. Then in time dt the moving point describes a distance $v_1 dt$, say the distance AD, along AB relative to the first frame, while the frame itself describes a distance $v_2 dt$, say AE, along AC relative to the second frame. Let AF be the diagonal of the parallelogram of which AD, AE are two edges; then AF will be the resultant motion of the point in time dt relative to the second frame. Since the moving point describes a distance AF in time dt, the resultant velocity will be $\dfrac{AF}{dt}$.

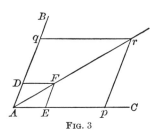

FIG. 3

Let us now agree that velocities are to be represented by straight lines, the direction of the line being parallel to that of the velocity and its length being proportional to the amount of the velocity, the lengths being drawn according to any scale we please; for example, we might agree that every inch of length is to represent a velocity of one foot per second, in which case a velocity of three feet a second will be represented by a line three inches long drawn parallel to the direction of motion.

In fig. 3 let Ap, Aq represent the velocities v_2, v_1 drawn on any scale we please. Since the scale is the same for both, we have

$$Ap : Aq = v_2 : v_1.$$

Now $AE = v_2 dt$, $AD = v_1 dt$, so that

$$AE : AD = v_2 : v_1,$$

and hence $\qquad Ap : Aq = AE : AD.$

If we complete the parallelogram $Aprq$, the diagonal Ar will pass through F, and we shall have

$$Ar : Ap = AF : AE.$$

If V is the resultant velocity, it has already been seen that

$$V = \frac{AF}{dt},$$

VELOCITY

so that
$$AF : AE = Vdt : v_2 dt$$
$$= V : v_2,$$
and hence
$$Ar : Ap = V : v_2.$$

Thus Ar represents the magnitude of the velocity V on the same scale as that on which Ap represents the velocity v_2. Also since Ar is in the direction of AF, the resultant motion, we see that Ar represents the velocity V both in magnitude and direction. We have accordingly proved the following theorem:

THEOREM. *If two velocities are represented in magnitude and direction by the two sides of a parallelogram which start from any point A, then their resultant is represented in magnitude and direction on the same scale by the diagonal of the parallelogram which starts from A.*

This theorem is known as the *parallelogram of velocities*. We may illustrate its meaning by two simple examples.

1. Suppose that a carriage is moving on a level road with velocity V. As a first frame of reference let us take the body of the carriage; as a second frame take the road itself. The velocity of frame 1 relative to frame 2 is then V. Relatively to frame 1, the center of any wheel P is fixed, so that any point on the rim describes a circle about P. Relatively to frame 1 the road is moving backward with velocity V, so that if there is to be no slipping between the rim and the road, the velocity of any point on the rim, relative to the first frame (the carriage), must be V. Thus the velocity of any point Q on the rim rela-

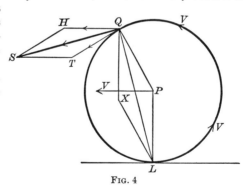

Fig. 4

tive to frame 1 will be a velocity V along the tangent QT. Representing this by the line QT, the velocity of the carriage relative to the road is represented by an equal line QH parallel to the road. Thus the resultant velocity of the point Q is represented by the diagonal QS of the parallelogram $QHST$. Clearly its direction bisects the angle HQT. Let L be the lowest point of the wheel, and let X complete the parallelogram $QPLX$. Obviously this parallelogram is similar to the parallelogram $QTSH$, corresponding lines in the two parallelograms being at right angles. Thus

$$QS : QT = QL : QP.$$

So that on a scale in which the velocity of the carriage is represented in magnitude by QP, the radius of the wheel, the velocity of the point Q will be represented by QL. Thus the velocities of the different points on the rim are proportional to their distances from L, their directions being in each case perpendicular to the line joining the point to L.

2. A battle ship is steaming at 18 knots, and its guns can fire projectiles with velocities of 2000 feet per second relative to the ship. How must the guns be pointed to hit an object the direction of which from the ship is perpendicular to that of the ship's motion?

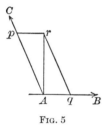

Fig. 5

Let AB be the direction of the ship's motion, and let us suppose the gun pointed in a direction AC. Then the velocity of the shot relative to the ship can be represented by a line Ap along AC, while that of the ship relative to the sea can be represented by a line Aq along AB. Completing the parallelogram $Aprq$, we find that the diagonal Ar will represent the velocity of the shot relative to the sea in magnitude and direction. Hence Ar must, from the data of the question, be at right angles to AB. If θ is the angle pAr through which the gun must be turned after sighting the object to be hit, we have

$$\sin \theta = \frac{pr}{Ap} = \frac{\text{velocity of ship}}{\text{velocity of firing of shot}}.$$

The velocity of the ship is 18 knots, or 18 nautical miles per hour. Now 1 nautical mile = 1.1515 ordinary miles = 6080 feet, so that a velocity of 18 knots is equal to 109,440 feet per hour, or 30.4 feet per second. Thus $\sin \theta = \dfrac{30.4}{2000} = .0152$, whence we find that $\theta = 0° 52' 16''$.

Triangle of Velocities

10. We can also compound velocities by a rule known as the *triangle of velocities*. In fig. 3 the two velocities were represented by Ap, Aq, and their resultant by Ar. The two velocities, however, might equally well have been represented by Ap, pr, and their resultant by Ar, from which we obtain the following rule:

VELOCITY

If two velocities are represented by the two sides of a triangle taken in order, their resultant will be represented by the third side, taken in the direction from the first side to the second side.

For example, let OP_1, OP_2 be two lines drawn through O to represent, on any scale, the velocities of a moving point at instants t_1, t_2. Then P_1P_2 will, on the same scale, represent the additional velocity acquired by the point in this interval.

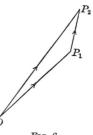

Fig. 6

For we can imagine a frame moving with the uniform velocity OP_1 of the particle at instant t_1. The velocity OP_2 at instant t_2 may be supposed compounded of the velocity OP_1 of the frame and a velocity P_1P_2 relative to the frame. Obviously this latter is the increase of velocity.

EXAMPLES

1. A car is running at 14 miles an hour, and a man jumps from it with a velocity of 8 feet per second in a direction making an angle of 30° with the direction of the car's motion. What is his velocity relative to the ground?

2. A railway train, moving at the rate of 60 miles an hour, is struck by a bullet, which is fired horizontally and at right angles to the train with a velocity of 440 feet a second. Find the magnitude and direction of the velocity with which the bullet appears to meet the train to a person inside.

3. A ship whose head points northeast is steaming at the rate of 12 knots in a current which flows southeast at the rate of 5 knots. How far will the ship have gone in $2\frac{1}{2}$ hours?

4. A train is traveling at the rate of 30 miles an hour, and rain falls with a velocity of 22 feet per second at an angle of 30° with the vertical in the same direction as the motion of the train. Find the direction of the splashes made on the windows by the raindrops.

5. A steamer's course is due south, and its speed is 20 knots; the wind is from the west, but the line of smoke from the steamer is observed to point in a direction 30° east of north. What is the velocity of the wind?

6. A man rows across a stream a mile wide, pointing his boat upstream at an angle of 30° with the bank. How long does he take to cross, if he rows with a velocity of 4 miles an hour and if the current has an equal velocity?

7. A stream has a current velocity a, and a man can row his boat with a velocity b. In what direction must he row, if he is to land at a point exactly opposite his starting point? And in what direction must he row so as to cross in the shortest time?

8. A ship whose head is pointing due south is steaming across a current running due west; at the end of two hours it is found that the ship has gone 36 miles in the direction 15° west of south. Find the velocities of the ship and current.

9. A person traveling eastward at the rate of 3 miles an hour finds that the wind seems to blow directly from the north; on doubling his speed it appears to come from the northeast. Find the direction of the wind and its velocity.

ACCELERATION

11. Acceleration is rate of increase of velocity. If we find that the velocity of a moving point increases by an amount f in a second, no matter which second is selected, we say that the motion of the point has a uniform acceleration f per second. For instance, a stone or other body falling under gravity is found to increase its velocity by a certain constant velocity f per second, where f denotes a velocity of about 32 feet per second. Thus we say that a falling stone has a uniform acceleration of f per second, or of about 32 feet per second per second.

Generally, however, an acceleration will not be uniform; the rate of increase of velocity will be different at different stages of the journey. To find the acceleration at any instant, we observe the change in velocity during an infinitesimal interval dt of time. If dv is the increase of velocity, we say that $\dfrac{dv}{dt}$ is the acceleration at the instant at which dt is taken. An acceleration will of course have sign as well as magnitude, for the velocity may be either increasing or decreasing. When the velocity is decreasing, the acceleration is reckoned with a negative sign. A negative acceleration is spoken of as a *retardation*. Thus a retardation f means that the velocity is diminished by an amount f per unit of time.

EXAMPLES

1. A workman fell from the top of a building and struck the ground in 4 seconds. With what velocity did he strike the ground, the acceleration due to gravity being 32 feet per second per second?

2. A train has at a given instant a velocity of 30 miles an hour, and moves with an acceleration of 1 foot per second per second. Find its velocity after 20 seconds.

ACCELERATION

3. A train comes to rest after the brakes have been applied for ten seconds. If the retardation was 8 feet per second per second, what was the velocity of the train when the brakes were first drawn?

4. How long does it take a body starting with a velocity of 22 feet per second and moving with an acceleration of 6 feet per second per second, to acquire a velocity of 60 miles an hour?

5. Two bodies start at the same instant with velocities u and v respectively; the motion of the first undergoes a retardation of f feet per second per second, while that of the second is uniform. How far will the second have gone by the time that the first comes to rest?

6. A body starting from rest moves for 4 seconds with a uniform acceleration of 8 feet per second per second. If the acceleration then ceases, how far will the body move in the next 5 seconds?

7. A train has its speed reduced from 40 miles an hour to 30 miles an hour in 5 seconds. If the retardation be uniform, for how much longer will it travel before coming to rest?

8. A body falling under gravity has an acceleration of 32.2 feet per second per second. Express this acceleration when the units are (*a*) centimeter, second; (*b*) mile, hour.

12. Parallelogram of accelerations. THEOREM. *Let the velocity of a point be compounded of two velocities v_1, v_2 along given directions, and let these velocities be variable, their accelerations being f_1, f_2. Then if two lines be drawn in the direction of the velocities, to represent f_1, f_2 on any scale, the resultant acceleration will be represented on the same scale by the diagonal of the parallelogram of which these lines are edges.*

To prove the theorem, we consider the motion during any small interval dt at which the component accelerations are f_1, f_2. In fig. 7 let AB, AC represent the two velocities v_1, v_2 at the beginning of this interval. Let BB', CC' represent, on the same scale, the infinitesimal increments in velocity in the interval dt,

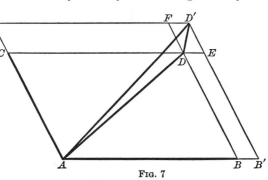

FIG. 7

namely $f_1 dt, f_2 dt$. Then AB', AC' will represent the velocities at the end of the interval dt.

In the figure the lines $BDF, B'ED', CDE, C'FD'$ are drawn parallel to AB and AC. Thus AD represents the resultant velocity at the beginning of the interval dt, and AD' that at the end of the interval. The velocity AD' can be regarded as compounded of the two velocities AD, DD', and, as in §10, DD' represents the increment in velocity in time dt. Thus, if F is the resultant acceleration, the line DD' will represent a velocity Fdt. On the same scale DE, DF represent velocities $f_1 dt, f_2 dt$, and $DED'F$ is a parallelogram.

If OF_1, OF_2 (fig. 8) represent the accelerations f_1, f_2 on any scale, and if OG is the diagonal of the completed parallelogram, we

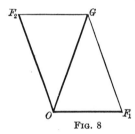

Fig. 8

clearly have $OF_1 : OF_2 = f_1 : f_2 = DE : DF$, so that the parallelograms $OF_1 GF_2$ (in fig. 8) and $DED'F$ (in fig. 7) will be similar and similarly situated. Thus

$$OG : OF_1 = DD' : DE = Fdt : f_1 dt = F : f_1,$$

so that OG represents the acceleration F on the same scale as that on which OF_1, OF_2 represent f_1, f_2; and OG, being parallel to DD', will also represent the direction of F, proving the theorem.

Clearly the acceleration at any instant need not be in the same direction as the velocity. In fig. 7 the directions AD, AD' represent velocities at the beginning and end of the interval dt. When in the limit we take $dt = 0$, these lines coincide, and the direction of the velocity at the instant at which dt is taken is that of AD. The direction of the acceleration at this instant is, however, DD'.

As an illustration of this, let us consider the motion of a particle moving uniformly in a circle; e.g. a point on the rim of a wheel, turning with uniform velocity V about its center.

Let A, B (fig. 9) be the positions of the point at two instants, let the tangents at A, B meet in C, and let D complete the parallelogram $ACBD$.

The velocity at the first instant is a velocity V along AC. Let us agree to represent this by the line AC itself. At the second instant the velocity is a velocity V along CB; this may, on the same scale, be represented by the line CB, or more conveniently by AD. Since AC, AD represent the

ACCELERATION

velocities at the two instants, the line CD will represent the change in velocity between these two instants.

Now let the two instants differ only by an infinitesimal interval dt, so that the points A, B coincide except for an infinitesimal arc Vdt. In the figure, CD passes through P wherever A, B are on the circle, so that when B is made to coincide with A, CD coincides with the radius through A. But if F is the acceleration of the moving point, the change in velocity produced in time dt must be Fdt. Thus CD represents the change of velocity Fdt in direction and magnitude, so that the change of velocity, and hence the acceleration at A, is along the radius at A.

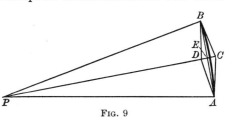

FIG. 9

Here, then, we have a case in which the acceleration is at right angles to the velocity.

To find the magnitude of the acceleration, we notice that $CD = 2\ CE$, and that, by similar triangles,

$$EC : CB = BE : BP.$$

Now EC, or $\frac{1}{2} CD$, represents the velocity $\frac{1}{2} Fdt$, while CB on the same scale represents the velocity V.

Thus $\qquad \frac{1}{2} Fdt : V = BE : BP.$

In the limit when BA is very small, BE, or $\frac{1}{2} BA$, becomes identical with half of the arc BA of the circle, and therefore with $\frac{1}{2} Vdt$. Thus, if a is the radius of the circle,
$$\tfrac{1}{2} Fdt : V = \tfrac{1}{2} Vdt : a,$$
giving $F = \dfrac{V^2}{a}$ as the amount of the acceleration.

EXAMPLES

1. A windmill has sails 20 feet in length, and turns once in ten seconds. Find the acceleration of a point at the end of a sail.

2. A wheel of radius 3 feet spins at the rate of 10 revolutions a second and is at the same time falling freely with an acceleration of 32 feet per second per second due to gravity. Find the resultant accelerations of the different points on the rim of the wheel.

3. Taking the earth to have an equatorial diameter of 7927 miles, find the acceleration towards the earth's center of (a) a point at rest, relative to the earth's surface, on the equator; (b) a body falling under gravity at the equator, with an acceleration, relative to the earth's surface, of 32.09 feet per second per second.

4. Supposing that the moon describes a circle of radius 240,000 miles round the earth in $29\frac{1}{2}$ days, find its acceleration towards the earth.

5. Assuming that the planets describe circles round the sun with different periodic times, such that the squares of the periodic times are proportional to the cubes of the radii of the circles, show that the accelerations of the planets are inversely proportional to the squares of their distances from the sun.

Vectors

13. We have found three kinds of quantities, — motion, velocity, and acceleration, — all of which can be compounded according to the parallelogram law.

Quantities which can be compounded according to the parallelogram law are called *vectors*. A vector must have magnitude and direction, and hence must be capable of representation, on an assigned scale, by a straight line. We have seen that motion, velocity, and acceleration are all vectors.

Composition and Resolution of Vectors in a Plane

14. By definition of a vector, two vectors can be compounded into one, by application of the parallelogram law. It also follows from the definition that any one vector may be regarded as equivalent to two, these two being represented by the edges of a parallelogram constructed so as to have the original vector represented by the diagonal; or, as we shall say, any vector can be *resolved* into two others.

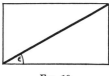

Fig. 10

In particular, if we construct a rectangular parallelogram so as to have a line which represents a vector R as its diagonal, we find that the vector R can be resolved into two vectors $R \cos \epsilon$ and $R \sin \epsilon$, at right angles to one another, and in directions such that R makes angles ϵ, $\dfrac{\pi}{2} - \epsilon$ with them.

If we take two fixed rectangular axes Ox, Oy in a plane, we see that any vector R can be resolved into two components $R \cos \epsilon$,

$R \sin \epsilon$ parallel to these axes, where ϵ is the angle which R makes with Ox. The components $R \cos \epsilon$, $R \sin \epsilon$ are spoken of as the *components of R* along Ox and Oy.

There are two ways of compounding a number of vectors R_1, R_2, $\cdots$, R_n. In the first place, we can construct a polygon $ABCDE \cdots N$, such that the sides AB, BC, CD, $\cdots$, MN represent the vectors R_1, R_2, R_3, $\cdots$, R_n. Then AN will represent the resultant. For R_1, R_2 can first be compounded into a vector R_{12} represented by AC. Combining R_3 with this vector, we obtain a vector represented by AD, and so on until finally AN is reached.

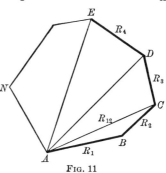

Fig. 11

As a second way, we can resolve each vector, such as R_s, into its two components

$$R_s \cos \epsilon_s, \quad R_s \sin \epsilon_s,$$

along rectangular axes Ox, Oy. The n vectors are now resolved into $2n$ vectors, of which n are parallel to Ox and n are parallel to Oy. The first set of n can be compounded into a single vector

$$X \equiv R_1 \cos \epsilon_1 + R_2 \cos \epsilon_2 + \cdots$$

parallel to Ox, while the second set can be compounded into a single vector

$$Y \equiv R_1 \sin \epsilon_1 + R_2 \sin \epsilon_2 + \cdots$$

parallel to Oy. We now have two vectors X, Y parallel to Ox, Oy.

If their resultant is a vector R making an angle ϵ with Ox, we have
$$R \cos \epsilon = X = R_1 \cos \epsilon_1 + R_2 \cos \epsilon_2 + \cdots. \tag{1}$$
$$R \sin \epsilon = Y = R_1 \sin \epsilon_1 + R_2 \sin \epsilon_2 + \cdots. \tag{2}$$

To find the numerical value of R, we square and add (1) and (2) and obtain

$$\begin{aligned}
R^2 &= X^2 + Y^2 \\
&= (R_1 \cos \epsilon_1 + R_2 \cos \epsilon_2 + \cdots)^2 + (R_1 \sin \epsilon_1 + R_2 \sin \epsilon_2 + \cdots)^2 \\
&= R_1^2 + R_2^2 + \cdots + 2 R_1 R_2 (\cos \epsilon_1 \cos \epsilon_2 + \sin \epsilon_1 \sin \epsilon_2) + \cdots \\
&= R_1^2 + R_2^2 + \cdots + 2 R_1 R_2 \cos (\epsilon_1 - \epsilon_2) + \cdots.
\end{aligned}$$

To find the direction of the resultant, we divide the corresponding sides of (1) and (2) and obtain

$$\tan \epsilon = \frac{Y}{X} = \frac{R_1 \sin \epsilon_1 + R_2 \sin \epsilon_2 + \cdots}{R_1 \cos \epsilon_1 + R_2 \cos \epsilon_2 + \cdots}.$$

If we have only two vectors R_1, R_2, making an angle θ with one another, we may put $\epsilon_1 - \epsilon_2 = \theta$, and obtain

$$R^2 = R_1^2 + R_2^2 + 2 R_1 R_2 \cos \theta.$$

Since R is obviously the diagonal of a parallelogram having two edges of lengths R_1, R_2, meeting at the angle θ, this result can be obtained directly from the geometry of the triangle ADC, in which the angle at C is evidently $\pi - \theta$. Thus

$$R^2 = R_1^2 + R_2^2 - 2 R_1 R_2 \cos (\pi - \theta),$$

which is clearly identical with the above expression.

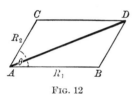
Fig. 12

We may take two examples to illustrate the method of resolving vectors into rectangular components in a plane.

1. In Ex. 2, p. 10, suppose that the direction of the ship (AB in fig. 5) is taken for axis Ox, and that that in which the shot is to travel is taken for axis Oy. Let the shot be fired with velocity V, making an angle θ with Ox, the velocity of the ship being v. The resultant velocity is to be along Oy, so that the velocity along Ox, say X, is to be nil. We have, however,

$$X = v + V \cos \theta,$$

so that we must have $\cos \theta = -\dfrac{v}{V}$, giving the result already obtained.

2. To find the acceleration of a point moving with uniform velocity V in a circle of radius a. Let A be the position of the particle at time $t = 0$, and take OA for axis of x. After time t the particle has described a length Vt of arc, so that if B is its position after time t, the angle BOA is $\dfrac{Vt}{a}$ in circular measure. The direction of velocity at B, namely the tangent at B, will accordingly

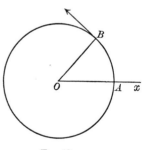
Fig. 13

make an angle $\dfrac{\pi}{2} + \dfrac{Vt}{a}$ with Ox, so that the components of the velocity along Ox, Oy, say v_1, v_2, will be

$$v_1 = V \cos\left(\dfrac{\pi}{2} + \dfrac{Vt}{a}\right) = -V \sin\dfrac{Vt}{a},$$

$$v_2 = V \sin\left(\dfrac{\pi}{2} + \dfrac{Vt}{a}\right) = V \cos\dfrac{Vt}{a}.$$

The acceleration along Ox is $\dfrac{dv_1}{dt}$, which, on differentiating v_1 with respect to t, is found to be

$$-\dfrac{V^2}{a} \cos\dfrac{Vt}{a},$$

while that along Oy is similarly found to be $\dfrac{dv_2}{dt}$, or

$$-\dfrac{V^2}{a} \sin\dfrac{Vt}{a}.$$

Compounding these, we obviously obtain an acceleration $\dfrac{V^2}{a}$ along BO, the result already obtained on page 15.

Composition and Resolution of Vectors in Space

15. It may be that the vectors to be compounded are not all in one plane. However, the method of determining the resultant is essentially the same. Thus we can construct a polygon in space $ABCD \cdots N$ such that the sides $AB, BC, \cdots, MN$ represent the vectors $R_1, \cdots, R_n$. As in the preceding case, it is readily shown that AN is the resultant.

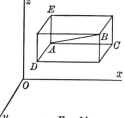

Fig. 14

It is usually more convenient to resolve each vector into three components parallel to rectangular axes in space. Given a vector AB, we draw through A, and likewise through B, three planes parallel to the coördinate planes. They inclose a rectangular parallelopiped of which AB is a diagonal. The edges AC, AD, AE represent three vectors by which AB can be replaced; they are the components parallel to the axes of the vector AB.

Suppose there are n vectors, and that the direction angles of the vector R_s are denoted by α_s, β_s, γ_s. As above, each vector R_s

can be replaced by three components parallel to the axes; these vectors are of amount

$$R_s \cos \alpha_s, \quad R_s \cos \beta_s, \quad R_s \cos \gamma_s.$$

Of the $3n$ vectors thus obtained, the n vectors parallel to the x-axis can be compounded into the single vector

$$X = R_1 \cos \alpha_1 + R_2 \cos \alpha_2 + \cdots + R_n \cos \alpha_n. \qquad (3)$$

The whole system of vectors can thus be replaced by this vector and two others parallel to the y and z axes respectively, namely

$$Y = R_1 \cos \beta_1 + R_2 \cos \beta_2 + \cdots + R_n \cos \beta_n. \qquad (4)$$
$$Z = R_1 \cos \gamma_1 + R_2 \cos \gamma_2 + \cdots + R_n \cos \gamma_n. \qquad (5)$$

Evidently the resultant of these three vectors, and consequently of the original n vectors, is a diagonal of a rectangular parallelepiped whose edges are of lengths X, Y, Z. If the length of the resultant be denoted by R, and the direction angles by α, β, γ, we have

$$R^2 = X^2 + Y^2 + Z^2,$$

and $\quad \cos \alpha = \dfrac{X}{R}, \quad \cos \beta = \dfrac{Y}{R}, \quad \cos \gamma = \dfrac{Z}{R}.$

Hence the resultant is completely determined in magnitude and direction.

· Centroids

16. Let a system of vectors be represented in direction by OA_1, $OA_2, \cdots, OA_n$, and let their magnitudes be $m_1 OA_1, \cdots, m_n OA_n$, where $m_1, m_2, \cdots, m_n$ are any quantities. Denote by x_r, y_r, z_r the coördinates of A_r with respect to axes through O; by $\alpha_r, \beta_r, \gamma_r$ the direction angles of OA_r with respect to these axes; and by R_r the magnitude of the vector $m_r OA_r$. The components of this vector along these axes are given by

$$R_r \cos \alpha_r = m_r OA_r \cos \alpha_r = m_r x_r,$$
$$R_r \cos \beta_r = m_r OA_r \cos \beta_r = m_r y_r,$$
$$R_r \cos \gamma_r = m_r OA_r \cos \gamma_r = m_r z_r.$$

Hence equations (3), (4), (5) can be written thus:

$$X = \sum_1^n m_r x_r, \quad Y = \sum_1^n m_r y_r, \quad Z = \sum_1^n m_r z_r. \qquad (6)$$

THE CENTROID

For the interpretation of this result, we make use of the idea of the *centroid of a system of points*. By definition the centroid of a system of points is the point such that its distance from any one of three coördinate planes is the average of the distances of all the points of the system from this plane, it being understood that each distance is measured with its proper algebraic sign.

From this definition, it follows that the distance of the centroid from any plane whatever is equal to the average of the distances of the n points from this plane. For if x_r, y_r, z_r are the coördinates of the rth point, the coördinates of the centroid, say $\bar{x}, \bar{y}, \bar{z}$, will be

$$\bar{x} = \frac{1}{n}\sum_1^n x_r, \quad \bar{y} = \frac{1}{n}\sum_1^n y_r, \quad \bar{z} = \frac{1}{n}\sum_1^n z_r, \tag{7}$$

and the perpendicular distance from the centroid to any plane

$$ax + by + cz + d = 0$$

is

$$\frac{1}{\sqrt{a^2+b^2+c^2}}(a\bar{x} + b\bar{y} + c\bar{z} + d)$$

$$= \frac{1}{\sqrt{a^2+b^2+c^2}} \frac{1}{n}\sum_1^n (ax_r + by_r + cz_r + d)$$

$$= \frac{1}{n}\sum_1^n \frac{ax_r + by_r + cz_r + d}{\sqrt{a^2+b^2+c^2}},$$

which proves the result.

Let us imagine that of the n points a number m_a all coincide at the point x_a, y_a, z_a, a number m_b at the point x_b, y_b, z_b, and so on. Then the centroid has coördinates (by equations (7)),

$$\left.\begin{array}{l} \bar{x} = \dfrac{1}{n}\sum m\, x_r = \dfrac{\sum m_r x_r}{\sum m_r} \\[1em] \bar{y} \qquad\quad = \dfrac{\sum m_r y_r}{\sum m_r} \\[1em] \bar{z} \qquad\quad = \dfrac{\sum m_r z_r}{\sum m_r} \end{array}\right\}. \tag{8}$$

where the summation is now taken over the various points in space at which the original points are accumulated. Calling these points in space $A, B, C, \cdots$, the point $\bar{x}, \bar{y}, \bar{z}$ is said to be the centroid of the points $A, B, C, \cdots$, corresponding to the multipliers $m_a, m_b, m_c, \cdots$.

By means of these results, equations (6) are reducible to

$$X = \bar{x} \cdot \sum_1^n m_r, \quad Y = \bar{y} \cdot \sum_1^n m_r, \quad Z = \bar{z} \cdot \sum_1^n m_r. \qquad (9)$$

Hence the resultant of the above set of vectors is directed along the line OC, and its magnitude is $\overline{OC} \cdot \sum_1^n m_r$. As defined by equations (9), the multipliers m_r can be any numbers whatever, positive or negative, so that the sum $\sum_1^n m_r$ may be positive, zero, or negative. In particular, when the vectors are represented in magnitude as well as direction by $OA_1, \cdots, OA_n$, the resultant is directed along OC_o, and its magnitude is $n \cdot OC_o$, where n is the number of vectors and the point C_o is the centroid, as defined above. Thus we have the theorem:

THEOREM. *If vectors of magnitude $m_1 OA_1, m_2 OA_2, \cdots$ act along the lines $OA_1, OA_2, \cdots$, then their resultant is of magnitude $(m_1 + m_2 + \cdots) OG$, and acts along OG, where G is the centroid of the points $A_1, A_2, \cdots$ for the multipliers $m_1, m_2, \cdots$.*

Obviously the parallelogram law is a particular case of this theorem.

EXAMPLES

1. Find the resultant of two vectors of magnitudes $5\,P$ and $12\,P$ which meet at right angles.

2. A vector P is the resultant of two vectors which make angles of $30°$ and $45°$ with it on opposite sides. How large are the latter vectors?

3. Show how to determine the directions of two vectors of given magnitude so that their resultant shall be of given magnitude and direction. When is this impossible?

4. Show that if the angle at which two given vectors are inclined to each other is increased, their resultant is diminished.

5. Under what conditions will the resultant of a system of vectors of magnitudes 7, 24, and 25 be equal to zero?

EXAMPLES 23

6. Three vectors of lengths P, P, and $P\sqrt{2}$ meet in a point and are mutually at right angles. Determine the magnitude of the resultant and the angles between its direction and that of each component.

7. Three vectors of lengths P, $2P$, $3P$ meet in a point and are directed along the diagonals of the three faces of a cube meeting at the point. Determine the magnitude of their resultant.

8. Show that the resultant of three vectors represented by the diagonals of three faces of a parallelepiped meeting in a vertex A is represented by twice the diagonal of the parallelepiped drawn from A.

9 D is a point in the plane of the triangle ABC, and I is the center of its inscribed circle. Show that the resultant of the vectors $a \cdot AD$, $b \cdot BD$, $c \cdot CD$ is $(a + b + c)\, ID$, where a, b, c are the lengths of the sides of the triangle.

10. $ABCD$, $A'B'C'D'$ are two parallelograms in the same plane. Find the resultant of vectors drawn from a point proportional to and in the same direction as AA', $B'B$, CC', $D'D$.

11. If O is the center of the circumscribed circle of the triangle ABC and P its orthocenter, show that OP is the resultant of the vectors OA, OB, and OC; also that $2\,PO$ is the resultant of PA, PB, PC.

12. The chords AOB and COD of a circle intersect at right angles. Show that the resultant of the vectors OA, OB, OC, OD is represented by twice the vector OP, where P is the center of the circle.

GENERAL EXAMPLES

(In these examples take the acceleration produced by gravity to be 32 feet per second per second)

1. A point possesses simultaneously velocities of 2, 3, 8 feet per second, in the directions of a point describing the three sides of an equilateral triangle in order. Find the magnitude of its velocity.

2. A point possesses simultaneously velocities, each equal to v, in the directions of lines drawn from the center of a regular hexagon to five of its angular points. Find the magnitude and direction of the resultant velocity.

3. When a steamer is in motion it is found that an awning 8 feet above the deck protects from rain the portion of the deck more than 4 feet behind the vertical projection of the edge of the awning; but when the steamer comes to rest the line of separation of the wet and dry parts is 6 feet in front of this projection. Find the velocity of the steamer, if that of the rain be 20 feet per second.

4. A ship sailing along the equator from east to west finds that from noon one day (local time) to noon the next day (local time) the distance covered is 420 miles. What would be the day's run, if the ship were sailing at the same rate from west to east?

5. A railroad runs due east and west in latitude λ. At what rate must a train travel along the road to keep the sun always directly south of it?

6. Determine the true course and velocity of a steamer going due north by compass at 10 knots through a 4-knot current setting southeast; and determine the alteration of direction by compass in order that the steamer should make a true northerly course.

7. A bicyclist rides faster than the velocity of the wind, and makes the error of judging the direction of the wind to be the direction in which it appears to meet him when he is in motion. Show that the wind will always appear to be against him, in whatever direction he rides.

8. One ship sailing east with a speed of 20 knots passes a lightship at 11 A.M.; a second ship sailing south at the same rate passes the same point at 1 P.M. At what time are they closest together, and what is then the distance between them?

9. Two particles move with velocities v and $2v$ respectively in opposite directions, in the circumference of a circle. In what positions is their relative velocity greatest and least, and what values has it then?

10. Find the relative motion of two particles moving with the same velocity v, one of which describes a circle of radius a while the other moves along a diameter.

11. Two particles move uniformly in straight lines. At a given time the distance between them is a and their relative velocity is V, the components of the latter in the direction of a and perpendicular to it being u and v. Show that, when they are nearest together, their distance is av/V, and that they arrive at this position after the interval au/V^2.

12. Three horses in a field are at a certain moment at the vertices of an equilateral triangle. Their motion relative to a person driving along a road is in a direction round the sides of the triangle (in the same sense), and in magnitude equal to the velocity of the carriage. Show that the three horses are moving along concurrent lines.

13. Two points describe concentric circles, of radii a and b, with speeds varying inversely as the radii. Show that the relative velocity is parallel to the line joining the points when the angle between the radii to these points is
$$\cos^{-1}\frac{2ab}{a^2+b^2}.$$

14. A stone dropped from a balloon moving horizontally is observed to be 4 seconds in the air, and to strike the earth in a direction making an angle of 15° with the vertical. Find the velocity of the balloon.

EXAMPLES

15. A ball is thrown from the top of a building with a velocity of 64 feet per second at an angle of 30° with the horizontal in an upward direction. Find the directions of its motion at the end of the first and second seconds, and also the velocities at these instants.

16. A ball is tossed into the air with a velocity of 20 feet a second, and at the end of a second is seen to be moving in a line at right angles to the direction of projection. What is its velocity at the instant?

17. If the velocity of a bullet is supposed to be a uniform horizontal velocity equal to n times that of sound, show that the points at which the sounds of the firing and of the bullet striking the target are heard simultaneously lie on a hyperbola of eccentricity n. Examine the case in which n is very nearly equal to unity.

18. Assuming that the earth moves in a circular orbit about the sun with a velocity 29.6 kilometers per second, and that the velocity of light is 300,000 kilometers per second, find the apparent displacement of the sun due to the earth's motion.

19. Assuming that the earth in a year describes a circle uniformly about the sun as center, that the distance between the centers is 220 radii of the sun, and that the radius of the sun is 108 times that of the earth, find the velocity of the vertex of the earth's shadow, taking the sun's radius as the unit of space and a year as the unit of time.

CHAPTER II

FORCE AND THE LAWS OF MOTION

Newton's Laws

17. The laws of motion, as we have said, form the material supplied by experimental mechanics for theoretical mechanics to work with. These laws have been stated in compact form by Newton:

LAW I. *Every body continues in its state of rest, or of uniform motion in a straight line, except in so far as it is compelled by impressed force to change that state.*

LAW II. *The rate of change of momentum is proportional to the impressed force, and takes place in the direction of the straight line in which the force acts.*

LAW III. *To every action there corresponds an equal and opposite reaction.*

18. These laws introduce several new terms, — "force," "momentum," "action," "reaction," — which must be explained before the laws can be fully understood.

The first law involves the idea of *motion*, which has already been discussed, and of *force*, which is new.

The word "force" is in common use. It is associated in the first instance with muscular effort; for example, we exert force in pushing against an obstacle. Scientifically, however, the word has a wider use; we say, for instance, that two railway trucks when they collide exert *force* on one another, and that the earth exerts *force* on all bodies, causing them to fall towards it unless they are supported in such a way that they resist this force.

The first law of motion, in fact, explains what is to be understood by force: it is that which tends to change the state of rest of a body, or of uniform motion in a straight line.

THE FIRST LAW

Consider, for instance, a railway truck standing at rest on a level line of rails. If a second truck runs into it the first truck will start into motion, so that force has been applied.

The first law, however, tells us more than this. It tells us that if a body is kept free from the action of forces, it will remain in its state of rest or of uniform motion in a straight line. Thus the normal state for a body to be in is one of rest or of uniform motion in a straight line, i.e. motion with uniform velocity; it is only the presence of force which can alter this normal state.

Consider again the case of the railway truck. Let us suppose it has been set in motion by collision, and that it starts off with a velocity of, say, 10 miles an hour. The first law tells us that unless forces act on the truck, it will continue its motion with an unaltered velocity of 10 miles an hour in the same straight line in which it started. When a truck is actually started into motion by collision, it may be taken for granted that it will not continue in uniform motion in a straight line, but will sooner or later be brought to rest. Thus forces must be at work. Let us consider the nature of these forces.

In the first place there is a force known as the resistance of the air. The air in front of the truck presses against it in such a way as to retard its motion. The air therefore exerts force on the truck just as a man might exert force on the truck by pressing against it with his hand. This force alone would stop the truck in time.

Let us suppose that the brakes are applied, and that the wheels are gripped so firmly as to be at rest relatively to the truck, so that they slide on the rails. There is then a large force applied to the truck by the rails, and this again tends to stop the motion of the truck. Even if the brakes are not applied, so that the wheels are left free to turn, there will still be a force applied by the rails, although this force will be smaller than before.

Suppose that the track is curved instead of straight. We can imagine the motion continuing for some time, but it will be motion along the curve, and not motion in a straight line, such as we are told by the first law would take place if it were not for the action of force. Force has therefore been applied, the force being that of the rails on the flanges of the wheels, tending to turn the truck round the curve. If the flanges were not present, this force could not act, so that the motion would continue in a straight line — the truck would run off the rails.

As another illustration of the meaning of the first law, let us examine the motion of a bullet fired from a gun. Here the forces which start the motion are supplied by the pressure of the powder. After the bullet has left the gun, the forces which act are small compared with those which have started the motion, so that we get an approximation to uniform motion in a straight line. The forces acting to alter this motion are (*a*) the resistance of the air and (*b*) the weight of the bullet. The former, as we have seen, tends to stop the motion by pressure on the ends and sides of the bullet; the latter tends to drag the bullet down to the earth, and so causes it, instead of describing a straight line, to describe a path which curves downward towards the earth.

19. The conception of uniform motion in a straight line, or of rest (the particular case of uniform motion in which the velocity is nil), as being the normal state of a body is due to Galileo (1564–1642). An interesting account of the discovery of this law will be found in Chapter II of Mach's *Science of Mechanics*,[1] or in Chapter IX of Cox's *Mechanics*.[2] Before the time of Galileo it was commonly supposed, on the authority of Aristotle, that every body had a *natural place*, and that its normal state was one of rest in this natural place. For instance, a stone was supposed to sink in water, not because the force of gravity was acting on it and setting it into downward motion, but because its natural place was at the bottom of the water; a cork was supposed to rise because its natural place was at the top. Thus Girard,[3] in 1634, speaks of " millions de matières, qui sont disposées chacunes en leurs lieux," and defines gravity as " la force qu'une matière démonstre à son obstacle, pour retourner en son lieu." Thus the effect of force, before Galileo, was supposed to be to keep a body out of its natural place. Galileo perceived that bodies had no *natural places* at all, but *natural states*, namely of rest or of uniform motion in a straight line, and the effect of force is not to move a body from its *natural place* but to disturb it from its *natural state*, — i.e. to alter its speed. This discovery of Galileo is what is expressed by Newton's first law of motion.

20. Having settled what is meant by the natural state of a body and also what is meant by force, — namely that which tends

---

[1] Ernest Mach, *Science of Mechanics* (Eng. trans. by McCormack).

[2] J. Cox, *Mechanics*, Cambridge, University Press, 1904.

[3] In the Elzevir edition of Stevin, Leyden, 1634. See Cox, *Mechanics, loc. cit. ante.*

THE SECOND LAW

to alter the natural state, — we next inquire as to what is the law which governs the effect produced by force. Given a force, by how much will this alter the natural state of uniform motion in a straight line? An answer to this is provided by the second law:

LAW II. *The rate of change of momentum is proportional to the impressed force, and takes place in the direction of the straight line in which the force acts.*

The force, then, produces change in a certain quantity, — the momentum of the body on which the force acts, — and the force is proportional to the rate of change of this momentum.

By momentum is meant the product of the velocity of the body by a quantity known as the *mass* of the body. The mass measures simply the quantity of matter of which a body is composed, and so does not depend on the motion of the body. Thus

$$\text{rate of change of momentum} = \text{mass} \times \text{rate of change of velocity}$$
$$= \text{mass} \times \text{acceleration},$$

by the definition of acceleration. We therefore see that the force is proportional to the product of two quantities, the mass of the body and its acceleration.

21. Measurement of mass. If we drop a body from our hand, it will, in general, be acted on by two forces, the resistance of the air and its weight. If we suspend the body in a vacuum, with an arrangement for letting it drop at any instant we please, we get rid of the resistance of the air, and the only force acting on the body will be its weight. Now if any two bodies are suspended side by side in a vacuum, and are let fall at precisely the same instant, it will be found that they remain side by side during the whole time they are falling towards the earth. Thus at any instant their accelerations are the same.

It follows from the second law of motion that the forces acting are proportional to their masses. These forces, as we have seen, are simply the weights of the bodies, so that, as the experimental result is true whatever the two bodies may be, we have the general law: *The masses of bodies are proportional to their weights.*

This gives us a means of comparing the masses of any two bodies. In every country a certain mass is taken as standard, and the mass of any other body is then compared with this standard, or with a copy of it, and in this way we get a knowledge of the actual mass of a body. For instance, in saying that the mass of a body is n pounds we mean that its mass (or weight) is equal to n times the mass (or weight) of a certain standard body kept at London.

22. Measurement of force. The weight of a unit mass is a force which may conveniently be taken to represent a unit force, and if this is done all other forces may be compared with this force. Thus a force of m pounds weight will mean a force m times as great as the weight of the standard pound.

This unit of force, however, is convenient rather than scientific, since it varies when the mass is moved about from place to place on the earth's surface. A unit pound mass will weigh more at London than at Washington; for instance, it will be found to extend or compress the spring of a spring balance more at London than at Washington, so that if the pound weight is taken as unit of force, we must remember that the unit of force is different at different parts of the earth's surface, and that a force of m pounds weight at London will be different from a force of m pounds weight at Washington.

For this reason a second unit of force is generally used for scientific purposes. This is called the absolute unit of force, and is chosen so as to be independent of position on the earth's surface. The second unit of force is defined to be one which produces unit acceleration in unit mass, whereas the former unit produced an acceleration equal to the value of gravity at the point. Thus, if g is the value of gravity, i.e. the acceleration of a body falling freely in a vacuum, the practical unit equals g times the absolute unit.

If unit force produces unit acceleration in unit mass, a force P will produce in mass m an acceleration $\frac{P}{m}$. Hence, denoting the acceleration by f, we have the fundamental equation

$$P = mf, \qquad (10)$$

which is the mathematical expression of Newton's second law. Here the force P must be measured in absolute units.

23. LAW III. *To every action there corresponds an equal and opposite reaction.*

It is a matter of common observation that a body A cannot exert force on a second body B without B at the same time exerting force on A. Thus an athlete trying to throw the hammer has to be on his guard that the hammer does not throw him; the force he exerts on the hammer is accompanied by the hammer exerting force on him, and he must steady himself against the effects of this force. So also when a gun fires a shot by exerting force on it, the shot exerts force on the gun, which is shown in the recoil of the gun. Thus all forces occur in pairs, which may conveniently be spoken of as action and reaction. The third law of motion tells us that the two forces which constitute such a pair are equal in magnitude and opposite in direction.

The meaning of the third law will be seen on examining the reaction corresponding to the forces which we have already used for illustrative purposes. The first illustration employed was that of a collision between two railway trucks. Truck A runs into truck B, exerting force on it and setting it in motion. The third law tells us that at the instant of collision B must exert force on A, this force being equal in amount to that exerted by A on B, and opposite in direction. The force of reaction will result in a change of velocity of A, lasting during the instant of collision only, and this may either merely check the motion of A, so that after the collision A proceeds with diminished velocity, or it may reverse the motion of A, so that truck A is seen to rebound from B and return in the direction in which it came.

After B has been set in motion we have imagined it to be acted on by three forces:

(*a*) the resistance of the air;
(*b*) the friction of the rails;
(*c*) the pressure of the rails on the flanges, turning the truck round a curve.

The reaction corresponding to the first force is a force exerted by the truck on the air in front of it and near it, tending to set the air in motion in the direction in which the car is moving; it is, in fact, this force which clears the air away from the space occupied at any instant by the truck.

The reaction corresponding to the second force is a force tending to drag the rails along with the truck. The rails are, of course, fastened down, so that this force cannot actually produce motion.

The reaction corresponding to the third force is a force exerted by the flanges of the truck wheels on the outer rail of the curve. The rails press on the flanges in a direction towards the center of the curve, so that the flanges press the rails outwards and away from the center of the curve. If the rails are not securely fixed, this pressure will cause them to move in the direction just mentioned; the rails will "spread" and the truck will run off the track.

In the illustration of the bullet we again had three forces operating on the bullet:

(*a*) the pressure of the powder before the shot leaves the barrel;

(*b*) the resistance of the air during the flight of the bullet;

(*c*) the weight of the bullet, dragging it downwards to the earth.

The reaction corresponding to the first force is the pressure of the shot driving the powder back. This in turn is transmitted to the gun, producing the "recoil" of the gun.

The reaction corresponding to the second force, just as with the truck, sets the air in motion, carving out a path for the bullet and producing the wind which accompanies its flight.

The reaction corresponding to the third force, the weight of the bullet, is more interesting, because we can obtain no direct evidence as to its existence. We merely infer from the principle of the uniformity of nature that as, in every case which has ever been tested, an action is accompanied by an equal and opposite reaction, therefore in this case, which is similar except in that it cannot be tested, we may suppose the action to be accompanied by its equal and opposite reaction.

The force which we can observe is the weight of the bullet, dragging it earthwards. This, we believe, represents a force exerted by the earth itself on the bullet, — the force of gravitation. This force must be accompanied by its reaction, so that the bullet must act on the earth with a force equal to the weight of the bullet, this force dragging the earth upwards to meet the bullet. The force exerted by the bullet on the earth is, by the third law, just as great as that exerted by the earth on the bullet. The upward acceleration produced in the earth by the bullet is, however, very much less than the downward acceleration produced in the bullet by the earth; for the force is jointly proportional to the mass and acceleration of the body acted on, and as the mass of the earth is very great compared with that of the bullet, its acceleration will be very small in comparison with that of the bullet.

Although for these reasons the acceleration produced in the earth by a bullet flying above it cannot be observed directly, yet in a very similar case the acceleration can be observed directly.

The moon, in describing a circle round the earth, is believed to be acted upon by the earth's gravitation in just the same way as the bullet. If no force acted on the moon, it would describe a straight line ; as it is, it is continually dragged down towards the earth, as we believe, by the same force of gravitation as the bullet. Just in the same way, then, the earth ought to experience an acceleration towards the moon. This acceleration is one which admits of astronomical observation.

24. In terms of ideas which have now been explained, the three laws may be restated as follows :

I. The normal state of a body is one of no acceleration. Departures from this normal state are produced by the action of force.

II. When a force acts so as to disturb the normal state of a body, the force is proportional to the product of the mass of the body by the acceleration produced.

III. Forces occur in pairs, every action being accompanied by a reaction, and each pair of forces being equal and opposite.

Frame of Reference

25. In stating the laws of motion we have spoken of the motion of a body without specifying the frame of reference relatively to which this motion is to be measured. In practice, motion is generally measured relatively to the surface of the earth, whereas Newton believed it to be possible to imagine a frame of reference actually fixed in space, and intended all motion to be measured relatively to this frame. Thus Newton's laws of motion apply to motion referred to axes fixed in space, whereas what we require to know, for all problems except those of astronomy, are the laws of motion referred to axes moving with the earth.

Let us first consider the effect of referring motion to a set of axes moving with uniform velocity in a straight line through space. A body under the action of no forces will have no acceleration in space, and, therefore, as the axes themselves have no acceleration in space, will have no acceleration relatively to the moving axes. Again, an acceleration has the same value whether referred to axes fixed in space or to the moving axes; for the acceleration referred to the moving axes is obtained by compounding the acceleration referred to axes fixed in space with that of the moving axes, and this latter acceleration is nil.

Thus it appears that the laws of motion retain exactly the same form when the motion is referred to axes which move in space, provided that these axes move with no acceleration.

This condition of no acceleration is not satisfied by a set of axes fixed in the earth's surface. A point on the earth's surface describes, on account of the earth's rotation, a circle about the earth's axis. If a is the radius of this circle, and v the velocity with which it is described, the point will have, by § 12, an acceleration $\dfrac{v^2}{a}$ towards the earth's axis of rotation. Thus a set of axes fixed in the earth's surface will have an acceleration of this amount, and this has to be borne in mind in applying the laws of motion. At a point on the equator $v = 46{,}510$ centimeters per second and $a = 637 \times 10^6$ centimeters, so that the acceleration is $\dfrac{v^2}{a} = 3.4$ centimeters per second per second. A body dropped at the equator will appear to have an acceleration of 978.1 centimeters per second per second, if the motion is referred to axes fixed in the earth; but will have a true acceleration of amount

$$978.1 + 3.4 = 981.5,$$

if the acceleration is referred to axes fixed in space.

This explains part of the reason why the force of gravity appears to vary from point to point at the earth's surface. The weight of a mass of one kilogramme will produce a certain extension of the spring of a spring balance at the North Pole. If taken to the equator, part of the weight goes towards producing the acceleration of the mass towards the earth's center, and it is only the remainder which extends the spring of the balance. The first part is the weight of about $3\frac{1}{2}$ grammes; the remainder is the weight of about $996\frac{1}{2}$ grammes. Thus we may say that, owing to the acceleration of the earth's surface towards its center, a mass of a kilogramme at the equator will appear to act on a spring balance with a force equal only to the earth's attraction on $996\frac{1}{2}$ grammes.

A second set of errors would be introduced by referring motion to axes in the earth's surface, these being caused by the change in the directions of the axes. For instance, if we use the laws of motion as though they were true for motion referred to axes fixed in the earth, and apply these laws to the fall of a stone, we shall find that the stone ought to strike the ground at a point vertically below that from which it is dropped. If we allow for the rotation of the earth, we shall find that the point at which the stone actually strikes must be somewhat to the east of the point vertically below that from which it started.

The errors introduced by treating motion on the earth as though it were motion with reference to axes fixed in space are, in general, either extremely small or very easily corrected. We shall, therefore, proceed at present by neglecting such errors altogether, and shall apply the laws of motion to motion with reference to the earth's surface.

Laws Applicable only to Motion of a Particle

26. There is a further limitation to the completeness of Newton's laws which ought to be noticed here. The second law would lead us to suppose that from a knowledge of the force acting on a body, and the mass of the body, we could deduce a definite acceleration of the body. But if the body is of finite size, the acceleration will be different at different points of the body; for example, we have seen that, as a consequence of the earth's rotation, the acceleration of a point at the equator of the earth is different from that of a point at the North Pole. Which acceleration, then, is it that is determined by the second law?

The answer to this difficulty is that the second law must be supposed to apply only to particles, i.e. to pieces of matter so small that they may be regarded as points. A moving particle has a single definite acceleration, just as a moving point has. From the law as applied to particles we shall be able to deduce laws which shall apply to bodies of finite size. This problem will be treated in a later chapter. Although, however, in strictness, the laws ought to be applied only to particles, it is obvious that there may be many problems in which we can treat bodies of finite size as particles without introducing any appreciable error. Such a case occurred when we discussed the flight of the bullet in § 18: the size of the bullet did not come into the question, as we could imagine all the points of the bullet to have the same acceleration. Many cases will occur in which a body of finite size may be treated as though it were a particle. In the next chapter we shall consider the application of forces to particles and to bodies which we find it is permissible to treat as particles.

CHAPTER III

FORCES ACTING ON A SINGLE PARTICLE

COMPOSITION AND RESOLUTION OF FORCES

27. The second law of motion enables us to find the acceleration produced when a particle of known mass is acted upon by a known force. In nature, however, forces do not generally act singly.

Consider, for example, the flight of the rifle bullet, discussed in § 18. While the bullet is in the air it is acted on by its weight and by the resistance of the air simultaneously. In addition to these, there may be a cross wind blowing and acting on the bullet with a horizontal pressure in a direction perpendicular to its motion. The resistance of the air retards the motion of the bullet, i.e. produces an acceleration in a direction opposite to that of the bullet's motion; the weight of the bullet drags it down, i.e. produces an acceleration towards the earth; while the cross wind will blow the bullet out of its course, i.e. will produce an acceleration in the direction in which the wind is blowing. Thus we can regard the three forces as each producing its own acceleration. The three accelerations can each be calculated from the second law of motion, and on compounding these three accelerations we shall have the resultant acceleration of the bullet. This resultant acceleration could have been produced by the action of a certain single force, so that we may say that this single force is equivalent, as regards the acceleration produced, to the combination of the three separate forces, or that the single force is the *resultant* of the three separate forces.

We must now put these ideas into exact mathematical form. As a preliminary, let us notice that a force has magnitude and direction, so that it can be represented by a straight line. We shall show that forces may be compounded according to the parallelogram law. Having proved this, it will follow that forces are vectors, and may be resolved and compounded according to the general rules already given.

28. Parallelogram of forces. THEOREM. *If two forces are represented in magnitude and direction by the two sides of a parallelogram, their resultant will be represented by the diagonal of the parallelogram.*

Let AB, AC represent the two forces, and let Ab, Ac represent the accelerations they would produce if they acted on any particle separately. Since, by the second law of motion, the acceleration is proportional to the force, we must have

$$Ab : Ac = AB : AC.$$

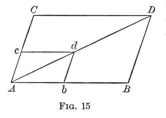

FIG. 15

Construct the parallelograms $Abdc$, $ABDC$. On account of the proportion just obtained, the two parallelograms will be similar, so that AdD will be a straight line, and we shall have

$$AD : Ad = AB : Ab.$$

But Ad, the diagonal of the parallelogram of edges Ab, Ac, represents the resultant acceleration. Since AB represents the force necessary to produce acceleration Ab, it follows from the proportion just obtained that AD will represent the force necessary to produce acceleration Ad. In other words, the acceleration of the particle is the same as if it were acted on by a single force represented by AD. Thus AD represents the resultant of the forces AB, AC.

It now follows that force is a vector, so that forces can be compounded according to the laws explained in §§ 14–16.

PARTICLE IN EQUILIBRIUM

29. In statics we are concerned only with particles, or systems of particles, at rest. The resultant force on each particle must accordingly be nil. It is therefore important to consider cases in which the resultant of a system of forces is nil.

CONDITIONS FOR EQUILIBRIUM

30. Polygon of forces. THEOREM. *If forces acting on a particle are represented by straight lines, the particle will be in equilibrium if the polygon formed by taking all these straight lines as edges is a closed polygon, i.e. if after putting all the lines end to end we come back to the starting point.*

Let AB, BC, CD, $\cdots$, MN represent in magnitude and direction any number of forces which act simultaneously on a particle. Since force is a vector the forces represented by AB and BC are equivalent to a single force represented by AC, and may therefore be replaced by this force.

Thus the system of forces may now be supposed to be forces represented by the lines AC, CD, $\cdots$, MN. The first two of these may again be replaced by a single force represented by AD, so that the system is reduced to forces represented by AD, DE, $\cdots$, MN. We can proceed

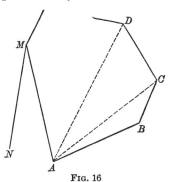

FIG. 16

in this way until we are left with only a single force represented by AN. This therefore represents the resultant of all the forces.

If the polygon is a closed polygon, the points A and N coincide, so that the resultant force represented by AN vanishes and the particle is in equilibrium. Conversely, if the particle is in equilibrium, AN vanishes, so that the polygon is a closed polygon. Thus the condition for equilibrium expressed by the theorem just proved is *necessary and sufficient*, — necessary because the condition must be satisfied if the particle is to be in equilibrium, and sufficient because equilibrium is insured as soon as the condition is satisfied.

31. Triangle of forces. If there are only three forces, the theorem reduces to a simpler theorem known as the *triangle of forces*. This is as follows:

THEOREM. *If a particle is acted on by three forces represented by straight lines, the particle will be in equilibrium if these three straight lines placed end to end form the sides of a triangle.*

40 FORCES ACTING ON A SINGLE PARTICLE

As this is a particular case of the *polygon of forces* no separate proof is needed. As before, the converse is also true, so that the condition is a necessary and sufficient condition for equilibrium.

When there are only three forces acting, the condition for equilibrium can be expressed in a still simpler form:

32. LAMI'S THEOREM. *When a particle is acted on by three forces, the necessary and sufficient condition for equilibrium is that the three forces shall be in one plane and that each force shall be proportional to the sine of the angle between the other two.*

Suppose that a particle is acted on by three forces P, Q, R. The necessary and sufficient condition for equilibrium is that we can

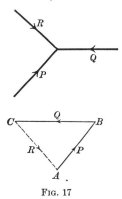

FIG. 17

form a triangle by placing end to end three lines which represent the forces P, Q, R in magnitude and direction.

Let us begin by taking AB to represent P, and placing against it at B a line BC to represent Q. Then CA must represent R, if the conditions for equilibrium are to be satisfied. Thus the three forces must be in one plane, namely, the plane parallel to ABC through the point of action of the forces.

Assume that there is equilibrium, so that the three forces are represented by the sides of the triangle ABC. Let us denote the angles of the triangle as usual by A, B, C, and its sides by a, b, c. Then, from a known property of the triangle,
$$\frac{a}{\sin A} = \frac{b}{\sin B} = \frac{c}{\sin C}.$$

By our construction, however, a, b, c are proportional to the magnitudes of the forces: we have
$$\frac{c}{P} = \frac{b}{R} = \frac{a}{Q}.$$

Thus
$$\frac{P}{\sin C} = \frac{Q}{\sin A} = \frac{R}{\sin B}.$$

CONDITIONS FOR EQUILIBRIUM 41

If pq denote the angle between the lines of action of the forces P and Q, we have $pq = \pi - B$, so that $\sin B = \sin pq$, and hence

$$\frac{P}{\sin qr} = \frac{Q}{\sin rp} = \frac{R}{\sin pq}. \quad (11)$$

The converse is true because, if the relation (11) is satisfied and if the lines of action of the forces are in one plane, we can construct a triangle of which the sides will represent the forces P, Q, R, so that there is equilibrium.

33. Analytical conditions for equilibrium. Expressed in an analytical form, the condition for equilibrium is that the resultant of all the forces acting shall be zero. If the individual forces are known, the resultant force can be obtained at once from the rules for the composition of vectors, which have already been given in §§ 14–16.

If the forces all act in a plane, let their magnitudes be $R_1, R_2, \cdots, R_n$, and let their lines of action make angles $\epsilon_1, \epsilon_2, \cdots, \epsilon_n$ with the axis of x. Then the resultant has components X, Y, where (cf. § 14)

$$X = R_1 \cos \epsilon_1 + R_2 \cos \epsilon_2 + \cdots,$$
$$Y = R_1 \sin \epsilon_1 + R_2 \sin \epsilon_2 + \cdots.$$

The magnitude of the resultant is $\sqrt{X^2 + Y^2}$, and this vanishes only if X and Y vanish separately. Thus the condition for equilibrium is that the components along the two axes shall vanish separately, i.e. that the sum of the components of the separate forces acting shall vanish when resolved along each axis.

Similarly, if the forces do not all act in one plane, the condition for equilibrium is that the sums of the components along three axes in space shall vanish separately.

EXAMPLES

1. Forces of 12 and 8 pounds weight act in two directions which are at right angles. Find the magnitude of their resultant.

2. Three forces each equal to F act along three rectangular axes. Find their resultant.

3. The resultant of two forces P_1 and P_2 acting at right angles is R; if P_1 and P_2 be each increased by 3 pounds, R is increased by 4 pounds and is now equal to the sum of the original values of P_1 and P_2. Find P_1 and P_2.

4. Forces acting at a point O are represented by $OA, OB, OC, \cdots, ON$. Show that if they are in equilibrium, O is the centroid of the points $A, B, C, \cdots, N$.

5. $ABCDEF$ is a regular hexagon. Find the resultant of the forces represented by AB, AC, AD, AE, AF.

6. $ABCDEF$ is a regular hexagon. Show that the resultant of forces represented by $AB, 2AC, 3AD, 4AE, 5AF$ is represented by $\sqrt{351} \cdot AB$, and find its direction.

7. ABC is a triangle, and P is any point in BC. If PQ represent the resultant of the forces represented by AP, PB, BC, show that the locus of Q is a straight line parallel to BC.

Types of Forces

Weight of a Particle

34. The weight of a particle acts always vertically downward; for at a given place on the earth it is found that the weights of all particles act in parallel directions, and this direction is called the *vertical* at the place in question. The weight is the gravitational force with which the particle is attracted by the earth, except for a small correction which has to be introduced on account of the fact that axes fixed in the earth do not move without acceleration. This correction we shall not discuss here. When the weight of a body is said to be W, it is meant that to keep the body at rest relatively to the earth's surface a force W is required to act vertically upward.

Tension of a String

35. A string or rope supplies a convenient means of applying force to a body, and this force is spoken of as the tension of the string. Let $ABCD \cdots$ be the string, and let P be a particle tied to the string at its end. Let the divisions AB, $BC, \cdots$ of the string be so small that each may be regarded as a particle.

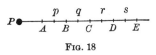

Fig. 18

There will be three forces acting on any particle such as BC: first, its weight; second, a force exerted on BC by the particle CD of the string; and third, a force exerted on BC by the particle AB.

Generally the weight of a string is very slight compared with the other weights in the problem. It is therefore convenient to regard a string as having no weight at all. In this case there are only two forces acting on the particle AB, so that for equilibrium these must be equal and opposite.

36. Flexibility. A string is said to be *perfectly flexible* when the force exerted by one particle on the next is in the direction joining the two particles. Thus, if the string now under discussion is perfectly flexible and weightless, the forces acting on the particle BC are along the directions pq, qr. To hold BC in equilibrium these must be equal in magnitude. Let T be taken as the magnitude of each. Also the two forces must be in opposite directions, so that pqr must be a straight line.

Since action and reaction, by the third law, are equal and opposite, the force exerted by BC on CD must also be T in the direction qr. This must, for equilibrium, be equal and opposite to the force exerted by DE on CD. This force must accordingly be of amount T, and qrs must be a straight line.

We can continue in this way, and find that all the particles must lie in a straight line $pqrs\cdots$, and that each acts on the next with the same force T. Also the particle A at the end of the string acts on P with this same force T in the direction of the string. The force T is called the tension. Thus we have the following:

The tension of a string at any point P is defined as the force with which the particle of the string on the one side of P acts on the particle on the other side of P.

The tension is the same in magnitude and direction at every point of a perfectly flexible, weightless string acted on by no external forces.

Hence it follows that

A perfectly flexible, weightless string acted on by no external forces must be in a straight line when in equilibrium.

If the tension vanishes, there is equilibrium whatever the direction of the elements of length $pq, qr, \cdots$. When the tension vanishes the string is said to be *unstretched*. Clearly an unstretched string can rest in equilibrium in any shape.

It will be proved later that when a perfectly flexible, weightless string passes over a smooth peg or pulley, the tension has the same magnitude at all points of the string, and at points of contact with the peg or pulley its direction is along the tangent to the peg or pulley.

37. If the string is not absolutely weightless, but is very light, any particle such as q will be acted on by three forces, — its weight vertically down, and the two forces from the adjacent particles acting along pq, rq. By Lami's theorem, each force must be proportional to the sine of the angle between the remaining two forces. Since the weight is small, sin pqr must be small; i.e. pqr must be very nearly a straight line. The line cannot be perfectly straight, however, unless the string is absolutely weightless; thus in a real string there must always be a certain "sag," due to the weight of the string, although this sag may be so slight as to be imperceptible.

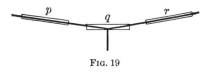

FIG. 19

38. Extensible and inextensible strings. The tension, as will have been seen, is a force acting at every point of the string, and tending to stretch the string in the direction of its length. The string either may or may not yield to this tendency to stretch. A string which stretches under tension is called *extensible;* a string which does not stretch at all, or which stretches so little that the amount of stretching is inappreciable, is called *inextensible.*

Thus an inextensible string remains of the same length whatever tension is applied to it, while the length of an extensible string depends on its tension.

In 1660 Hooke discovered a law which expressed a relation between the tension and the amount of stretching in a string: the one is proportional to the other.

DEFINITION. *The length of a string when the tension is zero is called the "natural length" of the string.*

DEFINITION. *The amount by which the length of a stretched string exceeds the natural length of the same string is called the "extension" of the string.*

HOOKE'S LAW. *The tension of a string is proportional to the extension.*

STRINGS

Although Hooke discovered this law in 1660, he did not publish it until 1676, and then only in the form of the anagram *ceiinosssttuv*.

In 1678 he explained that the letters of the anagram were those of the Latin words *ut tensio sic vis*, — "the power of any spring is in the same proportion with the tension thereof." By tension (*tensio*) Hooke meant the quantity which we have called the "extension"; by the power (*vis*) he meant the force tending to stretch the spring, i.e. the tension.

39. Hooke's law only enables us to compare the extensions produced by different tensions. To find the actual extension produced by a given tension we must know that produced by some other tension for comparison.

DEFINITION. *The force required to stretch a string to double its natural length is called the modulus of elasticity of the string.*

Thus if a is the natural length of a string, and λ the modulus of elasticity, we know that a tension λ produces an extension a, so that a tension T produces an extension Ta/λ

When we say that a string is inextensible, we mean that λ is so large that the extension Ta/λ may be neglected.

Hooke's law only holds within certain limits. If we go on increasing the tension in a string indefinitely, we find that, after a certain limit is passed, Hooke's law ceases to be true, and when a certain still greater tension is reached the string breaks in two parts.

EXAMPLES

1. A weight W hangs by a string and is pushed aside by a horizontal force until the string makes an angle of 45° with the vertical. Find the horizontal force and the tension of the string.

2. A weight suspended by a string is pushed sideways by a horizontal force. Show that as it is pushed farther from its position of rest, in which the string is vertical, the tension continually increases.

3. A weight of 100 pounds is suspended by two strings which make angles of 60° with the vertical. Find their tensions.

4. A weight of 30 pounds is tied to two extensible strings of natural length 2 feet, modulus of elasticity 100 pounds, and the other ends of the strings are tied to two points at a horizontal distance 4 feet apart. Find the position in which the weight can rest in equilibrium.

5. A weight W is suspended by three equal strings of length l from hooks which are the vertices of a horizontal equilateral triangle of side a. Find the tensions of the strings.

Reaction between Two Bodies

40. A second way in which force can be applied to a particle is by the pressure between the particle and the surface of a solid body. Such a force is commonly spoken of as a *reaction*.

A body standing on the floor of a room is acted on by its weight acting downwards, but is kept at rest by the action of a second force acting upwards from the floor; this is the reaction between the body and the floor. Clearly, in order that the body may rest in equilibrium, the reaction in this case must be equal to the weight of the body and must act vertically.

FRICTION

41. Imagine a small body standing on a plane of which the slope can be varied, such as the lid of a desk. If the plane is held horizontally, the body can stand at rest as already described. Let the plane be gradually tilted, and it will be found that as soon as the tilting reaches a certain angle the body will begin to slide down the plane. The angle at which sliding first occurs is found to be different for different pairs of substances; thus for wood sliding on wood it may vary from 10° to 25°, for iron on wood it varies from 10° to 30°, while for iron sliding on iron it is only about 10° or 15°.

When two substances are such that this angle is zero, — i.e. such that one can only rest on the other when the surface of contact is perfectly horizontal, — then the contact between them is said to be *perfectly smooth*. The nearest approximation to a perfectly smooth contact which we experience in actual life is probably that of steel on ice, as in skating.

It is found that the angle to which a plane made of one substance has to be tilted before a second substance begins to slide on it is independent both of the amount of the second substance and of the area in contact. Thus the angle depends only on the nature of the two substances in contact.

FRICTION

Further, when the two bodies are pressed together in any way it is found that the direction of the reaction can make any angle up to a certain limiting angle with the normal to the plane of separation without sliding taking place, but that as soon as this angle is reached sliding takes place. This angle is known as the *angle of friction*. It is clearly the same as the angle through which the plane before considered can be tilted, for the angle between the normal to the plane and the direction of the reaction (i.e. the vertical) is simply the slope of the plane.

42. In any case in which frictional forces act, let R denote the normal component of the reaction, and let F denote the component in the plane of the contact which is caused by friction. When slipping is just about to occur, the resultant must make an angle ϵ with the normal, where ϵ is the angle of friction. Thus, if S denotes the whole reaction, we must have

$$R = S \cos \epsilon, \quad F = S \sin \epsilon,$$

and hence $F = R \tan \epsilon.$

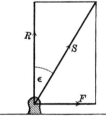

Fig. 20

The quantity $\tan \epsilon$ is called the *coefficient of friction* and is denoted by the single letter μ. Then, when slipping is just about to take place, we have
$$F = \mu R.$$

It must be clearly understood that this equation gives the true value of the frictional force *only when slipping is just about to take place*. It sets an upper limit to the value of the frictional force, but does not give the actual value of this force unless we know that the system is on the verge of sliding.

43. Consider, for instance, the experiment already discussed, in which a particle is placed on a horizontal plane which is gradually tilted up. When the plane is horizontal the particle is at rest, acted on only by gravity and the reaction with the plane. Thus the reaction is vertical, so that here $F = 0$. Consider next the state of things when the plane makes an angle α with the horizon.

48 FORCES ACTING ON A SINGLE PARTICLE

If slipping does not take place, the particle is in equilibrium under its weight W and the reaction between it and the plane. Thus the reaction must consist of a vertical force W. We can resolve this into components $W \cos \alpha$ and $W \sin \alpha$ perpendicular to and up the plane. The former is the normal component of the reaction, the latter is the frictional component. Thus in the notation already used we have

$$R = W \cos \alpha,$$
$$F = W \sin \alpha,$$

so that in this case we have

$$F = R \tan \alpha.$$

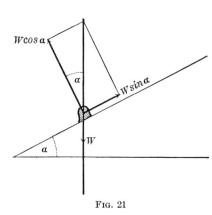

FIG. 21

As α increases, F and F/R both increase until, when α reaches the value ϵ, F/R reaches its limiting value μ, or $\tan \epsilon$, and after this slipping takes place.

EXAMPLES

1. A mass of 100 pounds placed on a rough horizontal plane is on the point of starting into motion when acted on horizontally by a force equal to the weight of 100 pounds. Find the angle of friction.

2. A body placed on an inclined plane which makes an angle of 30° with the horizontal is just on the point of moving down the plane when acted on by a horizontal force equal to the weight of the body. Find the coefficient of friction.

3. A man capable of exerting a pull of 200 pounds tries to drag a mass of 700 pounds over a horizontal road (coefficient of friction $\frac{1}{3}$). To help him, the chain from a crane is attached to the mass, the chain hanging vertically. How much tension must there be in the chain before the man can move the block?

4. An insect tries to crawl up the inside of a hemispherical bowl of radius a. How high can it get, if the coefficient of friction between its feet and the bowl is $\frac{1}{3}$?

5. A man trying to push a block of stone over ice pushes horizontally and finds that just as soon as the stone begins to move his feet begin to slip. Show that if he pushes upwards on the stone he will get it along without difficulty, but that if he pushes downwards he cannot possibly move it.

6. A smooth pulley is placed at the edge of a horizontal plane. A string passes over it, having at one end a weight w hanging freely, and at the other

ILLUSTRATIVE EXAMPLES 49

end a weight W resting on the plane. If the coefficient of friction μ is so large that motion does not take place, find through what angle the plane must be tilted before motion begins.

7. A tourist of mass M is roped to a guide of mass m on the side of a mountain, the side of which may be taken to be an inverted hemisphere. The length of the rope subtends an angle α at the center of the mountain, and the rope is not supposed to touch the mountain at any point. If the coefficient of friction between either man and the mountain is μ, how far can the tourist venture down the side of the mountain before both he and the guide fall to the bottom?

ILLUSTRATIVE EXAMPLES

1. *A heavy particle C rests on a smooth inclined plane, being supported by two strings of lengths l_1, l_2, which are attached to two points A, B in the plane, these points being in the same horizontal line and at a distance h apart. Find the tensions of the strings and the reaction with the plane.*

Let W be the weight of the particle and let α be the inclination of the plane to the horizon. The particle is in equilibrium, being acted on by the following forces:

(*a*) Its weight W, which acts vertically downwards.

(*b*) The reaction between the particle and the plane. Since the plane is smooth, this reaction acts at right angles to the plane. Let the amount of the reaction be R.

(*c*) The two tensions of which the amount is required. Let the amounts of these be denoted by T_1, T_2.

Since these four forces produce equilibrium, the sum of their resolved parts in any direction must vanish. The two tensions have no resolved parts at right angles to the plane; hence, by resolving at right angles to the plane, we shall get

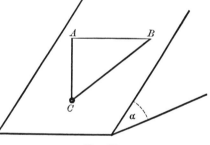

FIG. 22

an equation in which only two of the forces are involved.

The resolved part of the weight at right angles to the plane is $W\cos\alpha$. The reaction is wholly at right angles to the plane; hence the equation for which we are in search is

$$R - W\cos\alpha = 0.$$

This gives us the amount of the reaction at once.

Let us now consider the resolved parts of the forces in the inclined plane. The only forces which have components in this plane are the following:

(*a*) The weight, of which the component is $W\sin\alpha$, down the line of greatest slope.

(*b*) The tensions of the strings, which are entirely in the plane and which act along the strings CA, CB.

50 FORCES ACTING ON A SINGLE PARTICLE

The three forces $W\sin\alpha$, T_1, and T_2 must be in equilibrium; hence, by Lami's theorem, each must be proportional to the sine of the angle between the other two.

In fig. 23, CD, CA, CB are the lines of action of these three forces. The line CD, being the line of greatest slope through C, is at right angles to the line AB, which we are told is horizontal. Thus if DC is produced to meet AB in P, the angle APC is a right angle. Hence

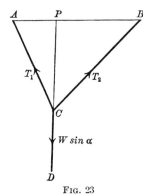

FIG. 23

$$\sin ACD = \sin ACP = \cos CAP,$$

and similarly

$$\sin BCD = \sin BCP = \cos CBP.$$

By Lami's theorem we have

$$\frac{W\sin\alpha}{\sin ACB} = \frac{T_1}{\sin BCD} = \frac{T_2}{\sin ACD}.$$

From the relations just obtained we can obtain these ratios in terms of the angles of the triangle ABC. We have

$$\frac{W\sin\alpha}{\sin C} = \frac{T_1}{\cos B} = \frac{T_2}{\cos A}.$$

We can now, if required, express $\cos A$, $\cos B$, and $\sin C$ in terms of the sides l_1, l_2, and h of the triangle, by means of the ordinary formulæ of trigonometry.

The student is advised to examine for himself the form assumed by the result in the two special cases

(a) $A = B = 0$, in which ACB are in a straight line;

(b) $C = 0$, $A = B = \dfrac{\pi}{2}$, in which the strings are parallel.

2. *Find the direction and magnitude of the smallest force which will start a body resting on an inclined plane into motion down the plane.*

Let α be the angle of the plane, and μ the coefficient of friction between it and the body to be moved. Let W be the weight of the body, and let a force F be applied in a direction making an angle θ with the line of greatest slope down the plane, this force being supposed to be just sufficient to move the body.

The forces acting on the body consist of

(a) its weight W;

(b) the applied force F;

(c) the reaction with the plane.

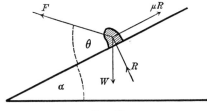

FIG. 24

Let the last force be resolved into two components along and perpendicular to the plane. Taking the latter to be R, the former will be μR acting up the plane, for, by hypothesis, the body is on the point of motion down the plane.

ILLUSTRATIVE EXAMPLES

The resultant of all these forces vanishes, so that the sum of their components in any direction vanishes. Resolving normal to the plane, we obtain

$$R + F \sin \theta - W \cos \alpha = 0,$$

and resolving along the plane,

$$F \cos \theta + W \sin \alpha - \mu R = 0.$$

Eliminating the unknown reaction R, we obtain

$$F(\mu \sin \theta + \cos \theta) - W(\mu \cos \alpha - \sin \alpha) = 0,$$

so that

$$F = \frac{W(\mu \cos \alpha - \sin \alpha)}{\mu \sin \theta + \cos \theta}.$$

Replacing μ by $\tan \epsilon$, we obtain

$$F = \frac{W(\cos \alpha \tan \epsilon - \sin \alpha)}{\cos \theta + \tan \epsilon \sin \theta} = \frac{W \sin(\epsilon - \alpha)}{\cos(\theta - \epsilon)}.$$

The value of F is a minimum when $\cos(\theta - \epsilon)$ is a maximum, and this occurs when $\cos(\theta - \epsilon) = 1$; i.e. when $\theta = \epsilon$. In this case the value of F is

$$F = W \sin(\epsilon - \alpha).$$

Thus this is the smallest force by which motion can be produced, and it must act so as to make an angle θ with the plane equal to the angle of friction ϵ. Since, by hypothesis, the weight rests without slipping when no force is applied, the angle ϵ must be greater than α. Thus the direction of the force F must always be inclined in an upward direction. The function performed by the force F is twofold: it supports part of the weight of the body (through its component normal to the plane), and so lessens the amount of friction to be overcome; and it also supplies (through its component in the inclined plane) the motive power for overcoming the frictional resistance. When these two parts of the force are balanced in the most advantageous way the value of F is a minimum, and this, as we have proved, occurs when $\theta = \epsilon$.

An interesting and instructive solution of this problem can also be obtained geometrically. For equilibrium, the three forces already enumerated must satisfy the condition of forming a triangle of forces.

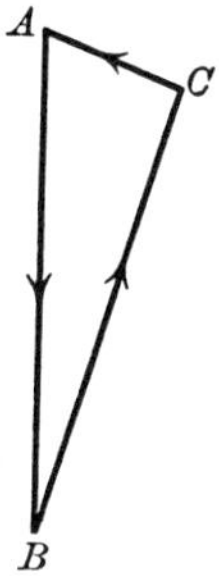

Let AB represent the weight and BC the reaction between the mass and the plane, then CA must represent the applied force F. If the body is on the point of motion, the reaction must make an angle ϵ with the normal to the plane, so that the angle ABC must be $\epsilon - \alpha$. Thus the line BC is fixed in direction, and the problem is that of finding the direction and magnitude of AC, when the length AC is a minimum. Obviously the minimum occurs when AC is perpendicular to BC, so that AC must be in a direction making an angle $\epsilon - \alpha$ with the horizontal, as already found, and since $AC = AB \sin ABC = AB \sin(\epsilon - \alpha)$, the magnitude of the force required will be $W \sin(\epsilon - \alpha)$.

Fig. 25

52 FORCES ACTING ON A SINGLE PARTICLE

3. *A particle is tied to an elastic string, the other end of which is fixed at a point in a rough inclined plane. Find the region of the plane within which the particle can rest.*

The forces acting on the particle are

(a) its weight, say W, vertically down;
(b) the tension of the string;
(c) the reaction with the rough plane.

Let the natural length of the string be l, the modulus of elasticity λ; then when the actual length of the string is r, where r is greater than l, the tension is $\dfrac{(r-l)\lambda}{l}$.

Let α be the inclination of the plane, and let μ be the coefficient of friction between the plane and the particle. Let the reaction with the plane be resolved into a normal component R, and a component F along the plane. The condition that the particle can remain at rest is that F shall be less than μR.

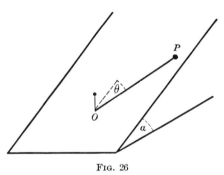

FIG. 26

Resolving at right angles to the plane, the only forces which have components in this direction are found to be the weight of the particle and its reaction with the plane. Thus
$$R - W \cos \alpha = 0.$$

Consider the equilibrium of the particle when at some point P, distant $r\,(>l)$ from O. The components in the inclined plane, of the forces acting on it, are

(a) $W \sin \alpha$ down the line of greatest slope through P;
(b) the tension $\dfrac{(r-l)\lambda}{l}$ along PO;
(c) the frictional component of the reaction, which we have called F.

Let OP make an angle θ with the line of greatest slope in the plane. Then, since the resultant of the first two forces must be of magnitude F, we must have

$$F^2 = W^2 \sin^2 \alpha + \frac{(r-l)^2 \lambda^2}{l^2} + \frac{2(r-l)\lambda}{l} W \sin \alpha \cos \theta,$$

giving the magnitude of the frictional force required to maintain equilibrium. If the particle is on the point of motion, $F = \mu R = \mu W \cos \alpha$, so that

$$W^2 (\sin^2 \alpha - \mu^2 \cos^2 \alpha) + \frac{(r-l)^2 \lambda^2}{l^2} + \frac{2(r-l)\lambda}{l} W \sin \alpha \cos \theta = 0. \qquad (a)$$

Since r, θ are polar coördinates of the point P, equation (a) is the polar equation of the boundary of the region within which the particle can remain at rest.

ILLUSTRATIVE EXAMPLES 53

The equation is most easily interpreted by noticing that if $r - l$ is replaced by r, the equation becomes

$$W^2(\sin^2\alpha - \mu^2\cos^2\alpha) + \frac{\lambda^2}{l^2}r^2 + 2\frac{\lambda}{l}W\sin\alpha \cdot r\cos\theta = 0, \qquad (b)$$

which is the polar equation of a circle. Thus the original locus represented by equation (a) can be drawn by first drawing the circle represented by equation (b), and then producing each radius vector through the origin to a distance l beyond the circumference of this circle.

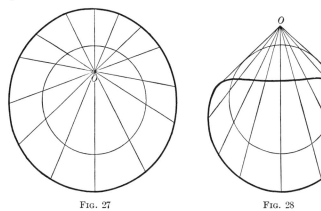

FIG. 27   FIG. 28

The same result can be obtained by a geometrical treatment of the problem. The particle is acted on by only three forces in the plane on which it rests, so that lines parallel and proportional to these forces must form a triangle of forces.

In fig. 29 let OP be the string, and let AP be a length l measured off from P, so that AO is the extension $r-l$ of the string. The tension is always proportional to AO and acts along AO. Let us then agree that in the triangle of forces the tension shall be represented by the actual line AO. On the same scale let the component of the weight, $W\sin\alpha$, be represented by the line OG, the direction of this being, of course, down the line of greatest slope through O. Then AOG must be the triangle of forces, so that GA must represent the frictional reaction between the particle and the plane. The maximum value possible for this is $\mu W\cos\alpha$, so that if slipping is just about to occur, GA will represent a force $\mu W\cos\alpha$. Thus corresponding to a position

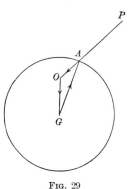

FIG. 29

of P in which slipping is just about to occur, the positions of A are such that GA represents the constant force $\mu W\cos\alpha$,— in other words, the locus of A is a circle of center G. This leads at once to the construction previously obtained.

54 FORCES ACTING ON A SINGLE PARTICLE

The region in which equilibrium is possible assumes two different forms according as the angle of the inclined plane α is less or greater than the angle of friction ϵ. In the former case the region of equilibrium is of the kind represented in fig. 27. On passing the value $\alpha = \epsilon$ the circle used in the construction passes through the point O, and for values of α greater than ϵ the region of equilibrium becomes an area of the kind drawn in fig. 28. On passing through the value $\alpha = \epsilon$ a sudden change takes place in the shape of the region of stability. For values of α which are greater, by however little, than ϵ, a circle of radius l, center O, is entirely outside the region of stability; while for values of α which are smaller, by however little, than ϵ, this circle is inclosed within the region of equilibrium. Clearly this circle maps out the region within which the weight can rest with the string unstretched, and this will be one of equilibrium or not according as $\alpha <$ or $> \epsilon$.

Thus this circle falls inside or outside the region of equilibrium in the way predicted by analysis. At the same time we could not have been sure, without a separate investigation, that the result given by analysis would be accurate as regards the region within a distance l of O. For the analysis began by assuming the string to be stretched, and so had no application except to the region at a distance greater than l from O.

4. *Two weights w, w' rest on a smooth sphere, being supported by a string which passes through a smooth ring at O, a point vertically above the center of the sphere. Find the configuration of equilibrium.*

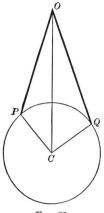

Fig. 30

Let P, Q, in fig. 30, be the positions of the two weights in a configuration of equilibrium. The weight w at P is acted on by the following forces:

(a) its weight w vertically downwards;

(b) the tension of the string along PO;

(c) the reaction between the sphere and the weight. Since the sphere is supposed smooth, the direction of this reaction is at right angles to the plane of contact between the particle and the sphere; i.e. along CP.

The three forces acting on the particle P are accordingly parallel to the three sides of the triangle OPC. Thus the triangle OPC may be regarded as a triangle of forces for these forces, so that the magnitudes of the forces must be proportional to the sides of this triangle. Denoting the tension and reaction by T and R, we obtain

$$\frac{w}{OC} = \frac{T}{OP} = \frac{R}{CP}. \qquad (a)$$

In the same way the triangle OCQ may be regarded as a triangle of forces for the particle Q. (This triangle does not represent force on the same scale as the former triangle OCP; for in the former case OC represented a weight w, whereas it now represents a weight w'.)

ILLUSTRATIVE EXAMPLES 55

From this second triangle of forces we obtain

$$\frac{w'}{OC} = \frac{T'}{OQ} = \frac{R'}{CQ}, \tag{b}$$

where T', R' represent the tension and reaction acting on Q.

Since the ring at O is supposed to be smooth, the tension in the string POQ is the same at all points. Thus $T = T'$. We now obtain, from equations (a) and (b),

$$w \cdot OP = w' \cdot OQ, \tag{c}$$

since each product is equal to $T \cdot OC$. If l is the whole length of the string, we have

$$\frac{w}{OQ} = \frac{w'}{OP} = \frac{w + w'}{l}, \tag{d}$$

showing that the string arranges itself so that it is divided by the ring at O in the inverse ratio of the two weights. We notice also from equations (a) and (b) that

$$\frac{R}{w} = \frac{R'}{w'},$$

for each ratio is that of the radius of the sphere to OC. Thus the reactions are in the direct ratio of the weights.

If the string is inextensible, the length l is known, so that equations (d) determine the lengths OP, OQ completely. Suppose, however, that the string is an extensible string, say of natural length a and modulus λ. Then, instead of l being a known quantity, we have one additional equation between unknown quantities, namely

$$T = \frac{l - a}{a} \lambda.$$

Remembering that equation (c) gives two values for the quantity $T \cdot OC$, we have

$$T \cdot OC = \frac{l-a}{a} \lambda \cdot OC = w \cdot OP = w' \cdot OQ$$

$$= \frac{OP}{\dfrac{1}{w}} = \frac{OQ}{\dfrac{1}{w'}} = \frac{l}{\dfrac{1}{w} + \dfrac{1}{w'}},$$

so that
$$\frac{l-a}{a} \lambda \cdot OC = \frac{l}{\dfrac{1}{w} + \dfrac{1}{w'}}.$$

This equation determines the value of l, and having found this we proceed as before.

5. *A weight W is supported by strings of which the tensions are $T_1, T_2, \cdots T_n$. The strings do not hang vertically, but the angles between the different pairs of strings are known, being ϵ_{12}, ϵ_{13}, etc. Find the weight W in terms of the tensions and of these angles.*

56 FORCES ACTING ON A SINGLE PARTICLE

Clearly the weight is equal to the resultant of the tensions $T_1, T_2, \cdots T_n$.

Let us take any three rectangular axes in space. Let the direction cosines of the first string be l_1, m_1, n_1; let those of the second string be l_2, m_2, n_2; and so on. Then, resolved along the axes, the components of the first tension will be

$$l_1 T_1, \quad m_1 T_1, \quad n_1 T_1.$$

The components of the other tensions are similar expressions, so that if X, Y, Z denote the three components of the resultant, we have

$$X = l_1 T_1 + l_2 T_2 + \cdots + l_n T_n,$$
$$Y = m_1 T_1 + m_2 T_2 + \cdots + m_n T_n,$$
$$Z = n_1 T_1 + n_2 T_2 + \cdots + n_n T_n.$$

Since the magnitude of the resultant is equal to W, we have

$$W^2 = X^2 + Y^2 + Z^2$$
$$= (l_1 T_1 + l_2 T_2 + \cdots + l_n T_n)^2 + (m_1 T_1 + m_2 T_2 + \cdots + m_n T_n)^2$$
$$\quad + (n_1 T_1 + n_2 T_2 + \cdots + n_n T_n)^2$$
$$= T_1^2 (l_1^2 + m_1^2 + n_1^2) + \cdots + 2\, T_1 T_2 (l_1 l_2 + m_1 m_2 + n_1 n_2) + \cdots$$
$$= T_1^2 + T_2^2 + \cdots + 2\, T_1 T_2 \cos \epsilon_{12} + \cdots,$$

which gives the result required.

GENERAL EXAMPLES

1. ABC is a triangle, with a right angle at A; AD is the perpendicular on BC. Prove that the resultant of forces, $\dfrac{1}{AB}$ acting along AB and $\dfrac{1}{AC}$ acting along AC, is $\dfrac{1}{AD}$ acting along AD.

2. At a point O there acts a force P, whose line of action is in the plane determined by two lines OA, OB, meeting at O. The resolved part of P in the direction OA is represented in magnitude and direction by OX, that in the direction OB by OY. Show that the force P is represented in magnitude by the diameter of the circle OXY, and find its direction.

3. Forces $P_1, P_2, \cdots, P_n$ acting in one plane at a point O are in equilibrium. Any transversal cuts their lines of action in points $L_1, L_2, \cdots, L_n$; and a length OL_i is considered positive when the direction from O to L_i is the same as to P_i. Prove that $\Sigma P_i / OL_i = 0$.

4. A body is sustained on a smooth inclined plane by two forces, each equal to half the weight, the one acting horizontally, and the other along the plane. Find the inclination of the plane.

5. The angle of a smooth inclined plane is $30°$, and a force P acting horizontally sustains a body. In what other direction can P act and support the body? Compare the pressure upon the plane in the two cases.

EXAMPLES

6. Two smooth planes, whose inclinations are α and β, meet in a horizontal line AB. At a point in AB is a small smooth ring through which passes a string with a weight at either end, resting one on each of the given planes, and in the same vertical plane with the ring. If the weights are in the equilibrium, find the tension of the string and the ratio of the weights.

7. Two smooth rings of weights W_1 and W_2 are connected by a string and rest in equilibrium on the convex side of a circular wire in the vertical plane. Show that, if the string subtends the angle α at the center of the circle, the angle of inclination θ of the string to the vertical is given by

$$\tan \theta = \frac{W_1 + W_2}{W_1 - W_2} \cot \frac{\alpha}{2}.$$

8. Two weights rest on a rough inclined plane and are connected by a string which passes over a smooth peg in the plane; if the angle of inclination α is greater than the angle of friction ϵ, show that the least ratio of the less to the greater is $\sin(\alpha - \epsilon)/\sin(\alpha + \epsilon)$.

9. Two weights support one another on a rough double inclined plane, by means of a fine string passing over the vertex, and both weights are about to move. Show that if the plane be tilted until both weights are again on the point of motion, the angle through which the plane will be turned is twice the angle of friction.

10. Two weights P, Q of similar material, resting on a double inclined plane, are connected by a fine string passing over the common vertex, and Q is on the point of motion down the plane. Prove that the greatest weight which can be added to P without disturbing the equilibrium is

$$\frac{P \sin 2\epsilon \sin (\alpha + \beta)}{\sin (\alpha - \epsilon) \sin (\beta - \epsilon)},$$

α, β being the angles of inclination of the planes, and ϵ the angle of friction.

11. A body is supported on a rough inclined plane by a force acting along it. If the least magnitude of the force, when the plane is inclined at an angle α to the horizon, be equal to the greatest magnitude, when the plane is inclined at an angle β, show that the angle of friction is $\frac{1}{2}(\alpha - \beta)$.

12. Two equal rings of weight W are movable along a curtain pole, the coefficient of friction being μ. The rings are connected by a loose string of length l, which supports by means of a smooth ring a weight W_1. How far apart must the rings be so that they will not come together?

13. Two weights P, Q of different material are laid on a rough plane, whose inclination is θ, and connected by a taut string inclined at 45° to the intersection of the plane with the horizon. Both weights are on the point of motion. Determine the coefficients of friction of P and Q, it being known that that of the upper weight is twice that of the lower.

14. A heavy ring is free to slide on a smooth elliptic wire of eccentricity e in a vertical plane, the major axis of the ellipse making an angle α with the horizontal, and a string fastened to the ring passes over a smooth peg at the center of the ellipse and supports a body of equal weight. Show that the angle ϕ which the tangent to the wire at the ring makes with the major axis is given by the equation

$$\tan(\phi + \alpha)(\sec^2\phi - e^2) = e^2 \tan\phi.$$

15. Two small smooth rings of weights W, W' are connected by a string, and slide on two fixed wires, the former of which is vertical and the latter inclined at an angle α to the horizontal. A weight P is tied to the string, and the two portions of it make angles θ, ϕ with the vertical. Prove that

$$\cot\theta : \cot\phi : \cot\alpha = W : P + W : P + W + W'.$$

16. Two particles of unequal mass are tied by fine inextensible strings to a third particle. They lie on a rough horizontal plane, with the strings stretched and making angles α, β with the horizontal line in the plane. Find the magnitude and direction of the least horizontal force which, on being applied to the third particle, will move all three.

17. A heavy particle is placed on a rough inclined plane of which the inclination α is equal to the angle of friction. A thread is attached to the particle, and passed through a hole in the plane which is lower than the particle, but is not in the line of greatest slope through it. Show that if the thread be gradually drawn through the hole, the particle will describe a straight line and a semicircle in succession.

CHAPTER IV

STATICS OF SYSTEMS OF PARTICLES

44. So far we have been considering the action of forces on a single particle. A different class of problems arises in considering the action of forces on a body composed of a great number of particles, to which forces are applied in such a way as to act on the different particles of the body.

Consider what happens when a force F is applied to one particle A of a body which is composed of a great number of particles $A, B, C, D, \cdots$. If the particle A were in no way influenced by the other particles $B, C, D, \cdots$ the particle A would start into motion under the action of the applied force, and would soon become separated from the other particles $B, C, D, \cdots$. If, however, the particles $A, B, C, D, \cdots$ constitute a single continuous body, this does not happen. What happens is that as

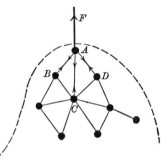

Fig. 31

soon as the particle A begins to move relatively to the other particles, systems of actions and reactions come into play between the particle A and the adjacent particles $B, C, D, \cdots$. Speaking loosely, we may say that the forces acting on A tend to check the motion of A, while the corresponding reactions tend to impart motion to $B, C, D, \cdots$. When $B, C, D, \cdots$ start into motion, further systems of forces begin to operate on the particles next beyond $B, C, D, \cdots$, and so on. Thus all the particles are set into motion, and instead of the particle A moving singly the complete body moves as a whole. We have now to discuss whether such a body, or system of bodies, will move or will remain at rest, when systems of forces act from outside on its different particles. We shall have to remember throughout that the forces applied from outside are not the only forces acting, but that these are accompanied by actions and reactions between the different particles.

45. One consequence of this last fact appears at once. Applying a force to one particle A of a body is not the same thing as applying an exactly similar force to another particle B. For the systems of internal actions and reactions will be different in the two cases. Any simple example will show that the resulting motion will, in general, also be different; e.g. a horizontal force applied to the middle point of the back of a chair will probably cause the chair to overturn. A similar force applied to one foot will drag it along the ground and also cause it to turn about a vertical axis.

The position occupied by the particle to which a force is applied is called the *point of application* of the force. The line drawn through this point in the direction of the force is called the *line of action* of the force.

Clearly, in order to have full data as to the action of a force, we must know

(a) its magnitude;
(b) its point of application;
(c) its line of action.

MOMENTS

46. DEFINITION. *The moment of a force about a line at right angles to the line of action of the force is defined to be the product of the force and of the shortest distance between the two lines.*

This moment, as we shall soon find, measures the *tendency to turn* around the line about which the moment is measured; e.g. if the arm of a balance is of length l, a weight w at its end has a moment lw about the pivot of the balance, and we shall find that this measures the tendency of the arm to turn.

DEFINITION. *The moment of a force about a line L which is not at right angles to the force is defined to be the same as the moment about L of the component of the force in a plane perpendicular to L.*

MOMENTS

Resolving the force into two components, one parallel to L and one perpendicular to L, it is clear that the former will not give any tendency to turn about L, so that the whole tendency to turn comes from the second component.

The two definitions which have now been given suffice to determine the moment of any force F about any line L. It may be noticed that the moment vanishes

(a) if the line of action of F is parallel to L;

(b) if the line of action of F intersects L.

Obviously, in either of these cases, the tendency to turn about L is zero.

47. Let the line L be at right angles to the plane of the paper, and intersect it in the point M. Let PA be the line of action of a force F in the plane of the paper, acting on a particle at A, and let MN be the perpendicular from M on to PA. Then, by definition, the moment of the force F about L is $F \times MN$.

Let the angle PAS be drawn equal to the angle NMA, say equal to θ, so that AS is perpendicular to MA. Then the moment of the force F about L

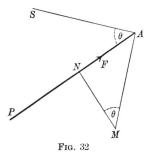

Fig. 32

$$= F \times MN$$
$$= F \times AM \cos \theta$$
$$= AM \times F \cos \theta$$
$$= AM \times \text{resolved part of } F \text{ along } SA.$$

Instead of F being the actual force acting at A, suppose that F is the resolved part, in the plane perpendicular to the line L, of some other force R. Then the moment of R about L is, by definition, the same as the moment of F, and the resolved part of R along $AS = F \cos \theta$. Hence what has just been proved may be put in the form

moment about L of any force R acting at A
$$= AM \times \text{resolved part of } R \text{ along } SA,$$

and SA is now determined as the direction which is perpendicular to L, and also to AM, the perpendicular from A on to L.

Thus we have a new definition of a moment, which is exactly equivalent to that previously given, namely:

The moment about a line L of a force R acting at A is equal to AM, the perpendicular from A on to L, multiplied by the component of R in a direction perpendicular to AM and to L.

48. From this conception of a moment we have at once the theorem:

The sum of the moments about any line L of any number of forces acting at a point A is equal to the moment of their resultant about L.

For, let R_1, R_2, $\cdots$ be the forces, and R their resultant. Let AM, as in § 47, be the perpendicular from A on to L, and let AS be a direction perpendicular to AM and to L. The theorem to be proved is that

$$AM \times \text{component of } R_1 \text{ along } AS$$
$$+ AM \times \text{component of } R_2 \text{ along } AS + \cdots$$
$$= AM \times \text{component of } R \text{ along } AS.$$

On dividing through by AM the theorem to be proved is seen to be simply that the component of R along AS is equal to the sum of the components of R_1, R_2, $\cdots$ along AS, which is known to be true.

We can now see more clearly how it is that the moment of a force, defined as we have defined it, gives a measure of the tendency to turn. In fig. 32 we are taking moments about a line L which is at right angles to the plane of the paper and meets this plane at M. The force whose moment is being considered is a force R acting at the point A. At A we have three directions mutually at right angles, namely

AS, AM, and the direction of a line through A parallel to L.

The moment of R about L has been defined to be

$$AM \times \text{component of } R \text{ along } AS.$$

MOMENTS

Now the component of R along AM is a force of which the line of action intersects L, and so can produce no tendency to turn a body about L, while the component of R along the line through A parallel to L can again produce no tendency to turn about L. Thus R can be resolved into three components, of which only the first, the component along AS, tends to set up rotation about L. We have defined the moment of the whole force R in such a way that it becomes identical with the moment of that one of its components which tends to set up rotation.

It will be noticed that a moment has sign as well as magnitude. In moving along the line of action of a force R, we may turn in either one direction or the other about a line L. We agree that when the turning is in one direction the moment of R about L is to be regarded as positive; when the turning is in the other direction the moment is taken to be negative.

49. If a particle is in equilibrium under the action of any number of forces, the resultant of all these forces must be nil. The sum of the moments of the separate forces, taken about any line whatever, is equal to the moment of the resultant and is therefore nil.

Hence we have the result:

When a particle is in equilibrium under the action of any forces, the sum of the moments of these forces about any line whatever must vanish.

System of Particles in Equilibrium

50. Consider a system of particles supposed to be in equilibrium under the action of any number of forces. As we have seen, the forces acting on any single particle will be of two kinds:

(a) *external forces*, forces applied to the particle from outside, as for instance the weight of the particle;

(b) *internal forces*, forces of interaction between the particle and the remaining particles of the system.

Now if the whole system of particles is in equilibrium, it follows that each particle separately must be in equilibrium. It follows from § 33, that

(a) the sum of the components in any direction, of all the forces acting on any single particle, must vanish;

and from the theorem just proved in § 48, that

(b) the sum of the moments about any line, of all the forces acting on any single particle, must vanish.

If, however, the sum of the components of the forces acting on each particle vanishes, it follows by addition that the sum of the components of all the forces acting on all the particles must vanish. The sum of the components of the internal forces, however, vanishes by itself, for the internal forces consist of pairs of actions and reactions, and the two components in any direction of such a pair of forces are equal and opposite.

Since the total sum vanishes, and the sum of the components of internal forces vanishes, it follows that the sum of the components of external forces vanishes.

A similar proposition is true of the moments of the external forces. The sum of the moments about any line L of all the internal forces is nil, for the moments of an action and reaction are equal and opposite. The sum of the moments of all the forces, internal and external, is zero, for each sum of the moments of the forces acting on each particle is zero separately. Thus the sum of the moments of the external forces is zero.

Thus we have proved the following theorems:

When a system of particles is in equilibrium under the action of any system of external forces,

(a) the sum of the components of all these forces in any direction is zero;

(b) the sum of the moments of all these forces about any line is zero.

Speaking loosely, we may say that these theorems express that there is no tendency to advance in any direction or to turn about any line.

ILLUSTRATIVE EXAMPLE

Wheel and axle. The apparatus known as the "wheel and axle" consists of a circular axle free to turn about its central axis, to which a circular wheel is rigidly attached, so that its center is on the center of the axle. A rope or string is wound round the axle, and has a weight attached to its end. A second rope or string is wound round the circumference of the circle in the opposite direction, and this again has a weight attached to its end. By a suitable choice of the ratio of these two weights, the apparatus may be balanced so that there is no tendency for it to turn about its axis.

Let us consider the equilibrium of the system consisting of the wheel and axle and of those parts of the strings or ropes which are wound round them. To simplify the problem, let us disregard altogether the weight of the system. Then the externally applied forces are

(a) the tension of the rope wound round the wheel;

(b) the tension of the rope wound round the axle;

(c) the action of the supports which keep the wheel and axle from falling.

Let the weights be denoted by P and Q, so that these are also the tensions of the strings, and let the radii of the wheel and axle be a, b respectively. Let us express mathematically that the sum of the moments of the externally applied forces about the axis is nil.

The moment of the tension of the string on the wheel is Pa, for P is the amount of the tension, which acts at right angles to the axis, and a is the shortest distance from the axis to the line of action of this tension.

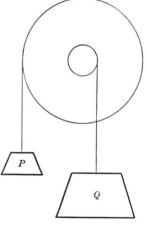

Fig. 33

Similarly the moment of force (b) is $-Qb$, the negative sign being taken because this tends to turn the system in the direction opposite to that in which the first tension tends to turn it.

If we imagine the system to be supported by forces acting on the axis itself, the moment of forces (c) vanishes, for the lines of action of these forces intersect the line about which we are taking moments. Thus the required equation is

$$Pa - Qb = 0.$$

This equation simply expresses that

[tendency of P to turn system] $-$ [tendency of Q to turn system] $= 0$.

Thus when the system is balanced so as to remain at rest we must have

$$P : Q = b : a,$$

so that the weights must be inversely as the radii. Practical examples of the principle of the wheel and axle are supplied by the windlass and capstan.

66 STATICS OF SYSTEMS OF PARTICLES

EXAMPLES

1. Eight sailors, each pressing on the arm of a capstan with a horizontal force of 100 lbs., at a distance of 8 feet from its center, can just raise the anchor. The radius of the axle of the capstan is 12 inches. Find the pull on the cable which raises the anchor.

2. In the apparatus of fig. 33 the weight P is disconnected, and the free end of the string is tied to the same point on Q as the other string. Show that in equilibrium this point is vertically below the axis.

3. A wheel is free to turn about a horizontal axis, and has fastened to it two strings which are wound round its circumference in opposite directions. The other ends are both tied to a small ring from which a weight is suspended. Show that when the system is at rest the two strings will make equal angles with the vertical.

4. A man finds that he can just move a lock gate against the pressure of the water, by pressing with a horizontal force of 150 lbs. at a distance of 8 feet from the pivot. What force must he exert if he presses at a distance of 9 feet from the pivot?

5. A wheel capable of turning freely about a horizontal axis, has a weight of 2 pounds fixed to the end of a spoke which makes an angle of $60°$ with the horizontal. What weight must be attached to the end of a horizontal spoke to prevent motion taking place?

6. A drawbridge is raised by a chain attached to the end farthest removed from the hinges. When the bridge is at rest in a horizontal position, the chain makes an angle of $60°$ with the bridge, and the pull on the chain necessary to move the bridge is equal to the weight of three tons. Find what additional pull is required in the chain when a weight of one ton is placed at the middle point of the bridge.

FORCES IN ONE PLANE

51. The simplest problems in statics are always those in which all the forces have their lines of action in one plane. In such a problem it is obviously most convenient to take moments about a line perpendicular to the plane in which the forces act. Let any such line intersect the plane in a point P. Each force is entirely perpendicular to the line about which moments are taken, so that the moment is equal to the product of the force and the shortest distance of the line of action of the force from P.

Taking moments about an axis which intersects the plane of the forces at right angles in a point P is often spoken of as *taking*

CO-PLANAR FORCES 67

moments about the point P, and the perpendicular from P to the line of action of a force is spoken of as the *arm* of the moment of this force.

52. THEOREM. *When three forces, acting in a plane, keep a body or system of bodies in equilibrium, these three forces must meet in a point.*

For let P, Q, R be the forces, and let P, Q intersect in the point A. Then the sum of the moments of P, Q, and R about A must vanish, and those of P and Q are already known to vanish. Thus the moment of R about A must vanish, — that is, R must pass through the point A, or, what is the same thing, the three forces must intersect in a single point.

An application of this principle is often sufficient in itself for the solution of statical problems in which the applied forces can be reduced to three.

ILLUSTRATIVE EXAMPLES

1. The seesaw. *Two persons of weights W_1, W_2 stand on a plank which rests on a rough support about which it is free to turn. Neglecting the weight of the plank, find how the persons must place themselves in order that the plank may balance.*

The forces may be supposed all to act in one plane, namely the vertical plane through the central line of the plank. The forces are

(a) the weight W_1 of the person at one end;

(b) the weight W_2 of the person at the other end;

(c) the reaction between the plank and its support.

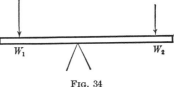

Fig. 34

Let a, b be the distances of the persons from the support; then, on taking moments about the point of support, we have

$$W_1 a - W_2 b = 0.$$

Thus the two persons should stand at distances from the support which are inversely proportional to their weights.

Notice that in this problem the system is acted on by three forces, which meet in a point, the point being at infinity.

STATICS OF SYSTEMS OF PARTICLES

2. The nutcracker. *It is found that a weight of 100 pounds placed on top of a nut will just crack it. How much force must be applied at the ends of the arms of a nutcracker 6 inches long to crack the nut when it is placed $\frac{1}{2}$ inch from the hinge?*

Let a force F applied at the extreme end of each arm be supposed just sufficient to crack the nut. Then when a force F is applied at the end of the arm, the pressure between the nut and the arm must be the weight of 100 pounds. Thus the forces acting from outside on either arm of the nutcracker will consist of

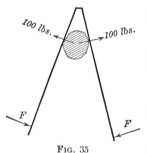

FIG. 35

(a) the force F applied at the end of the arm;

(b) the pressure of 100 pounds weight exerted by the nut on the arm at a distance of $\frac{1}{2}$ inch from the hinge;

(c) the reaction at the hinge.

The weight of the nutcracker is here supposed to be negligible.

Taking moments about the hinge, we obtain

$$6 \times F = \tfrac{1}{2} \times 100 \text{ pounds weight,}$$

so that $F = 8\tfrac{1}{3}$ pounds weight.

NOTE. When, as here, an unknown force neither enters in the data nor is required in the answer, we can always obtain equations in which the force does not occur, by taking moments about a point in its line of action. So again, if two such forces occur, we can obtain an equation into which neither force enters, by taking moments about the point of intersection of their two lines of action.

3. *A ladder stands on a rough horizontal plane, leaning against a rough vertical wall, the contacts at the two ends of the ladder being equally rough. Find how far a man can ascend the ladder without its slipping, it being supposed that the weight of the ladder may be neglected.*

The forces acting on the system composed of the man and ladder are three in number:

(a) the reaction with the horizontal plane;
(b) the reaction with the vertical wall;
(c) the weight of the man.

These forces are all in one plane; hence, by the theorem of § 52, their lines of action must meet in a point.

In the figure let AB be the ladder, C the position of the man, and P the point in which the three forces meet, so that PC is vertical, and AP, BP are the lines of action

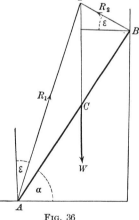

FIG. 36

of the reactions at A, B. When slipping is just about to begin, each of these reactions must make with the normal an angle equal to the angle of friction.

ILLUSTRATIVE EXAMPLES 69

Let ϵ be this angle of friction, and let α be the inclination of ladder to the horizontal. Then, from the geometry of the triangle ACP, we have

$$\frac{AC}{\sin \epsilon} = \frac{AP}{\sin\left(\dfrac{\pi}{2} + \alpha\right)},$$

and since APB is a right angle,

$$AP = AB \cos\left(\frac{\pi}{2} - \epsilon - \alpha\right).$$

Thus $\quad AC = AP \sin \epsilon \sec \alpha = AB \sin \epsilon \sin (\epsilon + \alpha) \sec \alpha.$

Thus slipping will begin as soon as the man has climbed a height equal to $\sin \epsilon \sin (\epsilon + \alpha) \sec \alpha$ times the whole height.

The condition that the man can reach the top without slipping is that $\sin \epsilon \sin (\epsilon + \alpha) \sec \alpha$ shall be greater than unity, or that

$$\sin \epsilon \sin (\epsilon + \alpha) > \cos\left[(\epsilon + \alpha) - \epsilon\right]$$
$$> \sin \epsilon \sin (\epsilon + \alpha) + \cos \epsilon \cos (\epsilon + \alpha).$$

Thus for the condition to be satisfied $\cos \epsilon \cos (\epsilon + \alpha)$ must be negative; i.e. $\epsilon + \alpha$ must be greater than $90°$. Thus the angle between the ladder and the vertical must be less than the angle of friction. This is also clear from the figure, for when the man reaches B, two of the forces, namely the reaction at B and the weight of the man, both pass through B, so that the third force must also pass through B; i.e. the reaction at A must have AB for its line of action, and if the ladder just slips here, the angle between AB and the vertical must be ϵ.

4. *If, in the last problem, the man has ascended to some point C without the ladder slipping, what are the reactions at A and B?*

Here it is not known what angles the reactions make with the normals: all that is known is that these angles are *less* than the angle of friction.

Let us resolve the reaction at A into two components N_1, F_1, and that at B into two components N_2, F_2, these being horizontal and vertical as in the figure. Then the forces acting on the system composed of the man and ladder are the five forces

$$N_1, F_1, N_2, F_2, \text{ and } W.$$

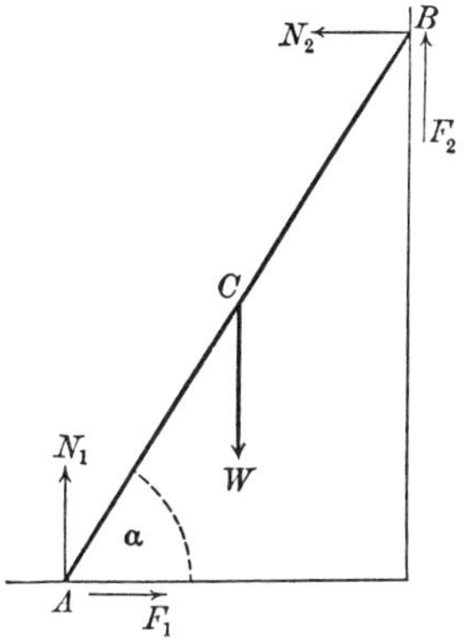

Fig. 37

Resolving vertically, $\qquad W - N_1 - F_2 = 0.$ $\qquad$ (a)
Resolving horizontally, $\qquad F_1 - N_2 = 0.$ $\qquad$ (b)
Taking moments about A,
$$W \cdot AC \cos \alpha - F_2 \cdot AB \cos \alpha - N_2 \cdot AB \sin \alpha = 0. \qquad (c)$$

70 STATICS OF SYSTEMS OF PARTICLES

There are four quantities which it is required to find. So far we have obtained only three equations. We can, of course, obtain other equations by resolving in other directions and by taking moments about other points, but it will be found that the equations so obtained will not be new equations, but simply equations of which the truth is already implied in the equations already obtained. Thus we cannot, by resolving and taking moments, obtain more than three independent equations, and these do not suffice to determine the four unknown quantities.

Here we have illustrated a problem which cannot be solved by the methods explained in this chapter, and which requires for its solution a consideration of the systems of forces set up between the separate particles of the bodies acted upon. It is important that the student should realize that such problems exist, although he may not yet be able to solve them.

5. *Force required to drag a car along.* To simplify the problem as far as possible, let us suppose that the car is mounted on four equal wheels, each of radius a and revolving round an axle of radius b, the coefficient of friction between wheel and axle being the same for each wheel. Let us suppose that a force P applied horizontally is found to be just sufficient to start the car into motion.

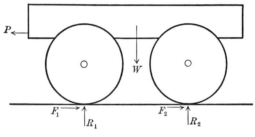

Fig. 38

Let us consider first the equilibrium of the whole car. We are most conveniently able to enumerate the forces acting on any system by taking a tour, in imagination, over the whole surface of a cover made just to fit the system, and noting the forces which act across this cover at its different points. These, together with the weight of the whole system, will give the whole system of forces. The forces acting on the car are in this way found to be

(a) its weight, say W;

(b) the horizontal applied force P;

(c) the reactions between the wheels and the ground. Let us resolve each reaction into a vertical component R and a horizontal component F; let us denote the components of the reaction between the first wheel and the ground by F_1, R_1; let the corresponding quantities for the second wheel be F_2, R_2; and so on.

ILLUSTRATIVE EXAMPLES

(As regards the frictional force acting between the wheel and the ground, we notice that although motion is about to take place, this motion is not one of *slipping* between the wheel and the ground, so that the ratio of F to R for any wheel is not the coefficient of friction between the wheel and the ground.)

The forces just enumerated hold the car in equilibrium. Thus the sum of their components in any direction must vanish and the sum of their moments about any line must vanish. Resolving horizontally and vertically, we obtain

$$P = F_1 + F_2 + F_3 + F_4. \qquad (a)$$
$$W = R_1 + R_2 + R_3 + R_4. \qquad (b)$$

There is nothing to be gained by taking moments about any line; as we do not know the line of action of P, we cannot know its moment.

Next let us consider the equilibrium of a single wheel. The wheel touches the ground and also touches the axle at some point C. (We may think of the axle as a circle of radius very slightly less than that of the inside of the hub of the wheel.) The forces acting on the wheel are accordingly

(a) its reaction with the ground;
(b) its reaction with the axle;
(c) its weight, which we shall neglect as being insignificant in comparison with that of the car.

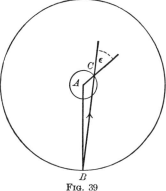

Fig. 39

Neglecting the third force, the two former forces must be equal and opposite. The line of action of each is accordingly the line joining B, the point of contact with the ground, to C the point of contact with the axle. Since slipping is just about to take place at C, the reaction at C will make an angle ϵ, equal to the angle of friction, with AC the normal at C. Thus the angle BCA is equal to ϵ.

In the triangle ACB we have $AC = b$, $AB = a$, and the angle $ACB = \epsilon$.

Thus
$$\frac{a}{\sin \epsilon} = \frac{b}{\sin ABC}.$$

Since, however, a force along BC has components R_1 and F_1 along and perpendicular to AB, we have

$$\tan ABC = \frac{F_1}{R_1}.$$

Thus
$$\frac{F_1}{R_1} = \tan ABC$$
$$= \frac{\sin ABC}{\sqrt{1 - \sin^2 ABC}}$$
$$= \frac{b \sin \epsilon}{\sqrt{a^2 - b^2 \sin^2 \epsilon}}.$$

Now ϵ, a, and b are supposed to be the same for each wheel, so that

$$\frac{b \sin \epsilon}{\sqrt{a^2 - b^2 \sin^2 \epsilon}} = \frac{F_1}{R_1} = \frac{F_2}{R_2} = \frac{F_3}{R_3} = \frac{F_4}{R_4}$$
$$= \frac{F_1 + F_2 + F_3 + F_4}{R_1 + R_2 + R_3 + R_4}$$
$$= \frac{P}{W},$$

by equations (a) and (b).

Thus
$$P = \frac{Wb \sin \epsilon}{\sqrt{a^2 - b^2 \sin^2 \epsilon}}, \qquad (c)$$

giving the horizontal pull required.

The value of b, the radius of the axle, will generally be small in comparison with a, the radius of the wheel. Thus, without serious error, we may neglect $b^2 \sin^2 \epsilon$ in comparison with a^2, and replace the denominator in equation (c) by a. The equation now becomes

$$P = \frac{Wb \sin \epsilon}{a}.$$

By making b/a very small, we see that the car can be made to run very smoothly. We notice also that even if there is so much friction between wheel and axle that the coefficient of friction may be regarded as infinite, we have $\sin \epsilon = 1$, and hence

$$P = \frac{Wb}{a},$$

so that the force required to drag the car along will still be small compared with that required to drag the same weight over a fairly smooth surface.

This analysis has assumed that the wheels may be supposed to touch the ground only at their lowest point. It applies pretty accurately to the case of steel wheels rolling on steel rails, but does not apply to the problem of an ordinary road carriage moving over a soft road, where the wheels are embedded to a small extent in the road. In fact, if the analysis just given took account of all the facts of the case, it is clear that the force required to haul a car would be independent of the state of the road.

EXAMPLES

1. A weight of 250 pounds is suspended from a light rod which is placed over the shoulders of two men and carried in a horizontal position. If the men walk 10 feet apart and the weight is 4 feet from the nearer of them, find the weight borne by each.

2. A weight is suspended from a light rod which passes over two fixed supports 6 feet apart. On moving the weight 6 inches nearer to one support, the pressure on that support is increased by 10 pounds. What is the amount of the weight?

EXAMPLES

3. A balance has two pans, each of weight 8 ounces, suspended from a beam, each at distance 7 inches from the pivot. A dishonest tradesman moves one pan half an inch nearer to the pivot, adding weight to the farther pan in order that the two pans may still balance. Find how much weight he must add, and by how much his profits will be increased by his dishonesty.

4. A balance has a weight of 20 ounces suspended from one end of the beam. A string is tied to the other end of the beam and at equal distance from the pivot, and this string makes an angle of 45° with the horizontal. With what force must it be pulled to maintain the beam of the balance in a horizontal position?

5. A crowbar 8 feet long is to be used to move a body, it being required to apply a force of 500 pounds weight vertically upwards to this body to move it. How near to the end of the crowbar must the fulcrum be placed in order that a man of 140 pounds weight may be able to apply the required force by pressing on the other end of the crowbar?

6. A table of negligible weight has any number of legs. A heavy particle is placed on the table. Show that the table will tilt over if the vertical through the particle meets the floor under the table, in a point *outside* the polygon formed by joining the points of contact of the feet with the floor.

7. A table of negligible weight has three legs, the feet forming an equilateral triangle. A heavy particle is placed on the table in a position such that the table does not tilt over. Find the proportion of weight which is carried by each foot.

8. A card is suspended in a horizontal position by three equal inextensible strings fastened to three points A, B, C in the card which form an equilateral triangle, and also to a point P above the card. A weight is placed on the card at any point Q inside the triangle ABC. Find the tensions of the strings.

9. A card is suspended by four equal inextensible strings which pass through the four points A, B, C, D of a square in the card and are tied to four points A', B', C', D' at equal heights h vertically above the points A, B, C, D. A weight is placed on the card at any point P inside the square $ABCD$. Show that the tensions in the strings cannot be determined without discussing the internal stresses in these strings.

10. If in the last question the internal stresses in the strings stretch the strings very slightly, so that Hooke's law is obeyed, show that the tensions can be found, and find them.

11. A seesaw rolls on a rough circular log of radius a, fixed horizontally. Two persons stand at distances b, c from the middle point of the seesaw, their weights being such that the seesaw is just balanced horizontally with the middle point resting on the log. The first person moves a distance d towards the center of the log. Through what angle will the seesaw turn? How far can the person advance before the seesaw slips off the log altogether?

12. A pair of wheels of radius a are connected by an axle of radius b, and run on horizontal rails. A string is wound round the axle and the end leaves the axle making an angle θ with the horizontal. If this string is pulled, show that the wheels will run toward or away from the person pulling according as $\cos \theta$ is greater or less than b/a. What happens when $\cos \theta = b/a$?

13. If the weight of the wheels and axles of a car is w, and if this may not be neglected in comparison with W, the total weight of the car, show that equation (c) of example 5, p. 72, must be replaced by

$$P = \frac{(W-w)b\sin\epsilon}{\sqrt{a^2 - b^2 \sin^2\epsilon}}.$$

14. A locomotive of weight 134 tons rests on a bogie, of which the wheels and axles weigh 4 tons, and two pairs of driving wheels, of which the wheels and axles weigh 10 tons. The weight taken on the axles of the bogie is 40 tons, that taken on the axles of the driving wheels being 80 tons. The diameters of the wheels are 2 feet 10 inches and 7 feet 1 inch respectively. Each axle, where it passes through the axle box, is of radius $2\frac{1}{8}$ inches, and the coefficient of friction is $\frac{1}{5}$. Find the horizontal force necessary to move the engine.

15. In question 14 the rails are greased so that the coefficient of friction between them and the wheels is less than $\frac{1}{100}$; show that the engine cannot be started without the wheels skidding on the greased rails, and explain the dynamical processes by which the engine is set in motion in this case.

STRINGS

53. Strings, ropes, and chains frequently form part of the systems of bodies with which statical problems are concerned, so that it is important to discuss the equilibrium of a string (or rope or chain). The first problem we shall consider is that of a string stretched over a surface,— as for example a pulley wheel,— it being supposed that the weight of the string may be neglected, and that the contact between the string and the surface is equally rough at all points. It will also be supposed that the string is all in one plane.

Let P, Q be two adjacent points of the string so near together that the portion PQ of the string may be treated as a particle.

The forces acting on this particle will be

(a) T_P, the tension at P, acting along the tangent to the string at P;

(b) T_Q, the tension at Q, acting along the tangent to the string at Q;

(c) the reaction with the surface.

By Lami's theorem, each force must be proportional to the sine of the angle between the remaining two forces.

STRINGS

Let A be the point of the surface at which the string leaves it. Let the normals to the surface be drawn at A, P, Q, and let the normal at P make an angle θ with the normal at A. If the points A, P, Q come in this order, as in fig. 40, the normal at Q will make with the normal at A an angle slightly greater than θ, — say $\theta + d\theta$, — so that $d\theta$ is the small angle between the normals at P and Q.

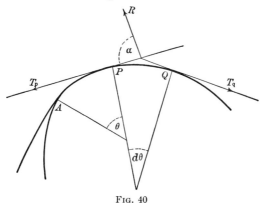

Fig. 40

With this notation the angle between the tensions T_P and T_Q is $\pi - d\theta$. Let the angle between the reaction R and the tension T_P be α, then the angle between the tension T_Q and R is $\pi - \alpha + d\theta$. Thus we have

$$\frac{R}{\sin(\pi - d\theta)} = \frac{T_P}{\sin(\pi - \alpha + d\theta)} = \frac{T_Q}{\sin \alpha}.$$

Since $\sin(\pi - \alpha + d\theta) = \sin(\alpha - d\theta)$, we have

$$\frac{T_Q}{\sin \alpha} = \frac{T_P}{\sin(\alpha - d\theta)},$$

and by a known theorem in algebra, each fraction is equal to

$$\frac{T_Q - T_P}{\sin \alpha - \sin(\alpha - d\theta)}.$$

Now $T_Q - T_P$ is the increase in T when θ changes from θ to $\theta + d\theta$, and this, in the notation of the differential calculus, may be written

$$\frac{dT}{d\theta} d\theta.$$

76   STATICS OF SYSTEMS OF PARTICLES

Also the denominator $\sin \alpha - \sin (\alpha - d\theta)$ is the increase in $\sin \alpha$ when α changes from $\alpha - d\theta$ to α, and this in the same way is equal to
$$\frac{d(\sin \alpha)}{d\alpha} d\theta \quad \text{or} \quad \cos \alpha\, d\theta.$$

Thus the original fraction is equal to
$$\frac{\frac{dT}{d\theta} d\theta}{\cos \alpha\, d\theta} \quad \text{or} \quad \sec \alpha \frac{dT}{d\theta}.$$

Thus
$$\frac{T_Q}{\sin \alpha} = \sec \alpha \frac{dT}{d\theta},$$

or
$$T_Q = \tan \alpha \frac{dT}{d\theta}. \tag{12}$$

When in the limit the particle PQ is supposed to become vanishingly small, T_P and T_Q become indistinguishable. Let us denote either by the single letter T, so that T is now simply the tension at a point at which the normal makes an angle θ with that at A. If the string is just on the point of slipping in the direction APQ, the angle between the reaction R and the normal of either Q or P will be ϵ, the angle of friction. Thus we shall have
$$\alpha = \frac{\pi}{2} - \epsilon,$$
so that $\tan \alpha = \cot \epsilon$, and equation (12) becomes
$$T = \cot \epsilon \frac{dT}{d\theta}. \tag{13}$$

54. If the contact between the surface and the string is perfectly smooth, $\epsilon = 0$, so that $\frac{dT}{d\theta} = 0$. It follows that T is a constant; i.e. the tension is the same at all points of the string. Thus the tension of a string is not altered by its passing over a *smooth* surface, — the result already given in § 36.

55. In the more general use in which the contact is not perfectly smooth, let μ be the coefficient of friction, so that $\mu = \tan \epsilon$, then equation (13) may be written

$$\frac{dT}{d\theta} = \mu T,$$

and integrating this, $\quad \dfrac{dT}{T} = d(\mu\theta),$

$$d(\log T) = d(\mu\theta),$$

or $\quad\quad\quad\quad \log T = \mu\theta + \text{a constant}.$

Let T_o be the tension at A, then we find, by putting $\theta = 0$, that the constant must be equal to $\log T_o$, so that

$$\log T - \log T_o = \mu\theta,$$

or $\quad\quad\quad\quad T = T_o e^{\mu\theta}. \quad\quad\quad\quad (14)$

If the string leaves the surface again at some point B at which the normal makes an angle ψ with the normal at A, we find for the tension at B

$$T = T_o e^{\mu\psi},$$

so that the tension is multiplied by $e^{\mu\psi}$ on passing over the surface from A to B.

If the string (or rope) is passed round and round a post or bollard, the tension is increased in the ratio $e^{2\mu\pi}$ for each complete turn. For a hemp rope on oak the coefficient of friction, according to Morin, is $\mu = 0.53$. Thus $2\mu\pi = 3.34$ and $e^{2\mu\pi} = 28.1$. The tension of a rope wound round an oak post is accordingly increased about twenty-eight fold for each complete turn.

EXAMPLES

1. A weight is suspended by a rope which, after being wound round a horizontal beam, leaves the beam horizontally, its end being controlled by a workman. If the rope makes $1\frac{1}{4}$ complete revolutions round the beam, what force must be exerted by the man
   (a) to keep the weight from slipping?
   (b) to raise the weight?
   (Take $\mu = \frac{1}{4}$.)

2. A weight of $2\frac{1}{2}$ pounds stands on a rough table. A string tied to the base of the weight hangs over the edge of the table and has attached to it a second weight which hangs freely. If the coefficients of friction between the weight and the table and the string and the table are $\frac{1}{2}$ and $\frac{1}{4}$ respectively, find how heavy the hanging weight must be to start the other weight into motion.

3. A weight of 2500 pounds is to be raised from the hold of a ship. A rope attached to the weight makes $3\frac{1}{4}$ turns round a steam windlass, its other end being held by a seaman. With what force must he pull his end of the rope so as to raise the weight when the windlass is in motion? ($\mu = \frac{1}{8}$.)

4. In the last question, find what pull would have to be exerted on the rope if the windlass were at rest.

5. It is found that two men can hold a weight on a rope by taking three turns about a post, and that one of them can do it alone by taking one half turn extra. If each can pull with a force of 220 pounds weight, find the weight sustained.

6. In a tug of war the rope is observed to rub against a post at the critical moment, in such a way that the two parts of the rope make an angle of 1° with one another. If the coefficient of friction between the rope and the post is $\frac{1}{6}$, show that this imposes a handicap on the winning side equal to about .0029 times its aggregate pull.

Suspension Bridge

56. An interesting problem is afforded by the kind of suspension bridge in which the weight of the bridge (supposed horizontal) is taken by a suspension cable by means of vertical chains connecting the bridge with the cable.

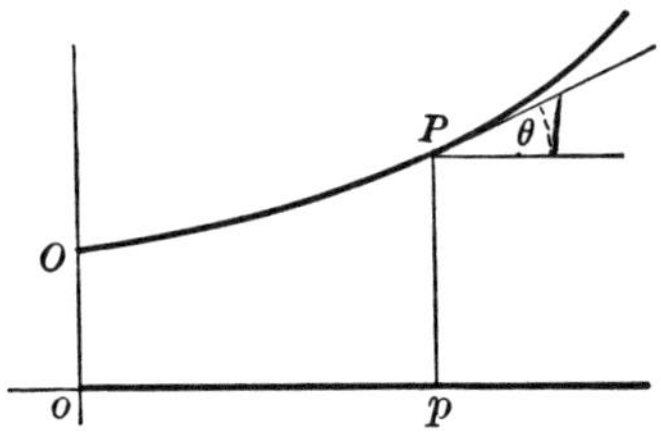

Let us, for simplicity, agree to neglect the weights of the chains and cable, and suppose the weight of the bridge to be distributed evenly along its length.

Fig. 41

Let O be the lowest point of the cable, and let P be any other point. Let o, p be the points of the bridge vertically below O, P, and let $op = x$. Let the tension at P be T, and let that at O be H. Let the direction of the cable at P make an angle θ with the horizontal.

SUSPENSION BRIDGE

The forces acting on the piece OP of the cable consist of

(a) the tension at O, of amount H acting horizontally;

(b) the tension at P, of amount T acting at an angle θ with the horizontal;

(c) the tensions of the vertical chains, all acting vertically.

Resolving horizontally, we obtain
$$H - T\cos\theta = 0. \tag{15}$$

Resolving vertically, we obtain
$$T\sin\theta - S = 0,$$
where S is the sum of the tensions of all the chains which leave the cable between O and P. These tensions support the portion op of the bridge, and if the weight of the bridge is w per unit length, the weight of op will be wx. Thus $S = wx$, and therefore
$$T\sin\theta = wx. \tag{16}$$
This and equation (15), $\quad T\cos\theta = H, \tag{17}$

will give us all the information we require.

To find the shape which the cable must have in order that the bridge may hang horizontally, we require to obtain a relation between θ and x. We accordingly eliminate T from equations (16) and (17), and obtain
$$\tan\theta = \frac{w}{H}x.$$

If y is the height of the cable above the bridge, x, y may be regarded as the Cartesian coördinates of a point P on the cable, and we have
$$\tan\theta = \frac{dy}{dx}.$$

Thus the coördinates x, y of P are related by
$$\frac{dy}{dx} = \frac{w}{H}x,$$
giving on integration $\quad y = \frac{1}{2}\frac{w}{H}x^2 + C,$

where C is a constant of integration.

80 STATICS OF SYSTEMS OF PARTICLES

This is the Cartesian equation of the cable. It is easily seen to represent a parabola of latus rectum $\dfrac{2H}{w}$. Thus the cable must hang in the form of a parabola. The greater the horizontal tension, the greater the latus rectum of the parabola, and therefore the flatter the curve of the cable. A perfectly straight cable is of course an impossibility — this would require infinite tension.

57. To find the tension at any point of the cable, we square equations (16) and (17), and add corresponding sides. Thus

$$T^2 = H^2 + w^2 x^2,$$

giving the tension at a point distant x from the center. If the bridge is of length $2a$, the tension at either pier must be

$$\sqrt{H^2 + w^2 a^2}.$$

The Catenary

58. In the problem of the suspension bridge, we neglected the weight of the cable. A second problem arises when the cable is supposed to be acted on by no external forces except its own weight. The problem here is simply that of a string of which the two ends are fastened to fixed points, and which hangs freely between these points.

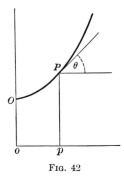

Fig. 42

As before, let O be the lowest point, and let P be any other point. The forces acting on the portion OP of the string are

(*a*) the tension at O, of amount H acting horizontally;

(*b*) the tension at P, of amount T acting at an angle θ with the horizontal;

(*c*) the weight of OP. If we take the string to be of weight w per unit length, and denote the distance OP by s, this weight is ws acting vertically.

THE CATENARY 81

Resolving horizontally, we obtain

$$H - T\cos\theta = 0. \tag{18}$$

Resolving vertically, we obtain

$$T\sin\theta - ws = 0. \tag{19}$$

To find the shape of curve in which the string will hang, we must obtain a relation between θ and s. Eliminating T, we obtain

$$H\tan\theta = ws,$$

or, if we replace H/w by the single constant c,

$$s = c\tan\theta. \tag{20}$$

This is one form of the equation of the curve, s and θ being taken as coördinates. The equation in this form is known as the *intrinsic equation* of the curve. We require, however, to deduce the equation in its Cartesian form.

59. If the point o in fig. 42 is taken as origin, the axes being horizontal and vertical, we have at once the relations

$$dx : dy : ds = \cos\theta : \sin\theta : 1, \tag{21}$$

Fig. 43

for dx and dy are the horizontal and vertical projections of the small element ds of length of the string. As a first step, let us use relations (21) to change the variables of equation (20) from s and θ to s and y. We have

$$c^2 = s^2\cot^2\theta = s^2\csc^2\theta - s^2 = s^2\left(\frac{ds}{dy}\right)^2 - s^2,$$

so that

$$s\frac{ds}{dy} = \sqrt{s^2 + c^2}.$$

Thus

$$dy = \frac{s\,ds}{\sqrt{s^2 + c^2}},$$

and integrating this, we obtain

$$y = \sqrt{s^2 + c^2} + \text{a constant}. \tag{22}$$

We can determine the constant of integration as soon as we decide where the origin is to be taken — so far we have not fixed the point o. Since s denotes the arc of the curve measured from O, we have $s = 0$ at the point O, and therefore the y coördinate of O (putting $s = 0$ in equation (22)) is

$$y = c + \text{a constant.}$$

Let us agree that Oo is to be made equal to c, so that $y = c$ at O. Then the unknown constant of integration must be zero. Thus equation (22) will be
$$y^2 = s^2 + c^2. \tag{23}$$

The last step is to transform the variables from y and s to y and x. The relation which enables us to do this is obtained by eliminating θ from relations (21), and is

$$(ds)^2 = (dy)^2 + (dx)^2. \tag{24}$$

The equation already obtained is

$$s = \sqrt{y^2 - c^2},$$

so that
$$ds = \frac{y\,dy}{\sqrt{y^2 - c^2}},$$

and on eliminating ds from this and relation (24), we obtain

$$\frac{y^2(dy)^2}{y^2 - c^2} = (dy)^2 + (dx)^2.$$

From this,
$$(dx)^2 = (dy)^2 \left[\frac{y^2}{y^2 - c^2} - 1 \right]$$

$$= \frac{c^2}{y^2 - c^2} (dy)^2,$$

so that
$$dx = \frac{c\,dy}{\sqrt{y^2 - c^2}}. \tag{25}$$

Integrating this, we obtain

$$\frac{y}{c} = \cosh \frac{x}{c}, \tag{26}$$

where
$$\cosh \frac{x}{c} = \tfrac{1}{2}(e^{\frac{x}{c}} + e^{-\frac{x}{c}}).$$

The student who is not familiar with the hyperbolic cosine (cosh) function will easily be able to verify equation (26) by differentiating it back into equation (25).

THE CATENARY

Equation (26) is the Cartesian equation of the curve formed by the string; this curve is known as the *catenary*.

From equation (23) we obtain the value of s in the form

$$s^2 = y^2 - c^2$$
$$= c^2\left(\cosh^2 \frac{x}{c} - 1\right)$$
$$= c^2 \sinh^2 \frac{x}{c},$$

so that
$$\frac{s}{c} = \sinh \frac{x}{c},$$

where
$$\sinh \frac{x}{c} = \tfrac{1}{2}(e^{\frac{x}{c}} - e^{-\frac{x}{c}}).$$

60. Expanding the exponentials $e^{\frac{x}{c}}$, $e^{-\frac{x}{c}}$, we obtain $\cosh \frac{x}{c}$ in the form

$$\cosh \frac{x}{c} = 1 + \frac{1}{2}\left(\frac{x}{c}\right)^2 + \frac{1}{24}\left(\frac{x}{c}\right)^4 + \cdots.$$

So long as x is small, we may neglect all the terms of this series beyond the second. Using the value obtained in this way, we obtain instead of equation (26)

$$y = c + \frac{x^2}{2c},$$

showing that so long as x is small the curve coincides very approximately with a *parabola* of latus rectum $2c$ or $2H/w$.

This parabola, it will be noticed, is one which would be formed by the cable of a suspension bridge of horizontal tension H, w being the weight per unit length of the bridge itself. Indeed, it is clear that when the cable is almost horizontal, it is a matter of indifference whether the cable itself possess weight w per unit length of its arc, or whether a weight w per unit length is hung from it so as to lie horizontally.

We can also obtain a simple approximation to the shape of the catenary when x is large, i.e. at points far removed from the lowest point. When x is very large, so that $\frac{x}{c}$ is very large, the value of $e^{\frac{x}{c}}$ becomes very large, while that of $e^{-\frac{x}{c}}$ becomes very small. Thus

84   STATICS OF SYSTEMS OF PARTICLES

the value of $\cosh\dfrac{x}{c}$ becomes approximately $\tfrac{1}{2}e^{\tfrac{x}{c}}$, and the equation of the catenary (equation (26)) becomes

$$\frac{y}{c} = \tfrac{1}{2}e^{\tfrac{x}{c}}.$$

Thus for large values of x the catenary coincides with the exponential curve.

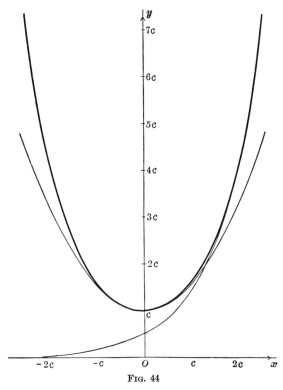

Fig. 44

Fig. 44 shows the form of the catenary. The thin curves are

(a) the parabola, to which the catenary approximates for small values of x;

(b) the exponential curve, to which the catenary approximates for large values of x.

THE CATENARY

61. Sag of a tightly stretched string. A string or wire stretched so as to be nearly horizontal all along its length — as for instance a telegraph wire — may, as we have seen, be supposed to form a parabola to within good approximation. Thus let A, B be two poles at equal height between which a wire is stretched; let C be the middle point of AB, and let D be the point of the wire vertically below C. Then,

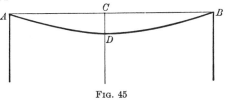

Fig. 45

from symmetry, D will be the lowest point of the wire and therefore will be the vertex of the parabola. Thus, from the equation of the parabola,
$$CB^2 = \frac{2H}{w} CD,$$
since, by § 60, its latus rectum is $2H/w$.

Thus if $h = AB$, the distance between the poles, the "dip" CD is given by
$$CD = \frac{w}{2H} CB^2$$
$$= \frac{1}{8} \frac{wh^2}{H}. \qquad (27)$$

To obtain the length of the wire we have to introduce a higher order of small quantities, and so are compelled to return to the equation of the catenary.

We have
$$s = \tfrac{1}{2} c (e^{\frac{x}{c}} - e^{-\frac{x}{c}})$$
$$= x + \frac{1}{6} \frac{x^3}{c^2} + \cdots.$$

The quantity we require is $s - x$, namely $DB - CB$ in fig. 45. When the string is tightly stretched c is very great, so that we may neglect the terms in s beyond those written down in the above equation, and obtain
$$s - x = \frac{1}{6} \frac{x^3}{c^2}, \text{ approximately,}$$
$$= \frac{1}{6} \frac{w^2 x^3}{H^2}.$$

Putting $x = \tfrac{1}{2} h$, we find as the total increase in a span of length h, caused by sagging,

$$2s - h = \frac{1}{24} \frac{w^2 h^3}{H^2}.$$

EXAMPLES

1. The entire load of a suspension bridge is 320 tons, the span is 640 feet, and the height is 50 feet. Find the tension at the points of support, and also the tension at the lowest point.

2. The weight of a freely suspended cable is 320 tons; the distance between the two points of support, which are in the same horizontal line, is 640 feet, and the height of these points above the lowest point of the cable is 50 feet. Find the tension at the points of support, and also the tension at the lowest point.

3. The wire for a telegraph line cannot sustain a weight of more than a mile of its own length without breaking. If the wire is stretched on poles at equal intervals of 88 yards, what is the least sag permissible?

4. In the last question how much wire is required for a mile of the line?

5. A telegraph line has to be built of a certain kind of wire, stretched over evenly spaced posts. Show that if the number of posts is very large, the line will be built most economically as regards the cost of wire and posts, if the cost of the posts is three times that of the additional length of wire required by "sagging."

GENERAL EXAMPLES

1. A block of stone weighing $\tfrac{1}{2}$ ton is raised by means of a rope which passes over a pulley vertically above it and is wound upon a windlass one foot in diameter. The windlass is worked by two men who turn cranks of length 2 feet. What force must each man exert perpendicular to the cranks?

2. A man sitting in one scale of a balance presses with a force of 60 pounds against the beam in a vertical direction and at a point halfway between the fulcrum and the end of the beam from which his scale is supported. If the beam is 5 feet long, find the additional weight which must be put in the other scale to maintain equilibrium.

3. The scales of a false balance hang at unequal distances a, b from the fulcrum, but balance when empty. A weight appears to have weights P, Q respectively when weighed in the two scales. Find its true weight and prove that

$$\frac{b}{a} = \sqrt{\frac{Q}{P}}.$$

EXAMPLES 87

4. A weightless string 24 inches in length is fastened to two points which are in the same horizontal line, and at a distance of 16 inches apart. Weights are fixed to two points at distances of 9 and 7 inches from the ends of the string and hang in such a way that the portion of the string between them is horizontal. Determine the ratio of the weights.

5. A light wire has a weight suspended from its middle point, and is itself supported by a string fastened to its two ends and passing over a smooth peg. Show that the wire can rest only in a horizontal or vertical position.

6. Three smooth pegs A, B, C stuck in a wall are the vertices of an equilateral triangle, A being the highest and the side BC horizontal; a light string passes once around the pegs and its ends are fastened to a weight W which hangs in equilibrium below BC. Find the pressure on each peg.

7. Two rings of weights P and Q respectively slide on a weightless string whose ends are fastened to the extremities of a straight rod inclined at an angle θ to the horizontal. On this rod slides a light ring through which the string passes, so that the heavy rings are on different sides of the light ring. All contacts are smooth and, in equilibrium, ϕ is the angle between the rod and those parts of the string which are close to the light ring. Prove that
$$\frac{\tan \theta}{\tan \phi} = \frac{P - Q}{P + Q}.$$

8. Two small heavy rings slide on a smooth wire, in the shape of a parabola with axis horizontal; they are connected by a light string which passes over a smooth peg at the focus. Show that their depths below the axis are proportional to their weights when they are in equilibrium.

9. Two equally heavy rings slide on a wire in the shape of an ellipse whose major axis is vertical, and are connected by a string which passes over a smooth peg at the upper focus. Show that there are an infinite number of positions of equilibrium.

10. $ABCD$ is a quadrilateral; forces act along the sides AB, BC, CD, DA measured by α, β, γ, δ times those sides respectively. Show that if these forces keep any system of particles in equilibrium, then
$$\alpha\gamma = \beta\delta.$$

11. A light rod rests wholly within a smooth hemispherical bowl of radius r, and a weight W is clamped on to the rod at a point whose distances from the ends are a and b. Show that θ, the inclination of the rod to the horizon in the position of equilibrium, is given by the equation
$$2\sqrt{r^2 - ab}\sin\theta = a - b.$$

88 STATICS OF SYSTEMS OF PARTICLES

12. A weightless rod, to which are fixed two rough beads of masses m and m', lies on an inclined plane and is free to turn about an axis through the rod perpendicular to the plane. If it be in a horizontal position, show that it will not begin to slip round unless

$$\mu < \frac{ma \sim m'b}{ma + m'b} \tan \alpha,$$

where α is the angle of the plane, and a and b the distances of m and m' respectively from the axis.

13. A bead of weight W, run on a smooth weightless string, rests on an inclined plane of angle α, the coefficient of friction between the bead and plane being μ. The ends of the string are tied to two points A, B in the plane at the same height. Show how to find the positions of limiting equilibrium for the bead, and show that in such a position P, the tension of the string is
$$\tfrac{1}{2} W \sec \tfrac{1}{2} APB \cdot (\tan^2 \alpha - \mu^2)^{\frac{1}{2}}.$$

14. A uniform string is placed on a rough sphere so as to lie on a horizontal small circle in altitude α. Prove that, if the string be on the point of slipping along the meridians, the tension is constant and equal to $W \cot (\alpha + \epsilon)$, where W is the weight of a length of the string equal to the radius of the circle, and ϵ is the angle of friction.

15. A weightless string is suspended from two fixed points and at given points on the string equal weights are attached. Prove that the tangents of the inclinations to the horizon of different portions of the string form an arithmetical progression.

16. A smooth semicircular tube is just filled with $2n$ equal smooth beads, each of weight W, that just fit the tube, and stands in a vertical plane with the two ends at equal height. If R_m is the pressure between the mth and $(m+1)$th beads from the top, show that

$$R_m = W \sin \frac{m\pi}{2n} \operatorname{cosec} \frac{\pi}{2n}.$$

17. In the last question, let the beads be indefinitely diminished in size. Prove that the pressure between any two beads is proportional to the depth below the top of the tube.

18. A heavy string hangs over two smooth pegs, at the same level and distance a apart, the two ends of the strings hanging freely and the central part hanging in a catenary. Show that for equilibrium to be possible, the total length of the string must not be less than ae.

EXAMPLES

19. A string of weight W is suspended from two points at the same level, and a weight W' is attached to its lowest point. If α, β are now the inclinations to the vertical of the tangents at the highest and lowest points, prove that
$$\frac{\tan \alpha}{\tan \beta} = 1 + \frac{W}{W'}.$$

20. A heavy string of length l is supported from two points, and at these points the string makes angles α, β with the vertical. Show that the height of one point above the other is
$$l \cos \tfrac{1}{2}(\alpha + \beta) \sec \tfrac{1}{2}(\alpha - \beta).$$

21. Prove that the direction of the least force required to draw a carriage is inclined at an angle θ to the ground, where $a \sin \theta = b \sin \epsilon$, a and b being the radii of wheels and axles respectively, and ϵ being the angle of friction.

CHAPTER V

STATICS OF RIGID BODIES

Rigidity

62. If we press a lump of wet clay or of soft putty with the finger, we find that a dent is left in the clay or putty; the force applied to the substance by the finger has caused it to change its shape. If we press a mass of jelly with the finger, we do not find any dent left in the jelly, but we notice that so long as the force is applied the shape of the jelly is altered, although it returns to its original shape as soon as the pressure is removed.

On the other hand, if we press a lead bullet or an ivory billiard ball with the finger, we do not notice any change of shape either while the pressure is applied or after. In ordinary language we say that the lead and ivory are *harder* than the clay and putty; in scientific language we say that they are *more rigid*.

63. A perfectly rigid body would be one which showed no change of shape under any force, no matter how great this force might be. A bullet and a billiard ball are not perfectly rigid. A billiard ball is pressed out of shape during the interval of collision with a second billiard ball, but regains its shape immediately, while a lead bullet is pressed permanently out of shape by striking a target. A perfectly rigid body does not exist in nature, although such bodies as a billiard ball or a bullet may be regarded as perfectly rigid, so long as the forces which act upon them are not too great.

We can give a mathematical definition of a perfectly rigid body as follows:

A body is perfectly rigid when the distance between any pair of particles in it remains unaltered, no matter what forces act on the body.

RIGIDITY

64. A rigid body can move about in space without changing the direction of any line in it — such a motion is called a *motion of translation*. It can also turn about any point P without the position of P altering — such a motion is called a *motion of rotation about P*. It can again have a motion compounded of a motion of translation and one of rotation, and this we shall show to be the most general motion which it can possibly have.

65. First we must notice that a rigid body is fixed when any three points in it are fixed, provided that these three points do not lie in a straight line. For let A, B, C be the three points. If we fix A and B, any motion which can take place must, since the body is by hypothesis perfectly rigid, be one in which the distances of any other point P from A and B remain unaltered. Thus P must describe a circle with AB as axis, and the motion of the body must be one of rotation about the line AB. Thus if A, B, C are not in a straight line, C must describe a circle about AB. But if C also is fixed, this cannot happen; in other words, no motion can take place, so that the body is fixed in position.

Thus the position of a rigid body is determined when the positions of three non-collinear points in it are determined.

66. We can now prove the theorem:

The most general motion of a rigid body is compounded of a motion of translation and one of rotation.

In fig. 46 let the figure on the left represent the body in its original position, and let the heavy curve on the right represent the body after it has been moved in any way. Let P be the position of any particle of the body in its original position, and let Q be the position of the same particle after the motion has taken place.

Imagine that the body is first moved from its original position in such a way that the point P moves to Q, while all lines in the body remain parallel to their original positions. This motion is one of pure translation. After this motion has taken place, we can turn the body about the point Q in such a way as to turn it into its final position. For let us take any two other particles R, S in the body (not in the same straight line with Q), and let R', S' be their final positions. Since the body is supposed to be perfectly

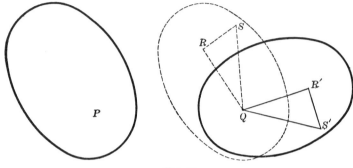

Fig. 46

rigid, it follows that all distances between particles of the body remain unaltered. Hence the distances QR, RS, SQ are respectively equal to QR', $R'S'$, $S'Q$. Thus the triangles QRS, $QR'S'$, being equal in all respects, can be superposed, and the motion of superposition of these triangles is the motion required and is a motion of pure rotation about Q, since Q does not move.

Since the position of a rigid body is fixed when any three points in it are fixed, it follows that the rigid body can only have one position in which the three points Q, R, S have given positions. But after the motion we have described, the three points Q, R, S are placed in their final positions. Hence the whole body must be in its final position, and this proves the theorem.

67. Axis of rotation. In a motion of rotation, let P be the point which remains fixed. Take any plane A through P, and let B be the position of the plane A after the rotation has occurred.

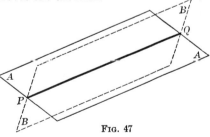

Fig. 47

These two planes both pass through P, and must therefore

CONDITIONS OF EQUILIBRIUM

intersect in some line PQ passing through P. This line is called the *axis of rotation*. The rotation can be imagined as a turning about an imaginary pivot running along the axis of rotation.

CONDITIONS OF EQUILIBRIUM FOR A RIGID BODY

68. We have seen that a rigid body is fixed when three non-collinear points in it are fixed. It follows that, whatever forces act on a rigid body, we can always hold it at rest by applying three suitably chosen forces at three points which are not in a straight line.

We can, however, select these forces in a special way.

Let us select any three points A, B, C, subject only to the condition that they are not in a straight line. By a suitably chosen force acting on the particle at A we shall always be able to keep the point A at rest.

When A is fixed B may or may not tend to move. If B tends to move, the direction of motion of B must be perpendicular to BA, since A cannot move. Hence after A is fixed it must be possible to fix B by applying at B a force perpendicular to BA.

When A and B are both fixed the only motion possible for the third point C is one perpendicular to both AC and BC, i.e. perpendicular to the plane ABC. Thus C can be held at rest by a force perpendicular to the plane ABC, and the whole body is now held at rest. Thus it has been proved that *a rigid body can be held at rest, in opposition to the action of any system of forces*, by the application of the following forces at three arbitrarily chosen points A, B, C not in a straight line:

(*a*) a force at A, direction unknown;

(*b*) a force at B, direction perpendicular to the line AB;

(*c*) a force at C, direction perpendicular to the plane ABC.

The condition that the original system of forces should hold the body in equilibrium is of course that no additional forces are required to fix the body, and hence that the three forces introduced at the points A, B, C should each vanish.

Transmissibility of Force

69. Consider a rigid body acted on by two forces W_A, W_B at two points A, B, these forces being equal in magnitude but acting in the opposite directions AB, BA.

Either the rigid body will be in equilibrium under the action of these two forces, or else it can be held at rest by three forces P_A, P_B, P_C at the points A, B, and any third point C not in the line AB, these forces being in the directions already mentioned, namely P_C being perpendicular to ABC, and P_B perpendicular to AB.

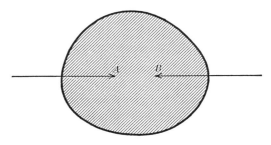

Fig. 48

Let these forces, if necessary, be put in so that the body is in equilibrium under the action of the forces W_A, W_B, P_A, P_B, P_C. The body being in equilibrium, the sum of the moments of these forces about any line, or of their components in any direction, must vanish, by § 50.

Let us consider what is the sum of the moments about the line AB. The forces W_A, W_B, P_A, P_B all meet this line, so that the moment of each of these forces vanishes. Thus the sum of the moments about the line AB consists of the moment of the single force P_C, and for the sum of these moments to vanish, the moment of P_C must vanish. Now P_C is perpendicular to the line AB, and does not intersect it, so that the moment of P_C can vanish only if the force P_C is itself equal to zero. This means that no force is required to keep the body from turning about AB as axis of rotation.

TRANSMISSIBILITY OF FORCE 95

The body is now held at rest by the two forces P_A, P_B, and is therefore in equilibrium under the forces P_A, P_B, W_A, W_B. Taking moments about a line through A perpendicular to AB, we find that the moments of P_A, W_A, W_B vanish, so that in order that the sum of the moments of the four forces taken about this line may vanish, we must have the moment of P_B equal to zero, and therefore P_B itself equal to zero. Thus the only force required to keep the body at rest is the force P_A at A.

A condition for equilibrium is now that the sum of the components of W_A, W_B, and P_A shall vanish in any direction. The components of W_A and W_B are, however, equal and opposite, so that the component of P_A must vanish in every direction. That is to say, we must have $P_A = 0$.

It has now been proved that the rigid body is in equilibrium under the action of the two forces W_A, W_B.

70. This establishes at once a principle known as the *transmissibility of force*.

The effect of a force acting on a rigid body depends on its magnitude and on the line along which it acts, but not on the particular particle in this line to which it is applied.

For, let the same force be applied at any two points Q, R of its line of action. An

FIG. 49

equal and opposite force at R can neutralize either of the two forces, which are therefore equivalent.

COMPOSITION OF FORCES ACTING IN A PLANE

71. Suppose that we have two forces P, Q acting at two points A, B of a rigid body, it being supposed that the two lines of action of these forces lie in one plane. Then the two lines of action, produced if necessary beyond the points A, B, will meet in some point C.

By the principle of the transmissibility of force it is immaterial whether the force P acts at A or at C; let us suppose it to act at C. In the same way let us suppose the force Q to act at C

instead of at B. Then we have the body acted on by two forces P, Q which act on the same particle C. These may be compounded, according to the rules explained in Chapter III, into a single force

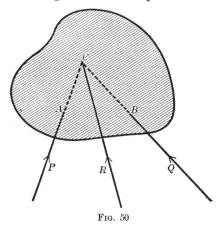

FIG. 50

R acting at C. Thus we can compound forces of which the lines of action intersect, even though they do not act on the same particle.

Having compounded two forces into a single force, we may compound this resultant with any third force which lies in the same plane as the two original forces, and in this way obtain a resultant of three forces, and so on.

Thus any number of forces which all lie in one plane may be compounded into a single force. This force is called the *resultant* of the original force.

72. An exception arises when we attempt to compound two parallel forces, for their lines of action do not meet. This difficulty, however, is easily surmounted. Let P, Q be the two forces to be compounded, and let AB be any line cutting their lines of action in A, B. Let us add to the system P, Q two forces:

(*a*) a force R acting along BA;

(*b*) a force R acting along AB.

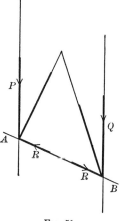

FIG. 51

These two forces being equal and opposite can be introduced without producing any effect. On compounding the first with P we obtain a resultant P' acting at A, and on compounding the second with Q we obtain a resultant Q' acting at B. Thus the original forces P, Q have been replaced by the new forces P', Q'. The lines of action of these forces, however, will not in general be parallel, so that they may be compounded into a single resultant force acting through their point of intersection.

73. Let us suppose that the forces originally to be compounded were R_1, R_2, $\cdots$, and that these have been compounded into a single resultant R. Let us take axes x, y in the plane in which these forces act, and let the components of R_1 along these axes be X_1, Y_1, those of R_2 being X_2, Y_2, and so on. Finally let the components of R be X, Y.

The system of forces which consists of the original forces R_1, R_2, $\cdots$, together with the resultant R reversed, constitutes a system in equilibrium. Resolving parallel to the axes, we obtain

$$X_1 + X_2 + X_3 + \cdots - X = 0,$$
$$Y_1 + Y_2 + Y_3 + \cdots - Y = 0.$$

Thus the components of R are given by the equations

$$X = X_1 + X_2 + X_3 + \cdots,$$
$$Y = Y_1 + Y_2 + Y_3 + \cdots.$$

The magnitude of R can be found from the equation

$$R^2 = X^2 + Y^2,$$

while the angle θ which the line of action of R makes with the axis of x can be found from the equation

$$\tan \theta = \frac{Y}{X}.$$

To obtain the position of the line of action of R we use the fact that the sum of the moments of

$$R_1, R_2, R_3, \cdots, \text{ and } - R$$

taken about any point in the plane must vanish. This gives us the moment of R about any point, and hence, since we know the magnitude and direction of R, we can find the position of its line of action.

STATICS OF RIGID BODIES

ILLUSTRATIVE EXAMPLE

Forces P, Q, R act on a rigid body, all the forces being in one plane, and their lines of action forming a right-angled isosceles triangle of sides a, a, $\sqrt{2}\,a$. Find their resultant.

Let ABC be the triangle, the forces P, Q, R acting along BC, CA, AB respectively. Take C for origin and CB, CA for axes of x, y. Let the resultant have components X, Y. Then, resolving along Cx, we obtain

$$X = -P + \frac{R}{\sqrt{2}},$$

and similarly resolving along Cy,

$$Y = Q - \frac{R}{\sqrt{2}}.$$

Thus the resultant is of magnitude **R** given by

$$\mathbf{R}^2 = X^2 + Y^2 = \left(-P + \frac{R}{\sqrt{2}}\right)^2 + \left(Q - \frac{R}{\sqrt{2}}\right)^2$$
$$= P^2 + Q^2 + R^2 - \sqrt{2}\,R(P+Q).$$

The angle θ which it makes with Cx is given by

$$\tan\theta = \frac{Y}{X} = -\frac{R - \sqrt{2}\,Q}{R - \sqrt{2}\,P}.$$

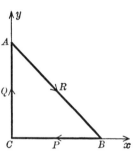

Fig. 52

To find the line of action, we express that the moment of **R** about C must be equal to the sum of the moments of P, Q, and R. If p is the perpendicular from C on to the line of action of the resultant **R**, this gives

$$\mathbf{R}p = R\frac{a}{\sqrt{2}},$$

so that

$$p = \frac{R}{\mathbf{R}}\frac{a}{\sqrt{2}},$$

and this determines the line of action of **R**.

EXAMPLES

(All forces are supposed to be acting on rigid bodies)

1. $ABCD$ is a square, and along the sides AB, BC, CD there act forces of 1, 2, 3 pounds respectively. Determine the magnitude and line of action of the resultant.

2. $ABCD$ is a square and forces P, Q, R, S act along the sides AB, BC, CD, DA respectively. What is the condition that their resultant shall pass through the center of the square?

3. In question 2, what is the condition

(a) that the resultant shall pass through A?

(b) that it shall pass through B?

(c) that the four forces shall be in equilibrium?

4. Forces P, Q, R act along the sides of a triangle ABC, and their resultant passes through the centers of the inscribed and circumscribed circles. Prove that
$$\frac{P}{\cos B - \cos C} = \frac{Q}{\cos C - \cos A} = \frac{R}{\cos A - \cos B}.$$

5. If four forces acting along the sides of a quadrilateral are in equilibrium, prove that the quadrilateral must be plane.

6. $ABCD$ is a plane quadrilateral and forces represented by AB, CB, CD, AD act along these sides, respectively, of the quadrilateral. Show that if there is equilibrium, the quadrilateral must be a parallelogram.

7. If a quadrilateral can be inscribed in a circle, prove that forces acting along the four sides and proportional to the opposite sides will keep it in equilibrium. Show also that the converse is true, namely that for equilibrium, the forces must be proportional to the opposite sides.

8. A quadrilateral is inscribed in a circle, and four forces act along the sides, and are inversely proportional to the lengths of these sides. Show that the resultant has for line of action the line through the intersections of pairs of opposite sides.

9. Forces act along the four sides of a quadrilateral, equal respectively to a, b, c, and d times the lengths of those sides. Show that if there is equilibrium,
$$ac = bd,$$
and that the further conditions necessary to insure equilibrium are that the ratios $a:b$ and $b:c$ shall be the ratios in which the diagonals are divided at their points of intersection.

10. In the last question show that the perpendicular distances to the first side, from the two points of the quadrilateral which are not on that side, are in the ratio
$$a(c-b) : d(b-a).$$

Parallel Forces

74. Let us use the method just explained, to determine the resultant of two parallel forces P, Q.

Take any point O on the line of action of P as origin, and take this line of action of P for axis Oy, as in fig. 53. Let the resultant be R, components X, Y. Then resolving we obtain
$$X = 0,$$
$$Y = P + Q,$$
so that the resultant force is of magnitude $P + Q$ and acts parallel

to Oy. Let us suppose that its distance from Oy is b, that of Q being denoted by a. Then taking moments about O we have

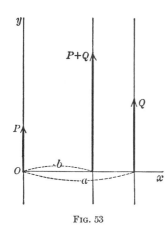

FIG. 53

$$(P + Q)b = Qa,$$

so that $\dfrac{b}{Q} = \dfrac{a}{P + Q} = \dfrac{a - b}{P}$,

showing that the line of action divides the distance between P and Q in the ratio $Q : P$.

Thus we have shown that *the resultant of two parallel forces P, Q is a force of magnitude $P + Q$, parallel to these forces, of which the line of action divides the distance between the lines of action of P and Q, in the ratio $Q : P$.*

75. Alternative treatment of parallel forces. We can prove directly from § 68 that parallel forces P, Q will be in equilibrium with a force $-(P + Q)$ parallel to them and acting along a line which divides the distance between them in the ratio $Q : P$.

Take two points A, B on the line of action of P, Q, and a third point C not on the line AB. Then the body, acted on by the forces P, Q, and $-(P + Q)$, can be kept in equilibrium by the further application of

(a) a force R_C at C, perpendicular to ABC;
(b) a force R_B at B, perpendicular to AB;
(c) a force R_A at A.

Thus the system of forces

$$P, \ Q, \ -(P + Q), \ R_C, \ R_B, \ R_A$$

will be in equilibrium.

Taking moments about the line AB, we find that $R_C = 0$. Taking moments about A, we find that $R_B = 0$, or else that it acts along BA, in which case it can be absorbed into R_A. Resolving perpendicular to the plane of the forces P, Q, we find that R_A can have no component perpendicular to the plane. Thus the four remaining forces

$$P, \ Q, \ -(P + Q), \ R_A,$$

are all in one plane.

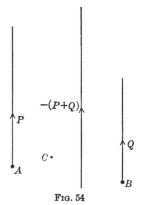

FIG. 54

Next, resolving parallel and perpendicular to the line of action of P, we find that both components of R_A vanish, and hence that $R_A = 0$. Thus the original forces were in equilibrium.

76. Clearly these methods of compounding forces can be extended, so that any number of parallel forces can be compounded into a single resultant force. We see at once, on reversing the resultant and resolving, that the resultant is parallel to the lines of action of the original forces, while its magnitude is equal to their algebraic sum.

This result is of importance in connection with the weights of bodies. It shows that the effect of gravity on any rigid body — i.e. the resultant of the weights of the individual particles of which the body is composed — may be regarded as a single force acting vertically along a single line.

In the next chapter it will be shown that, whatever position the rigid body is in, this line always passes through a definite point, fixed relatively to the body, known as its *center of gravity*.

77. Without assuming this, we can find the line of action in a number of simple cases. Suppose, for instance, we are dealing with a uniform rod. The weights of two equal particles equidistant from the center may be compounded into a single force acting through the middle point of the rod. Treating the weights of all the particles in this manner, we find that the weight of a uniform rod may be supposed to act at its middle point.

In the same way we can see that the weight of a circular disk, of a circular ring, or of a sphere may be supposed to act at its center; the weight of a cube or a parallelepiped at the intersection of its diagonals, and so on.

COUPLES

78. If we try to compound two parallel forces which are equal in magnitude but opposite in sign, we obtain as the resultant a force of zero amount of which the line of action is at infinity. Although such a force is of zero amount, its effect cannot be neglected: its moment does not vanish, being equal to the sum of the moments of the component forces. If, in fig. 55, AA', BB' are the parallel lines of action of two opposite forces each equal to R, and if PAB is a line at right angles to their direction, then the sum of their moments about a line through P, at right angles to the plane in which the forces act,

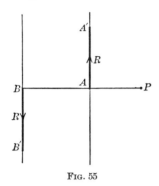

Fig. 55

$$= R \cdot PB - R \cdot PA$$
$$= Rd,$$

where d is the distance between the line of action of the forces. A pair of forces, equal in magnitude and opposite in direction, but not acting in the same line, is called a *couple*. Their moment about any point P in the plane containing their lines of action is independent of the position of the point P, and is spoken of as the *moment of the couple*.

Condition of Equilibrium

79. Since the resultant of a system of forces in a plane may be either a single force or a couple, the condition for there being no resultant will be that the resultant single force shall be nil, and that there shall be no couple. The component of the resultant force in any direction vanishes if the sum of the components vanishes. Thus, in order that the resultant force may vanish, it is necessary that the resolved parts in two different directions should vanish. If this condition is satisfied, there can be no resultant except a couple, and since the moment of a couple is the same, no matter about what point the moment is taken, it appears that there can be no couple if the moment about any one point is zero. Thus, as a necessary and sufficient condition of equilibrium for a system of forces in a plane, we have found the following:

A system of forces in a plane will be in equilibrium if the sum of the resolved parts in two directions each vanishes, and if the sum of the moments about any point also vanishes.

CONDITION OF EQUILIBRIUM

We can express the condition for equilibrium in a different form:

A system of forces in a plane will be in equilibrium if the sums of the moments about any three non-collinear points are each zero.

For, if the moment about any one point is zero, the resultant cannot be a couple. It must, therefore, be a single force. If the moments about each of two points A, B vanish, the line of action of this force must in general be AB, but if the moment about some third point C, not in the line AB, also vanishes, then the force itself must vanish.

EXAMPLES

1. Parallel forces of 5, 12, and 7 pounds act at the two ends and middle point, respectively, of a line 2 feet in length. Find the magnitude and line of action of their resultant.

2. Find the resultant of the forces in the last question when their magnitudes are respectively 5, -12, and 7 pounds.

3. Find the resultant of three forces, each of amount P, acting along the sides of an equilateral triangle, taken in order.

4. Prove that a system of forces acting along and represented by the sides of a plane polygon taken in order, is equivalent to a couple, whose moment is represented by twice the area of the polygon.

5. If the sums of the moments of any co-planar forces about three points which are not in a straight line are equal, and not each zero, prove that the system is equivalent to a couple.

6. A uniform rod is of length 3 feet and weight 24 pounds. Weights of 16 and 18 pounds are clamped to its two ends. Find at what point the rod must be supported so as just to balance.

7. A uniform beam weighing 20 pounds is suspended at its two ends, and has a weight of 50 pounds suspended from a point distant 7 feet and 3 feet from the two ends. Find the pressures at the points of suspension of the beam.

8. A uniform rod of weight 50 pounds and length 18 feet is carried on the shoulders of two men who walk at distances of 2 feet and 3 feet respectively from the two ends. A weight of 50 pounds is suspended from the middle point of the beam. Find the total weight carried by each man.

9. A dumb-bell weighing 32 pounds is formed of two equal spheres, each of radius 3 inches, connected by a bar of iron so that the centers of the spheres are 16 inches apart. One of the spheres is now removed, and the remainder of the dumb-bell is found to weigh 20 pounds. Find where this remaining part must be supported in order that it may just balance.

COUPLES IN PARALLEL PLANES

80. The result of § 79 shows that two couples acting in the same plane produce the same effect if their moments are equal. For, on reversing one of them, all the conditions of equilibrium are satisfied.

Thus we can determine the effect of a couple in any plane by knowing its moment only. We shall now show that the actual plane in which the couple acts is immaterial, the direction of this plane alone being of importance. In other words:

Couples of equal moments acting in parallel planes produce the same effect.

To prove this, we reverse one couple and show that the two couples are then in equilibrium. Let the first couple consist of two forces each equal to R, and let a common perpendicular to their lines of action meet the latter in points A, B. Let $A'B'$ be a line equal and parallel to AB in the plane of the second couple, and let the second couple reversed be represented by two forces

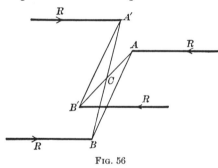
FIG. 56

R, R at $A'B'$. We can regard this couple as representing the second couple reversed, for its moment is equal and opposite to that of the second couple, while it is acting in the same plane as the second couple.

We have now to show that the four forces each equal to R acting at A, B, A', B' are in equilibrium. By construction, $ABB'A'$ is a parallelogram, so that C, the point of intersection of its diagonals, is also the middle point of each diagonal.

The two parallel forces R, R acting at A, B' may be compounded into a single force $2R$ acting at C, the middle point of AB', and similarly the two forces R, R at B, A' may be compounded into

COMPOSITION OF COUPLES 105

a force $2R$ at C, the middle point of $A'B$. These two forces $2R$, $2R$ are equal and act in opposite directions at the same point C. There is therefore equilibrium, proving that two couples are equivalent if they have equal moments and if the planes in which they act are parallel.

81. The direction perpendicular to the plane in which a couple acts is called the axis of a couple. Thus:

Two couples having the same axis and the same moment are equivalent.

COUPLES COMPOUNDED ACCORDING TO THE PARALLELOGRAM LAW

82. A couple, as we have just seen, is determined by a *quantity* (its moment) and a *direction* (its axis). Thus it may be fully represented by a straight line, the direction of this line being that of the axis, and the length representing, on any scale we please, the magnitude of the moment of the couple.

We shall now show that couples can be compounded according to the parallelogram law.

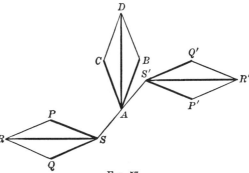

FIG. 57

THEOREM. *Two couples, represented in magnitude and direction by two lines AB, AC, will have as resultant a couple represented in magnitude and direction by AD, the diagonal of the parallelogram of which AB, AC are edges.*

Let AB, AC be lines representing by their direction and magnitude the axes and moments of two couples. Let SAS' be a line perpendicular to the plane ABC, A being its middle point. Let us draw planes through S, S' parallel to the plane ABC, and let the couple AB be replaced by two forces PS, $P'S'$ in these two planes, the lines PS, $P'S'$ being both perpendicular to AB. In the same way let the couple AC be replaced by two forces QS, $Q'S'$ in these same two planes.

The two couples have now been replaced by the four forces PS, QS, $P'S'$, $Q'S'$.

Let us complete the parallelograms $PSQR$, $BACD$, $P'S'Q'R'$. Obviously these parallelograms are all similar to one another, and corresponding lines in the first and second parallelograms are at right angles to one another. Thus a couple represented by AD may be replaced by forces RS, $R'S'$. But these two forces are exactly equivalent to the four forces PS, QS, $P'S'$, $Q'S'$ to which, as we have seen, the couples AB, AC may be reduced, and this proves the theorem.

FORCES IN SPACE

83. When the forces acting on a body are not all in one plane, their resultant will not in general be a single force.

THEOREM. *Any system of forces acting on a rigid body can be replaced by a force acting at an arbitrarily chosen point, and a couple.*

Let G be the chosen point, and let R be any force of which the line of action does not pass through G. At G let us introduce two equal and opposite forces, each equal to R and parallel to the line of action of R. By combining one of these forces with the original force R, we get a couple, so that the original force R can be replaced by a force parallel and equal to the original force but acting at G, and a couple.

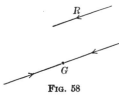
FIG. 58

Treating all the forces of the system in this way, we find that the original system of forces may be replaced by

(a) a number of forces acting at the chosen point G;

(b) a number of couples.

The forces acting at G can be combined into a single force at G, and the couples into a single couple, proving the result.

84. Theorem. *Any system of forces acting on a rigid body can be replaced by a force and a couple of which the axis is parallel to the line of action of the force.*

By the theorem just proved, the system may first be replaced by a force acting at any point O, and a couple. Let the force be of amount R, having OP as its line of action, and let the couple be of moment G, having OQ for its axis. If the angle POQ is denoted by θ, we may resolve the couple into two couples:

(*a*) a couple of moment $G \cos \theta$, having OP for axis;

(*b*) a couple of moment $G \sin \theta$, having its axis perpendicular to OP.

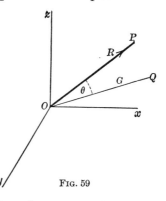

Fig. 59

The second of these couples may be replaced by any two forces provided these are chosen so as to be equivalent to the couple. Let us choose the first force to be a force $-R$ acting along OP, i.e. the force which will exactly neutralize the force R which we already have acting along OP. The second force of the couple must then be a force R acting along a line parallel to OP but at a distance from it equal to $G \sin \theta / R$.

The system has now been replaced by

(*a*) forces $+R, -R$ acting along OP;

(*b*) a force R parallel to OP;

(*c*) a couple $G \cos \theta$ of which the axis is parallel to OP.

The two forces (*a*) neutralize and we are left with a force R and a couple $G \cos \theta$ of which the axis is parallel to the line of action of the force. This proves the theorem.

The line of action of the force, which is now also the axis of the couple, is called the *central axis* of the system of forces. A system of forces is most simply specified by a knowledge of the magnitude of the force and couple, and of the position and direction of the central axis. Such a system is called a "wrench."

ILLUSTRATIVE EXAMPLES

1. *Two equal uniform planks are hinged at one end, and stand with their free ends on a smooth horizontal plane, being prevented from slipping by a rope which is tied to each plank at the same height up. Find the tension in this rope and the action at the hinge.*

In the figure let AB, AC represent the two planks, hinged at A, and let PQ be the rope. The forces acting on the plank AB will consist of

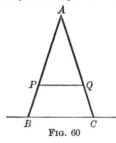

FIG. 60

(a) the action at the hinge A;
(b) the tension of the rope acting along PQ;
(c) the reaction at the foot B;
(d) the weight.

Of these four forces, (a) and (b) are the forces which it is required to find. Force (c) also is at present unknown. Force (d), as explained in § 77, can be regarded as a single force W, the total weight of the plank, and since we are told that the plank is uniform, this must be supposed to act through its middle point.

There is a simple way of finding force (c), the reaction at B. Since we are told that the contact at B is smooth, the direction of the reaction must be vertically upwards. Let its amount be R. From symmetry, there must be an exactly similar reaction at the foot C of the second plank. Now consider the equilibrium of the whole system which consists of the two planks and the rope. The only external forces which act on this system consist of

(a) the weight;
(b) the two reactions at B and C.

If we resolve vertically, we obtain, since this system is in equilibrium,

$$2W - 2R = 0,$$

so that $R = W$; each reaction is just equal to the weight of one plank, as we might have anticipated.

Of the four forces acting on the plank AB, the last two are now known, while the first two remain unknown. If we take moments about A, we shall get an equation between forces (b), (c), and (d), and this will enable us to find the unknown force (b), the tension.

If we denote the tension by T, and the angle BAC by 2θ, the equation obtained on taking moments about A is

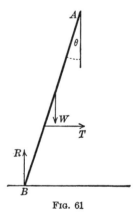

FIG. 61

$$R \cdot AB \sin\theta - W \cdot \tfrac{1}{2} AB \sin\theta - T \cdot AP \cos\theta = 0,$$

so that, remembering that $R = W$, we have

$$T = \frac{1}{2} \frac{AB}{AP} \tan\theta.$$

ILLUSTRATIVE EXAMPLES 109

Also on resolving horizontally and vertically, it is evident that the action at A must consist of a horizontal force of amount T and of direction opposite to that of T.

2. *A ring (e.g. a dinner napkin ring) stands on a table, and a gradually increasing pressure is applied by a finger to one point on the ring. Having given the coefficients of friction at the two contacts, examine how equilibrium will first be broken.*

Let A be the point of contact of the ring and table, and let B be the point of contact of the ring and finger. Let ϵ, ϵ' be the angles of friction at A and B respectively. Let the line BA make an angle α with the vertical.

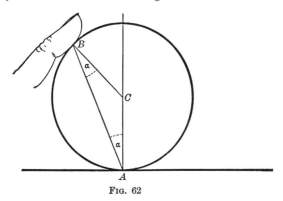

Fig. 62

The forces applied to the ring from outside are
(a) the reaction at A;
(b) the reaction at B;
(c) the weight of the ring.

Regarding the latter as a single force W acting along the vertical diameter CA of the ring, we see that so long as the ring remains at rest it is in equilibrium under the action of three forces.

Hence, by the theorem of § 52, the lines of action of the three forces must meet in a point.

The line of action of the weight is already known to be the vertical CA, and the line of action of the reaction at A must pass through A. Hence either

(α) the point in which the three lines of action meet must be A; or

(β) the reaction at A must act along CA, so that the point in which the three lines of action meet will be some point in CA, other than A.

The second alternative may be dismissed at once. For if the reaction at A acts along CA, this and the weight may be combined into a single force, and there must now be equilibrium under this force and the reaction at B. This requires that each force should vanish, i.e. there must be no pressure at B, and the reaction at A must be just equal to the weight of the ring. This obviously gives a state of equilibrium — the ring is standing at rest on the table, acted on solely by its weight — but this state of equilibrium is not the one with which we are concerned in the present problem.

Let us now consider the meaning of alternative (α). If the three lines of action meet in A, the reaction at B must act along BA, and this must be true no matter how great the pressure at B. Hence the reaction at B will always make an angle α with the normal.

If α is less than ϵ', the angle of friction at B, this will be a possible line of action for the reaction, and no slipping can take place at B, no matter how great the pressure applied at B may be.

On the other hand, if α is greater than ϵ', equilibrium is impossible, no matter how small the pressure applied at B may be. Thus, when there is equilibrium, the pressure at B must vanish, and we are led to the same state of equilibrium as was reached from case (β). As soon as the pressure at B becomes appreciable, equilibrium is obviously broken by slipping taking place at B, since for equilibrium to be maintained at B, the reaction would have to act at an angle greater than the actual angle of friction.

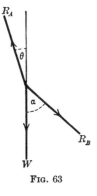

FIG. 63

Thus the solution resolves itself into two different cases:

CASE I. If $\alpha > \epsilon'$, as soon as pressure is applied at B, motion takes place. The ring slips at B, and consequently rolls at A.

CASE II. If $\alpha < \epsilon'$, we have seen that no matter how great a pressure is applied at B, there can never be slipping at B. It remains to examine whether there can be slipping at A.

To settle this question, we have to determine whether the reaction at A can ever be made to act at an angle with the vertical which is as great as the angle of friction at A, namely ϵ. Now the ring is acted on by three forces, the reactions at A and B, say R_A and R_B, and its weight W. The lines of action of these forces meet in the point A, and from Lami's theorem we can connect the magnitudes of the forces with the angles between them.

The lines of action of the three forces are represented in fig. 63. The angle between W and R_B is, as we have seen, always equal to α. Let the angle between R_A and the vertical be supposed to be θ. Then, by Lami's theorem,

$$\frac{R_A}{\sin \alpha} = \frac{R_B}{\sin \theta} = \frac{W}{\sin (\alpha - \theta)}.$$

The value of R_A is not given, but on equating the last two fractions we have

$$\frac{W}{R_B} = \frac{\sin (\alpha - \theta)}{\sin \theta} = \sin \alpha \cot \theta - \cos \alpha,$$

so that

$$\cot \theta = \left(\cos \alpha + \frac{W}{R_B}\right) \cosec \alpha.$$

EXAMPLES

This equation enables us to trace the changes in the value of the angle θ as R_B is gradually increased. We find that when $R_B = 0$ the value of θ is $\theta = 0$, and that as R_B increases θ increases continually, but never exceeds the value $\theta = \alpha$, which is reached when $R_B = \infty$.

If $\epsilon < \alpha$, the value of θ will pass through the value ϵ when R_B reaches a certain value, namely

$$R_B = \frac{\sin \epsilon}{W \sin(\alpha - \epsilon)},$$

and slipping at A will take place at this point.

If $\epsilon > \alpha$, the value of θ will never reach the value ϵ, so that slipping at A can never occur. Thus equilibrium is never broken, and the harder we press at B the more firmly the ring is held between the finger and the table.

We can now summarize the results which have been obtained, as follows:

If $\alpha > \epsilon'$, the ring rolls along the table as soon as we begin to press at B.

If $\alpha < \epsilon'$, there are two cases:

(a) $\alpha > \epsilon$, the ring will slip at A as soon as sufficient pressure is applied;

(b) $\alpha < \epsilon$, the ring cannot be made to move under any amount of pressure.

To make the ring shoot out from under the finger by slipping at A (in which case it returns to the hand, as in the well-known trick), it is necessary to press at a point on the ring at which α is greater than ϵ, while being less than ϵ'. We notice that if ϵ is greater than ϵ', it is impossible to project the ring in this way; this can only be done if the contact with the finger is rougher than the contact with the table.

GENERAL EXAMPLES

1. A pair of steps is formed by two uniform ladders each of length 12 feet and weight 20 pounds, jointed at the top, and having their points at distances 5 feet from the ground connected by a rope. The steps stand on a smooth horizontal plane, and a man of weight 160 pounds ascends to a height of 9 feet on one side. Find the tension in the rope.

2. A heavy uniform rod is supported by two strings of lengths a, b. The upper ends of the strings are tied to the same point, the lower ends being tied to the two ends of the rod. Show that the tensions of the strings are proportional to a and b respectively.

3. Two small fixed pegs are in a line inclined at an angle θ to the horizon. A rough thin rod passes under the lower and rests on the higher, this latter being lower than the center of gravity of the rod. The distances of the center of gravity from the two pegs are a and b respectively, and the coefficient of friction is μ. Show that if the rod is on the point of motion,

$$\mu = \frac{b - a}{b + a} \tan \theta.$$

4. Two heavy uniform rods have their ends connected by two light strings, and the whole system is suspended by the middle point of one rod. Prove that in equilibrium either the rods or the strings are parallel.

5. Two rods AB, CD lying on a smooth table are connected by stretched strings AC, BD. If the system is kept in equilibrium, by forces acting at the middle points of the rods, prove that

(a) the rods must be parallel;

(b) the tensions must be proportional to the strings.

6. $ABCD$ is a parallelogram and E is the intersection of the diagonals AC, BD. Show that parallel forces 7, 5, 16, 4 at A, B, C, D respectively are equivalent to other parallel forces, 8 at the middle point of CD, 10 at the middle point of BC, and 14 at E.

7. A solid cube is placed on a rough inclined plane of angle α with two edges of its base along lines of greatest slope. The angle of friction is ϵ. Prove that if $\alpha > 45°$ it will at once topple over, while if $\epsilon < \alpha < 45°$ it will slide down the plane. If α is less than either ϵ or $45°$, find the friction brought into action.

8. A uniform rod of length $2l$ and weight W rests over a smooth peg at distance $h(<l)$ from a smooth vertical wall at an angle θ with the horizontal, its lower end pressing against the wall, and its upper being held by a vertical string. Find the tension of the string and show that it vanishes if

$$\theta = \cos^{-1} \sqrt[3]{\left(\frac{h}{l}\right)}.$$

9. Two equal uniform spheres, each of weight W and radius a, rest in a smooth hemispherical bowl of radius b. Find the pressure between the two spheres and also the pressure of each on the bowl.

10. A uniform rod rests with its two ends on smooth inclined planes, inclined to the horizontal at angles α and β. Find the inclination of the rod to the horizontal.

11. In the last question a weight equal to that of the rod is clamped to it. At what point must it be clamped in order that the rod may rest horizontally?

12. A uniform circular ring of weight W has a bead of weight w fixed on it and hangs on a rough peg. Show that if $\sin \epsilon > \dfrac{w}{W+w}$, then the ring can rest without slipping, whatever point of it rests on the peg, ϵ being the angle of friction.

13. A pentagon $ABCDE$, formed of equal uniform heavy rods connected by smooth joints at their ends, is supported symmetrically in a vertical plane with A uppermost, and AB, AE in contact with two smooth pegs in the same horizontal line. Prove that if the pentagon is regular, the pegs must divide AB and AE each in the ratio

$$1 + \sin \tfrac{1}{10}\pi : 3 \sin \tfrac{1}{10}\pi.$$

EXAMPLES

14. A uniform beam of length l leans against the horizontal rim of a hemispherical bowl of radius a, with its lower end resting upon the smooth concave surface. Find its inclination to the vertical.

15. A bowl in the shape of a paraboloid of revolution is placed with its axis vertical. A uniform rod rests on a peg at the focus and has its lower end resting on the inner surface. Both contacts are perfectly smooth. Find the inclination of the rod to the vertical.

16. A uniform beam of weight W rests against a vertical wall and a horizontal plane with which it makes the angle α. Both contacts are perfectly smooth. The lower end of the beam is attached by a string to the foot of the wall. Find the tension of the string.

17. One end of a straight uniform heavy rod rests on a rough horizontal plane, the other end being connected with a fixed point by a string. If θ, ϕ, ψ be the inclinations of the string, the rod, and the total reaction of the horizontal plane respectively to the vertical, show that
$$\cot \theta \pm 2 \cot \phi - \cot \psi = 0.$$

18. Two uniform rods AB, BC of the same material but of different lengths are jointed freely at B and fixed to a vertical wall at A and C. Show that the direction of the reaction at B bisects the angle ABC.

19. A uniform regular-hexagonal board $ABCDEF$ of given weight W is supported in a horizontal position on three pegs, placed at the corners A, B and the middle point of DE. Find the pressures on the pegs.

20. Two spheres of radii a, b and weights W, W' respectively are suspended freely by strings of lengths l, l' respectively from the same hook in the ceiling. If $l' > l + 2a$, show that the angle which the first string makes with the vertical is
$$\sin^{-1} \frac{Wa}{(W+W')(a+l)}.$$

21. A uniform rod hangs by two strings of lengths l, l' fastened to its ends and to two hooks in the same horizontal line at the distance a. If the strings cross one another and make the respective angles α, α', with the horizontal, show that when the rod is in equilibrium
$$\sin(\alpha + \alpha')(l' \cos \alpha' - l \cos \alpha) = a \sin(\alpha - \alpha').$$

22. A uniform plank of length $2b$ rests with one end on a rough horizontal plane, touches a smooth fixed cylinder of radius a lying on the plane, and makes an angle 2α with the plane, the angle of friction being ϵ. Show that equilibrium is possible if
$$a \sin \epsilon > b \tan \alpha \cos 2\alpha \sin(2\alpha + \epsilon).$$

23. Two equal and similar isosceles wedges, each of weight W and vertical angle 2α, are placed side by side with their bases on a rough horizontal table so as to be just in contact along an edge. A smooth sphere of weight w and radius r is supported between them, being in contact with a face of each. Prove that for equilibrium it is necessary that

$$\mu > \frac{w \cot \alpha}{2W + w}, \quad r < 2a \sin \alpha \tan \alpha \left(1 + \frac{W}{w}\right),$$

where μ denotes the coefficient of friction and $2a$ is the length of either base.

24. A seesaw consists of a plank of weight w laid across a fixed rough log whose shape is that of a horizontal circular cylinder. The inclination to the horizontal at which it balances is increased to α when loads W, W' are placed at the lower and higher ends respectively; and the inclination is reduced to β when these loads are interchanged. Show that the inclination of the plane when unloaded is

$$\frac{w'(W + W' + w)(W'\alpha - W\beta)}{w(W + W' - w')(W - W')},$$

w' being the weight which, placed at the higher end, would balance the plank horizontally.

25. A chain is formed of $2n$ exactly similar links, the contacts between consecutive links being perfectly smooth. The two end links can slide on a horizontal wire, the contact here being rough, coefficient of friction μ. Show that in the limiting position of equilibrium, the inclination of either of the upper links to the vertical is

$$\tan^{-1} \frac{2n\mu}{2n - 1}.$$

26. Two equal circular disks of radius r, with smooth edges, are placed on their flat sides in the corner between two smooth vertical planes inclined at an angle 2α, and touch each other in the line bisecting the angle. Prove that the smallest disk which can be pressed between them without causing them to separate is one of radius $r(\sec \alpha - 1)$.

27. How is the result of the last question modified if all the contacts are rough, the angle of friction at each being ϵ?

28. Two uniform ladders are jointed at one end and stand with their other ends on a rough horizontal plane. A man whose weight is equal to that of one of the ladders ascends one of them. Prove that the other will slip first.

If it begins to slip when he has ascended a distance x, prove that the coefficient of friction is $\dfrac{a + x}{2a + x} \tan \alpha$, a being the length of each ladder and α the angle each makes with the vertical.

29. A weightless ladder rests against a smooth cube of weight W, standing on smooth ground, with the foot of the ladder tied to the middle

point of one of the lowest edges of the cube; a man of weight w ascends the ladder. Prove that, if the ladder projects above the top of the cube, the cube will tilt before he reaches the top of the cube unless
$$W > 2w \cos \alpha (\sin \alpha - \cos \alpha),$$
where α is the inclination of the ladder to the horizontal.

30. Four equal spheres rest in contact at the bottom of a smooth spherical bowl, their centers being in a horizontal plane. Show that if another equal sphere be placed upon them the lower spheres will separate if the radius of the bowl be greater than $(2\sqrt{13} + 1)$ times the radius of a sphere.

31. Three equal spheres rest in contact on a smooth horizontal plane, so that their centers form an equilateral triangle, and are bound together by a fine string passing around them on the level with their centers. If another equal sphere be placed symmetrically on them, show that the tension of the string is increased by $\dfrac{1}{3\sqrt{6}} w$, where w is the weight of the upper sphere.

32. A right circular cone of vertical angle 2α rests with its base on a rough horizontal plane. A string is attached to the vertex, and is pulled in a horizontal direction with a gradually increasing force. Find in what way equilibrium will first be broken.

33. A heavy particle is placed on a rough inclined plane whose inclination is exactly equal to the angle of friction. A thread is attached to the particle and is passed through a hole in the plane which is lower than the particle, but is not in the line of greatest slope through it. Show that if the thread be gradually drawn through the hole, the particle will describe a straight line and a semicircle in succession.

34. A uniform cubical block of weight W rests with one edge horizontal on a rough inclined plane, and against the block rests a rough sphere of weight W' of radius less than an edge of the cube. The inclination of the plane is gradually increased. Examine the different ways in which equilibrium may be broken, and determine which will actually occur in a given case.

35. A rough uniform rod is placed on a horizontal plane and is acted on, at one of the points of trisection of its length, by a horizontal force in a direction perpendicular to its length. Find about what point the rod will begin to turn.

36. A heavy bar AB is suspended by two equal strings of length l, which are originally parallel. Find the couple which must be applied to the bar to keep it at rest after it has been twisted through an angle θ in the horizontal plane.

37. The line of hinges of a door is inclined at an angle α to the vertical. Show that the couple required to keep it in a position inclined at an angle β to that of equilibrium is proportional to $\sin\alpha \sin\beta$.

38. Show that any system of forces acting on a rigid body can be reduced to two equal forces equally inclined to the central axis.

39. Prove that the central axis of two forces P and Q intersects the shortest distance c between their lines of action and divides it in the ratio
$$Q(Q + P\cos\theta) : P(P + Q\cos\theta),$$
θ being the angle between their directions. Also prove that the moment of the principal couple is
$$\frac{cPQ\sin\theta}{\sqrt{P^2 + Q^2 + 2PQ\cos\theta}}.$$

40. Prove that the axis of the resultant of two given wrenches (R_1, H_1) and (R_2, H_2), the axes of which are inclined to each other at an angle θ, intersects the shortest distance, $2c$, between their axes at a point the distance of which from the middle point is
$$\frac{(R_1^2 - R_2^2)c + (H_1 R_2 - H_2 R_1)\sin\theta}{R_1^2 + R_2^2 + 2R_1 R_2 \cos\theta}.$$

CHAPTER VI

CENTER OF GRAVITY

85. As we have seen, the action of gravity on a system of masses may be represented by a system of parallel forces, these forces consisting of a force acting on each particle equal to the weight of the particle, its direction being vertically downwards. By the rules explained in the last chapter, these forces may be compounded into a single force. The magnitude of this force is the sum of all the component forces, and is therefore the total weight of the body, while the direction of the force, being parallel to the component forces, is itself vertically downwards. The problem discussed in the present chapter is that of determining the position of the line of action of this force.

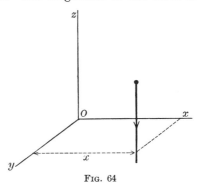

FIG. 64

86. Let the particles be of masses $m_1, m_2, \cdots$. Let rectangular axes be taken, the axis of z being vertical, and let the coördinates of the first particle be x_1, y_1, z_1, the coördinates of the second be x_2, y_2, z_2, and so on.

The weight of the first particle is $m_1 g$, and its line of action cuts the plane Oxy in a point of which the coördinates are $x_1, y_1, 0$. Hence the moment of the force about the axis Oy is $m_1 g x_1$. Let the line of action of the resultant cut the plane Oxy in the point $\bar{x}, \bar{y}, 0$. Then the moment of the resultant about the axis Oy is $\left(\sum m_1\right) g \bar{x}$, where $\sum m_1$ is the sum of the masses of all the particles.

CENTER OF GRAVITY

Since the moment of the resultant is equal to the sum of the moments of the separate forces, we must have

$$\left(\sum m_1\right) g\bar{x} = \sum (m_1 g x_1),$$

so that
$$\bar{x} = \frac{\sum m_1 x_1}{\sum m_1}.$$

Similarly
$$\bar{y} = \frac{\sum m_1 y_1}{\sum m_1}.$$

These equations determine the coördinates $\bar{x}, \bar{y}$ of the point in which the line of action of the resultant meets the plane Oxy.

We have, however, seen that the coördinates of the centroid of masses m_1 at x_1, y_1, z_1, m_2 at x_2, y_2, z_2, etc., are

$$x = \frac{\sum m_1 x_1}{\sum m_1}, \quad y = \frac{\sum m_1 y_1}{\sum m_1}, \quad z = \frac{\sum m_1 z_1}{\sum m_1},$$

so that the point in which a vertical through the centroid will meet the plane Oxy must be

$$\frac{\sum m_1 x_1}{\sum m_1}, \quad \frac{\sum m_1 y_1}{\sum m_1}, \quad 0,$$

i.e. the point must be the point $\bar{x}, \bar{y}, 0$ in which the line of action of the resultant force meets the plane Oxy. Thus

The line of action of the resultant force of gravity is the vertical line through the centroid of the particles.

For this reason the centroid of a number of points, weighted according to the masses of the particles which occupy these points, is called the *center of gravity* of the particles. The effect of gravity acting on a rigid body is, as we have now seen, represented by a single force acting vertically downwards through the center of gravity of the body, the amount of the force being equal to the total weight of the body. The action of gravity is, accordingly, the same as if the whole mass of the body were concentrated in a single particle placed at the center of gravity.

SYSTEM OF MASSES 119

87. It is clear that if we suspend a rigid body or system of bodies by a string, the center of gravity must be vertically below the string. For all the forces acting on the system reduce to two, — the tension of the string and the weight acting at the center of gravity, — and in equilibrium these two must act along the same line.

In the same way it will be seen that if a body is placed on a point in such a way as to balance in equilibrium on this point, then the center of gravity must be vertically above the point.

88. A few simple instances of the position of the center of gravity have been mentioned in § 77. These were as follows:

(a) the center of gravity of a uniform rod is at its middle point;

(b) the center of gravity of a uniform circular disk, circular ring, or sphere is at the center;

(c) the center of gravity of a cube or parallelepiped is at the center (i.e. the intersection of the diagonals).

89. It is easy to find the center of gravity of a system of bodies when the center of gravity of each of the component parts is known. For, regarding the weight of each of the bodies as a single force acting through its center of gravity, we have a number of parallel forces in action, and on compounding these according to the rules already explained, the line of action of the resultant determines the line along which the total weight will act. Thus the center of gravity of the whole system of bodies will be the centroid of the centers of gravity of the separate bodies, weighted according to the masses of these bodies.

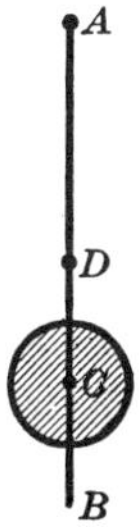

90. For instance, let us suppose we require to find the center of gravity of a pendulum which consists of a wire of length l and weight w, to which is affixed a circular bob of weight W, the center of the circle being at a distance a from the end of the wire. Let AB be the wire, C the center of the bob, and D the middle point of the wire. The center of gravity of the wire will be at D and that of the bob at C, so that the center of gravity of the whole will be at the centroid

FIG. 65

of the points D and C, these being weighted in the ratio $w:W$. Denoting this center of gravity by G, we have, from the formula

$$x = \frac{\sum m_1 x_1}{\sum m_1},$$

on treating the line $ADCB$ as the axis of x and taking A as origin,

$$AG = \frac{W \cdot AC + w \cdot AD}{W + w}$$

$$= \frac{W(l - a) + \tfrac{1}{2} wl}{W + w}.$$

EXAMPLES

1. Weights of 3 pounds are placed at each of three corners of a square, and a weight of 5 pounds at the fourth corner. Find their center of gravity.

2. From one corner of a cardboard square of edge 6 inches, a square of edge 3 inches is cut. Find the center of gravity of the remainder.

3. A thin rod of weight 6 ounces and length 6 inches is nailed on to a circle of cardboard of weight 6 ounces and radius 6 inches so that its two ends are on the circumference of the circle. Find the center of gravity of the whole.

4. A bicycle wheel of diameter 26 inches weighs 3 pounds. Each spoke is of length 11 inches, and starts from the hub at a distance of $\tfrac{1}{2}$ inch from the central axis of the wheel. If one spoke is taken out, find the center of gravity of the wheel.

5. A hammer has for handle a wooden cylinder, length 8 inches, radius $\tfrac{3}{4}$ inch, weight 8 ounces, and for head an iron cylinder, from which a hollow is cut into which the handle exactly fits, the handle coming through so as to be exactly flush with the iron. The head is of length 3 inches, radius $1\tfrac{1}{4}$ inch, and weight 3 pounds. Find the approximate position of the center of gravity.

6. A box, without lid, is made of 1-inch wood so as to have internal dimensions $12 \times 12 \times 12$ inches. Find the position of its center of gravity.

7. A uniform thin rod 28 inches in length is bent so that the two parts, of lengths 12 and 16 inches, are at right angles to one another. Find the center of gravity.

8. A uniform wire is bent into the form of a triangle. Show that the center of gravity of the wire coincides with the center of the circle inscribed in the triangle formed by joining the middle points of the sides.

9. A T-square is made of cedar of uniform density, the crosspiece being of dimensions $6 \times 2 \times \tfrac{1}{4}$ inches, and the arm being of dimensions $8 \times 1\tfrac{1}{2} \times \tfrac{1}{8}$ inches. The crosspiece is cut away so that the under surface of the instrument is plane. Find the position of the center of gravity of the whole.

TRIANGLE

10. Three beads of weights W_a, W_b, W_c are placed on a circular wire, and when the beads are at the points A, B, C on the circle, the center of gravity of the whole is found to coincide with O, the center of the circle. Show that

$$\frac{W_a}{\sin BOC} = \frac{W_b}{\sin COA} = \frac{W_c}{\sin AOB}.$$

Center of Gravity of a Lamina

91. It is often of importance to be able to find the position of the center of gravity of a lamina, i.e. of a thin, plane shell of uniform thickness and density, such, for instance, as is obtained by cutting a figure out of a sheet of cardboard.

92. Center of gravity of a triangle. Let ABC represent a triangular lamina of which it is required to find the position of the center of gravity. Let us imagine the triangle divided by lines parallel to the base BC into a very great number of infinitely narrow strips. Let pq be any single strip. Since, by hypothesis, we may regard this strip as of vanishingly small width and thickness, we may treat it as a thin uniform rod. The center of gravity of a thin uniform rod is at its middle point, so that the weight of the strip pq may be supposed to act at r, the middle point of pq.

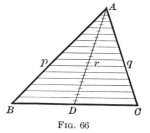

Fig. 66

The weights of the other strips may be treated in the same way, so that the weight of the whole triangle may be replaced by the weights of a system of particles situated at the middle points of these strips.

Now if D is the middle point of the base BC, the middle points of all the strips lie in the line AD. Thus the weight of the triangle is replaced by the weights of a number of particles, all of which are situated in the line AD. It follows that the center of gravity of the whole triangle must lie in the line AD.

We might equally well have supposed the triangle divided into strips parallel to the side AC. We should then have found that the center of gravity must lie in the line BE, where E is the middle point of AC.

These two results fully fix the position of the center of gravity; it must be at the intersection of the lines AD, BE.

Join DE. Then the triangles DCE, BCA are two similar triangles, the former being just half the size of the latter. Thus DE must be parallel to AB, and of half the length of AB.

It now follows that DGE, AGB are similar triangles, of which the former is half the size of the latter. Hence GD is half of AG.

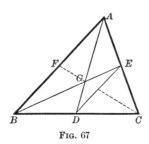

Fig. 67

Thus G divides AD in the ratio $2:1$. If we join C to F, the middle point of AB, we can in the same way show that CF must divide AD in the ratio $2:1$. Thus CF must also pass through G.

The three lines AD, BE, CF, which join the vertices of the triangle to the middle points of the opposite sides, are called the *medians* of the triangle. We have shown that the three medians meet in the same point G, and that this point is the center of gravity of the triangle. We have also shown that the center of gravity divides any median in the ratio $2:1$, i.e. that it is one third of the way up the median, starting from the base.

93. Center of gravity of any polygon. The center of gravity of any rectilinear polygon can be found by dividing it up into triangles and replacing each triangle by a particle at its center of gravity.

EXAMPLES

1. Show that the center of gravity of a triangle coincides with that of three equal particles placed at its angular points.

2. Prove that if the center of gravity of a triangle coincides with its orthocenter the triangle is equilateral.

3. A cardboard square is bent along a diagonal until the two parts are at right angles. Find the position of its center of gravity.

4. A quarter of a triangular lamina is cut off by a line parallel to its base. Where is the center of gravity of the remainder?

ROD OF VARYING DENSITY

5. A right-angled isosceles triangle is cut out from a lamina in the shape of an equilateral triangle so as to have the same base as the original triangle. Find the center of gravity of the V-shaped piece left over.

6. The center of gravity of a quadrilateral lies on one of its diagonals. Show that this diagonal bisects the other diagonal.

Centers of Gravity obtained by Integration

94. Center of gravity of a rod of varying density. Let AB be a rod of which the weight per unit length varies from point to point, and let ρ denote the weight per unit length at any point.

Let P, Q be two adjacent points, the distances of P, Q from the point A being x and $x + dx$ respectively. Then the length PQ is dx, and its mass is $\rho\, dx$, where ρ is the mass per unit length at this point. When dx is made vanishingly small, the distance of the center of gravity of PQ from A may be taken to be x. Hence if $\bar{x}$ denotes the distance of the center of gravity of the whole rod from A,

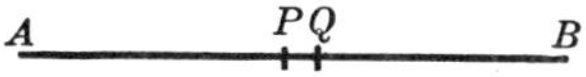

$$\bar{x} = \frac{\sum(mx)}{\sum m},$$

where m is the mass of any element such as PQ, and the summation is taken over all the particles of which the rod is formed. Putting $m = \rho\, dx$, this becomes

$$\bar{x} = \frac{\sum(\rho x\, dx)}{\sum(\rho\, dx)},$$

or, in the notation of the integral calculus,

$$\bar{x} = \frac{\int \rho x\, dx}{\int \rho\, dx}, \qquad (28)$$

where the integration extends in each case over the whole rod. The variable ρ will be a function of x, and the integrations cannot be performed until the exact form of this function is known.

Fig. 68

95. To take a definite instance, let us suppose that the density increases uniformly from one end to the other. Let the density at A be 0, and that at B be k. If the rod is of length a, the density at a distance x from A will be $k\left(\dfrac{x}{a}\right)$. Thus we must put

$$\rho = k\left(\dfrac{x}{a}\right)$$

in formula (28), and so obtain

$$\bar{x} = \dfrac{\int k\left(\dfrac{x}{a}\right) x\,dx}{\int k\left(\dfrac{x}{a}\right) dx},$$

where the integration is from $x = 0$ to $x = a$. Dividing numerator and denominator by $\dfrac{k}{a}$, we obtain

$$\bar{x} = \dfrac{\int_0^a x^2\,dx}{\int_0^a x\,dx}$$

$$= \dfrac{\tfrac{1}{3} a^3}{\tfrac{1}{2} a^2} = \dfrac{2}{3} a,$$

showing that the center of gravity is two thirds of the way along the rod.

96. We can use this result to obtain the center of gravity of a triangle. As in § 92, we divide the triangle into parallel strips, and replace each strip by a particle at its middle point. The mass of each particle must be that of the strip which it replaces, and this is jointly proportional to the width and the length of the strip. If x is the distance of any particle from A measured along the median AD, the width of a strip is proportional to dx, the length intercepted on the median, while the length of the strip is proportional to x, the distance from a. Thus $\rho\,dx$ must be simply

CIRCULAR ARC 125

proportional to $x\,dx$, and as we have just found, this at once leads to the result
$$\bar{x} = \tfrac{2}{3}\,a,$$
where a is the length of the median. This is exactly the result previously obtained.

97. Center of gravity of a circular arc. The same method can be used to find the center of gravity of a wire bent into the form of a circular arc PQ. Let O be the center of the circle, and A the middle point of the arc, and let the whole arc subtend an angle 2α at the center. Consider a small element cd of the half PA of the wire. Let the angle dOA be θ, and cOA be $\theta + d\theta$, so that the element subtends the angle $d\theta$ at the center. If a is the radius of the circle, the length of this element is $a\,d\theta$, so that if w is the mass of the wire per unit length, the mass of the element will be $wa\,d\theta$. This and the similar element $c'd'$ in the half AQ of the wire form a pair of equal particles equidistant from the central line OA. They may be replaced by a single particle of mass $2wa\,d\theta$ at their center of gravity. This center of gravity is in OA, at the point at which the line joining the two elements meets OA, and hence at a distance $a\cos\theta$ from O. Denoting this by x, and the mass $2wa\,d\theta$ by m, we have, for the distance $\bar{x}$ of the center of gravity of the whole wire from O,

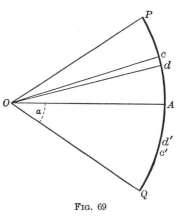

FIG. 69

$$\bar{x} = \frac{\sum(mx)}{\sum m}$$
$$= \frac{\int a\cos\theta\,(2wa\,d\theta)}{\int 2wa\,d\theta},$$

126   CENTER OF GRAVITY

where the integration is from $\theta = 0$ to $\theta = \alpha$. Simplifying, we find

$$\bar{x} = \frac{a \int_{\theta=0}^{\theta=\alpha} \cos\theta \, d\theta}{\int_{\theta=0}^{\theta=\alpha} d\theta}$$

$$= \frac{a \sin \alpha}{\alpha}, \qquad (29)$$

giving the position of the center of gravity.

When α is very small, $\sin \alpha$ and α become equal, so that for very small values of α, formula (29) reduces to $x = a$, as it ought. This simply expresses that as the curvature of the arc decreases, the center of gravity approximates more and more closely to the middle point of the arc. Finally, when $\alpha = 0$, the arc becomes a straight rod, and the center of gravity is, of course, found to be exactly at the middle point.

For an arc bent into a semicircle, we take $\alpha = \dfrac{\pi}{2}$, and obtain

$$\bar{x} = \frac{a \sin \dfrac{\pi}{2}}{\sin \dfrac{\pi}{2}} = \frac{2a}{\pi} = .6366\, a.$$

98. The center of gravity of a circular arc PQ can be found in an interesting manner, without the use of the integral calculus.

From symmetry it is clear that the center of gravity of the arc AP must lie in the radius which bisects the angle AOP. Let p be this center of gravity, and let q be the center of gravity of the arc AQ. Then the center of gravity or the whole arc PQ must be N, the middle point of pq.

Now since the angle $pON = \tfrac{1}{2}\alpha$, we have

$$ON = Op \cos \tfrac{1}{2}\alpha.$$

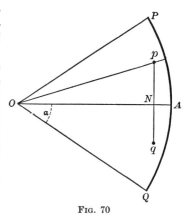

Fig. 70

CIRCULAR ARC

This relation shows that

(the distance of c. g. of arc 2α from center)

$$= \cos\frac{\alpha}{2} \times \text{(the distance of c. g. of arc } \alpha \text{ from center)}.$$

Similarly

(the distance of c. g. of arc α from center)

$$= \cos\frac{\alpha}{4} \times \text{(the distance of c. g. of arc } \frac{\alpha}{2} \text{ from center)},$$

and so on. Continuing in this way, and substituting, we obtain

(the distance of c. g. of arc 2α from center)

$$= \cos\frac{\alpha}{2} \cdot \cos\frac{\alpha}{4} \cdot \cos\frac{\alpha}{8} \cdots \cos\frac{\alpha}{2^{n+1}}$$

$$\times \text{(the distance of c. g. of arc } \frac{\alpha}{2^n} \text{ from center)}.$$

If we make n very great, the value of $\frac{\alpha}{2^n}$ becomes zero. Thus the distance of the c. g. of an arc $\frac{\alpha}{2^n}$ from the center becomes equal to a, the radius of the circle. Making n infinite, we have

(the distance of c. g. of arc 2α from center)

$$= a \cos\frac{\alpha}{2} \cos\frac{\alpha}{4} \cos\frac{\alpha}{8} \cdots \text{ to infinity}.$$

Now
$$\cos\frac{\alpha}{2} = \frac{\sin\alpha}{2\sin\frac{\alpha}{2}},$$

$$\cos\frac{\alpha}{4} = \frac{2\sin\frac{\alpha}{2}}{4\sin\frac{\alpha}{4}}, \text{ etc.,}$$

so that
$$\cos\frac{\alpha}{2} \cdot \cos\frac{\alpha}{4} \cdot \cos\frac{\alpha}{8} \cdots \cos\frac{\alpha}{2^n} = \frac{\sin\alpha}{2^n \sin\frac{\alpha}{2^n}}.$$

Making n infinite, the value of $\sin\frac{\alpha}{2^n}$ becomes identical with $\frac{\alpha}{2^n}$, so that $2^n \sin\frac{\alpha}{2^n}$ becomes identical with α, and we have

$$\cos\frac{\alpha}{2}\cos\frac{\alpha}{4}\cos\frac{\alpha}{8} \cdots \text{ to infinity} = \frac{\sin\alpha}{\alpha},$$

so that the distance of the c. g. of the arc 2α from the center is found to be $a\frac{\sin\alpha}{\alpha}$, as before.

99. Center of gravity of a segment of a circle. Suppose next that we require to find the center of gravity of a segment $PAQN$ of a circle, cut off by a chord PNQ, which subtends an angle 2α at the center O of the circle. Let us divide the whole segment into thin strips parallel to the chord, and let $cc'dd'$ in fig. 71 be a typical strip bounded by chords cc', dd'. Let the angle cOA be θ, and let dOA be $\theta + d\theta$. Then the width of the strip is $cd \sin \theta$ or $a \sin \theta \, d\theta$, while its length is $2\,cn$ or $2\,a \sin \theta$. Thus the area is $2\,a \sin^2 \theta \, d\theta$. Its mass may be supposed to be all concentrated at n, of which the distance from O is $a \cos \theta$.

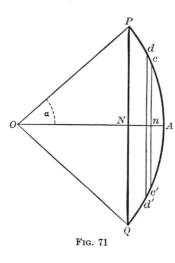

FIG. 71

Thus if $\bar{x}$ is the distance from O of the center of gravity of the whole segment, we shall have

$$\bar{x} = \frac{\int (a \cos \theta)(2\,a \sin^2 \theta \, d\theta)}{\int (2\,a \sin^2 \theta \, d\theta)},$$

and the integration has to be taken from $\theta = 0$ to $\theta = \alpha$. Simplifying, we have

$$\bar{x} = a \frac{\int_0^\alpha \sin^2 \theta \cos \theta \, d\theta}{\int_0^\alpha \sin^2 \theta \, d\theta}$$

$$= a \frac{\frac{1}{3} \sin^3 \alpha}{\frac{1}{2}(\alpha - \sin \alpha \cos \alpha)}$$

$$= \frac{2}{3} a \frac{\sin^3 \alpha}{\alpha - \sin \alpha \cos \alpha}.$$

We find, on putting $\alpha = \dfrac{\pi}{2}$, that the center of gravity of a semi-circle is at a distance $\dfrac{4}{3\pi} a$ from the center.

100. Center of gravity of a sector of a circle. The center of gravity of a sector of a circle can be found by regarding the sector as made up of a triangle and a segment. The center of gravity of the triangle and of the segment both being known, it is easy to find the center of gravity of the whole figure.

A simpler way is the following: We can divide the sector by a series of radii into a great number of very narrow triangles. The weight of each triangle may be replaced by the weight of a particle placed at its center of gravity. Now, in the limit, when the triangles become of infinitesimal width, the center of

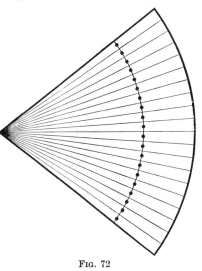

FIG. 72

gravity of each is on its median at a distance from the center of the circle equal to $\tfrac{2}{3} a$, where a is the radius of the circle. Thus all the particles lie on a circle of radius $\tfrac{2}{3} a$.

The weight of any particle must be equal to the weight of the triangle OPQ which it replaces. It must, therefore, be proportional to the base PQ of the triangle, and this again is proportional

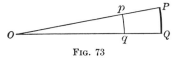

FIG. 73

to pq, the piece of the circle of radius $\tfrac{2}{3} a$ which is inclosed by the triangle. Thus the weight of the particle which has to be placed in the small element pq of this circle is proportional to the length pq. On passing to the limit, and making the number of triangles infinite, we find that the string of particles may be replaced by a wire of uniform density. The center of gravity of this wire has already been determined. If 2α is the angle of the wire, the center of gravity lies on the radius to the middle point of the wire at a distance $\frac{2}{3} a \frac{\sin \alpha}{\alpha}$ from the center.

Thus the center of gravity of the original sector of a circle of radius a and angle 2α is found to lie on the central radius of the sector at a distance $\frac{2}{3} a \frac{\sin \alpha}{\alpha}$ from the center.

101. Center of gravity of a spherical cap. The piece cut off from a spherical shell by a plane is called a spherical cap.

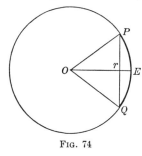

Fig. 74

The center of gravity of a spherical cap cut from a uniform shell can easily be found by the methods already explained.

Let PQ be the spherical cap, O being the center of the sphere from which it is cut. Let OE be the radius perpendicular to the plane PQ by which the cap is bounded, and let a denote the radius of the sphere.

Any plane parallel to PQ will cut the sphere in a circle of which the center will lie on OE. Hence by taking a great number of planes parallel to PQ, we can divide the spherical cap into a number of narrow circular rings, each having its center on the line OE. Let us consider a single circular ring cut off by the planes AaA', BbB'. Let the angles AOE, BOE be equal to θ and $\theta + d\theta$ respectively, so that the ring itself subtends an angle $d\theta$ at the center. The width AB of the ring is $a\,d\theta$. Its circumference may, in the limit, be supposed equal to the circumference of the circle AaA'. Since $Aa = a \sin \theta$,

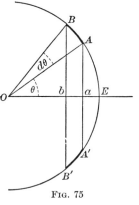

Fig. 75

this circumference is $2\pi a \sin \theta$. Thus the ring under consideration

SPHERICAL CAP AND BELT

may be regarded as a narrow strip of length $2\pi a \sin\theta$ and of width $a\,d\theta$. Its area is accordingly $2\pi a^2 \sin\theta\,d\theta$.

When $d\theta$ is made very small, the arc BA may be regarded as a straight line of length $a\,d\theta$, making an angle $\dfrac{\pi}{2}-\theta$ with OE. Thus the length of ba, the projection of BA on OE, is $a\,d\theta \cos\left(\dfrac{\pi}{2}-\theta\right)$ or $a\sin\theta\,d\theta$. The area of the ring BA is now seen to be

$$= 2\pi a^2 \sin\theta\,d\theta$$
$$= 2\pi a \cdot ba.$$

Thus the mass of the ring is the same as the mass of the element ba of a rod OE, if this rod is of uniform density such that its mass per unit length is that of an area $2\pi a$ of the shell. The center of gravity of the ring we have been considering clearly lies on the axis OE, so that in finding the center of gravity of the spherical cap this ring may obviously be replaced by the element ba of this rod.

In the same way each small ring may be replaced by the corresponding element of the rod. Thus the whole cap may be replaced by the length rE of the rod (fig. 74) which is intercepted between the boundary-plane PQ and the sphere. Since the rod is uniform, the center of gravity of the portion rE of the rod is at its middle point. This point is therefore the center of gravity of the spherical cap.

102. Center of gravity of a belt cut from a spherical shell by two parallel planes. In the same way we can find the center of gravity of the belt cut off from a uniform spherical shell by two parallel planes. In fig. 76 let PQ, $P'Q'$ be the two planes. Then we can divide the belt into narrow rings by planes parallel to PQ. Each ring, as before, may be replaced by the corresponding element of a uniform rod along the axis OE, so that the whole belt may be replaced by the portion rr' of this rod, the portion intercepted between the two planes PQ,

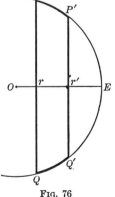

Fig. 76

$P'Q'$. The center of gravity is now seen to be the middle point of rr'.

CENTER OF GRAVITY OF A SOLID

103. Center of gravity of a pyramid on a plane base. Let a pyramid be formed having any plane figure $OPQR$ as base and any point A as vertex. We can find the center of gravity of a homogeneous pyramid by dividing it into thin layers parallel to its base, by a series of parallel planes.

Let $opqr$ be any such layer, this layer being regarded as an infinitely thin lamina. Let G be the center of gravity of a uniform lamina coinciding with the base $OPQR$, and let the line AG meet the lamina $opqr$ in g. Then, from the geometry of similar figures, it is clear that g occupies a position in the lamina $opqr$ which corresponds exactly with that occupied by the point G in the lamina $OPQR$. Thus g will be the center of gravity of the lamina $opqr$. The mass of this lamina may, accordingly, be replaced by the mass of a single particle at g.

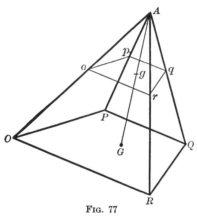

FIG. 77

In the same way each of the laminas into which we are supposing the pyramid to be divided may be replaced by a single particle at the point at which the lamina intersects the line AG. Thus the whole pyramid may be supposed replaced by a series of particles lying along AG. These form a rod of varying density, and the center of gravity of the pyramid will coincide with that of this rod.

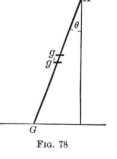

FIG. 78

The center of gravity of the rod may be found by the method already explained in § 94. Consider the lamina which lies between two adjacent

PYRAMID

parallel planes meeting AG in g, g' respectively. Let $Ag = x$ and $Ag' = x + dx$, so that the lamina intercepts a length dx on AG.

Let θ be the angle between AG and the perpendicular from A on to the base of the lamina. Then the thickness of the lamina
$$= gg' \cos \theta = dx \cos \theta.$$

If S is the area of the base of the pyramid, the area of the lamina under discussion is
$$\frac{x^2}{AG^2} S,$$
for the areas of the different laminas are proportional to the squares of their linear dimensions. Thus the volume of the lamina we are considering
$$= \frac{x^2}{AG^2} S\, dx \cos \theta.$$

If this is to be replaced by a particle occupying the length dx of the rod AG, the density of the rod must be
$$\rho = x^2 \frac{S \cos \theta}{AG^2}.$$

Thus the rod AG must be of a density which varies as the square of the distance (x) from the end (A).

The distance of the center of gravity of this rod from A is now, by the formula of § 94,
$$\bar{x} = \frac{\int_A^G \rho x\, dx}{\int_A^G \rho\, dx}$$
$$= \frac{\int_A^G x^3\, dx}{\int_A^G x^2\, dx}$$
$$= \frac{\frac{1}{4} AG^4}{\frac{1}{3} AG^3}$$
$$= \tfrac{3}{4} AG.$$

Thus the center of gravity of the pyramid is in the line AG, three quarters of the way down from A.

104. Center of gravity of the sector of a sphere. We can now find the center of gravity of the sector of a sphere, — the volume cut out of a solid sphere by a right circular cone having its vertex at the center of the sphere. To do this we divide the base PQ of the sector into a number of small elements of area, and then divide the volume of the sector into a number of pyramids of small cross section by taking these elements of area as bases and joining them to the common vertex O. These pyramids are all of the same height, so that their masses are proportional to their bases. The center of gravity of each pyramid is three quarters of the distance down from O to its base, and is, therefore, at a distance from O equal to three quarters of the radius of the sphere. Thus, if we construct a second sphere having O as its center and of radius equal to three quarters of the radius of the original sphere, the center of gravity of each small pyramid will lie on this new sphere. Each pyramid may be replaced by a particle at its center of gravity, so that the whole spherical sector may be replaced by a series of particles lying on this sphere and forming the spherical cap peq (fig. 79).

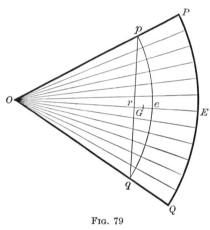

Fig. 79

The mass of each pyramid is proportional to the base, and this again is proportional to the part of the spherical shell peq which is intercepted by the pyramid. Thus the spherical shell peq which is to replace the original volume must be supposed to be of uniform density.

SECTOR OF A SPHERE

The sector of a sphere OPQ has now been replaced by the uniform spherical shell pq, and the center of gravity of this shell is known to be G, the middle point of re in fig. 79. This point G is, accordingly, the center of gravity required.

If the semivertical angle of the cone by which the sector is bounded is α, and if a is the radius of the sphere, we have

$$Oe = \tfrac{3}{4} a, \ Or = \tfrac{3}{4} a \cos \alpha,$$

so that $\quad\quad OG = \tfrac{3}{8} a (1 + \cos \alpha).$

In particular, if $\alpha = \dfrac{\pi}{2}$, the sector becomes a hemisphere, and

$$OG = \tfrac{3}{8} a.$$

Thus the center of gravity of a hemisphere is three eighths of the way along the radius which is perpendicular to its base.

Center of Gravity of Areas and Volumes obtained by Direct Integration

105. Center of gravity of a lamina. To find the center of gravity of a lamina of any shape by integration, we take any convenient set of axes Ox, Oy in the plane of the lamina, and imagine the lamina divided into small elements by two series of lines, one parallel to the axis Ox, and the other parallel to the axis Oy.

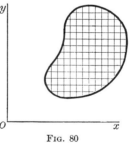

Fig. 80

Consider the small rectangular element for which the values of x for the two edges parallel to Oy are x and $x + dx$, and the values of y for the two other edges are y and $y + dy$. The area of this element is $dxdy$, so that if ρ is the mass of the lamina per unit area at this point, the mass of the element will be $\rho \, dxdy$. Moreover, when dx, dy are made vanishingly small in the limit, the mass may be treated as a particle. Thus the whole mass of the lamina may be regarded as the masses of a number of particles.

In § 86 we obtained for the center of gravity of a number of particles the formulæ

$$\bar{x} = \frac{\sum mx}{\sum m}, \quad \bar{y} = \frac{\sum my}{\sum m}.$$

In the present instance these become

$$\bar{x} = \frac{\iint \rho x \, dxdy}{\iint \rho \, dxdy}, \quad \bar{y} = \frac{\iint \rho y \, dxdy}{\iint \rho \, dxdy}, \tag{30}$$

the sign of summation being replaced by an integration which is to extend over the whole area of the lamina.

If the lamina is uniform, the value of ρ is constant, so that

$$\iint \rho x \, dxdy = \rho \iint x \, dxdy,$$

and so on, and on dividing throughout by ρ, the formulæ reduce to

$$\bar{x} = \frac{\iint x \, dxdy}{\iint dxdy}, \quad \bar{y} = \frac{\iint y \, dxdy}{\iint dxdy}.$$

106. Center of gravity of a solid. To find the center of gravity of a solid we divide it into small solid elements by three systems of planes parallel to the three coördinate planes. The volume of any small element is then $dxdydz$, and its mass is $\rho \, dxdydz$. The formulæ of § 86 now give the coördinates of the center of gravity in the form

$$\bar{x} = \frac{\iiint \rho x \, dxdydz}{\iiint \rho \, dxdydz}, \quad \bar{y} = \frac{\iiint \rho y \, dxdydz}{\iiint \rho \, dxdydz}, \text{ etc.} \tag{31}$$

If the solid is homogeneous, ρ is constant, and the formulæ become

$$\bar{x} = \frac{\iiint x\, dxdydz}{\iiint dxdydz}, \quad \bar{y} = \frac{\iiint y\, dxdydz}{\iiint dxdydz}, \text{ etc.}$$

107. Use of polar coördinates. Any other system of coördinates can, of course, be used for finding a center of gravity by integration. The only coördinates besides Cartesians which are of much use for this purpose are polar coördinates.

We can find the center of gravity of a lamina in polar coördinates by supposing the Cartesian coördinates x, y connected with the polar coördinates r, θ by the usual transformation

$$x = r \cos \theta, \quad y = r \sin \theta.$$

Formulæ (31) then become

$$\bar{r} \cos \bar{\theta} = \frac{\iint \rho (r \cos \theta)(r\, drd\theta)}{\iint \rho (r\, drd\theta)} = \frac{\iint \rho r^2 \cos \theta\, drd\theta}{\iint \rho r\, drd\theta},$$

$$\bar{r} \sin \bar{\theta} = \frac{\iint \rho (r \sin \theta)(r\, drd\theta)}{\iint \rho (r\, drd\theta)} = \frac{\iint \rho r^2 \sin \theta\, drd\theta}{\iint \rho r\, drd\theta}.$$

in which $\bar{r}$, $\bar{\theta}$ are the polar coördinates of the center of gravity. On dividing corresponding sides of these equations, we can obtain an equation giving the $\bar{\theta}$ coördinate alone, namely

$$\tan \bar{\theta} = \frac{\iint \rho r^2 \sin \theta\, drd\theta}{\iint \rho r^2 \cos \theta\, drd\theta}.$$

Similarly we can find the center of gravity of a solid in three-dimensional polars by supposing the polar coördinates r, θ, ϕ connected with x, y, z by the usual transformation

$$x = r \sin \theta \cos \phi, \quad y = r \sin \theta \sin \phi, \quad z = r \cos \theta.$$

Using this transformation, the first of formulæ (31) becomes

$$\bar{r} \sin \bar{\theta} \cos \bar{\phi} = \frac{\iiint \rho (r \sin \theta \cos \phi)(r^2 \sin \theta \, dr d\theta d\phi)}{\iiint \rho (r^2 \sin \theta \, dr d\theta d\phi)}$$

$$= \frac{\iiint \rho r^3 \sin^2 \theta \cos \phi \, dr d\theta d\phi}{\iiint \rho r^2 \sin \theta \, dr d\theta d\phi}, \qquad (32)$$

while similarly we have, from the remaining two formulæ,

$$\bar{r} \sin \bar{\theta} \sin \bar{\phi} = \frac{\iiint \rho r^3 \sin^2 \theta \sin \phi \, dr d\theta d\phi}{\iiint \rho r^2 \sin \theta \, dr d\theta d\phi}, \qquad (33)$$

$$\bar{r} \cos \bar{\theta} = \frac{\iiint \rho r^3 \sin \theta \cos \theta \, dr d\theta d\phi}{\iiint \rho r^2 \sin \theta \, dr d\theta d\phi}. \qquad (34)$$

108. An exactly similar method will lead to formulæ giving the position of the center of gravity in any system of coördinates.

The methods which have already been employed, or a combination of them, will suffice to determine any center of gravity. As illustrations of the use and combination of these methods, we shall find the center of gravity of the same solid figure in three different ways.

ILLUSTRATIVE EXAMPLE

A right circular cone OPQ is scooped out of a solid homogeneous sphere, the vertex of the cone O being on the surface of the sphere, and its axis being a diameter of the sphere. It is required to find the center of gravity of the remainder.

Method I. *Polar coördinates.* First let us use polar coördinates, taking the vertex O of the cone as origin, and the axis of the cone as initial line. If α is the semivertical angle of the cone, the equation of the cone is $\theta = \alpha$. If a is the radius of the sphere, the equation of the sphere is $r = 2\,a \cos\theta$. The center of gravity must from symmetry lie on the axis $\theta = 0$, so that $\bar{\theta} = 0$, and equation (34) becomes

$$\bar{r} = \frac{\iiint \rho r^3 \sin\theta \cos\theta \, dr d\theta d\phi}{\iiint \rho r^2 \sin\theta \, dr d\theta d\phi}.$$

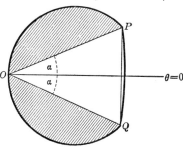

Fig. 81

The solid is supposed to be homogeneous, so that ρ is a constant, and may, therefore, be taken outside the sign of integration in both numerator and denominator. The limits of integration for ϕ are from $\phi = 0$ to $\phi = 2\pi$, so that this integration may be performed in each case. Doing this, and dividing out by $2\pi\rho$, we are left with

$$\bar{r} = \frac{\iint r^3 \sin\theta \cos\theta \, dr d\theta}{\iint r^2 \sin\theta \, dr d\theta}.$$

We may next integrate with respect to r, the limits being $r = 0$ to $r = 2\,a \cos\theta$, and obtain

$$\bar{r} = \frac{\int \tfrac{1}{4}(2\,a\cos\theta)^4 \sin\theta \cos\theta \, d\theta}{\int \tfrac{1}{3}(2\,a\cos\theta)^3 \sin\theta \, d\theta}$$

$$= \tfrac{3}{2} a \frac{\int \cos^5\theta \sin\theta \, d\theta}{\int \cos^3\theta \sin\theta \, d\theta}.$$

The limits of integration for θ are obviously from $\theta = \alpha$ (the cone) to $\theta = \dfrac{\pi}{2}$ (the tangent plane to the sphere). We have

$$\int_\alpha^{\frac{\pi}{2}} \cos^5\theta \sin\theta \, d\theta = -\tfrac{1}{6}\left[\cos^6\theta\right]_\alpha^{\frac{\pi}{2}} = \tfrac{1}{6}\cos^6\alpha,$$

$$\int_\alpha^{\frac{\pi}{2}} \cos^3\theta \sin\theta \, d\theta = -\tfrac{1}{4}\left[\cos^4\theta\right]_\alpha^{\frac{\pi}{2}} = \tfrac{1}{4}\cos^4\alpha.$$

CENTER OF GRAVITY

Substituting these values, we find

$$\bar{r} = \tfrac{3}{2} a \cdot \frac{\tfrac{1}{6} \cos^6 \alpha}{\tfrac{1}{4} \cos^4 \alpha} = a \cos^2 \alpha.$$

Thus the center of gravity is on the axis of the cone at a distance $a \cos^2 \alpha$ from the vertex.

METHOD II. *Cartesian coördinates.* We may next employ Cartesian coördinates, taking O as origin and the axis of the cone as axis of x. The equation of the cone is now

$$y^2 + z^2 = x^2 \tan^2 \alpha,$$

while that of the sphere is

$$x^2 + y^2 + z^2 - 2ax = 0.$$

From § 106, we have

$$\bar{x} = \frac{\iiint x\, dx\, dy\, dz}{\iiint dx\, dy\, dz}.$$

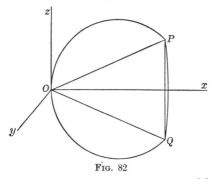

FIG. 82

In each integral we may integrate first with respect to y and z together. We have to evaluate the same integral in both cases, namely $\iint dy\, dz$, the limits being given by

$$y^2 + z^2 = x^2 \tan^2 \alpha$$
and
$$y^2 + z^2 = 2ax - x^2.$$

The problem is the same as that of finding the area of a circular ring of inner and outer radii $x \tan \alpha$ and $\sqrt{2ax - x^2}$ respectively. (This ring is, of course, the intercept of the solid on the plane parallel to the yz plane.) The area of the ring is

$$\pi (2ax - x^2) - \pi (x^2 \tan^2 \alpha) = \pi (2ax - x^2 \sec^2 \alpha),$$

and on substituting this value for $\iint dy\, dz$, the formula becomes

$$\bar{x} = \frac{\int \pi x (2ax - x^2 \sec^2 \alpha)\, dx}{\int \pi (2ax - x^2 \sec^2 \alpha)\, dx}.$$

The limits of integration are now from $x = 0$, the origin, to $x = 2a \cos^2 \alpha$, the value of x on the plane PQ. Evaluating the integrals, and substituting these limits, we obtain

$$\bar{x} = \frac{2 a \pi \tfrac{1}{3} (2 a \cos^2 \alpha)^3 - \pi \sec^2 \alpha \tfrac{1}{4} (2 a \cos^2 \alpha)^4}{2 a \pi \tfrac{1}{2} (2 a \cos^2 \alpha)^2 - \pi \sec^2 \alpha \tfrac{1}{3} (2 a \cos^2 \alpha)^3}$$
$$= a \cos^2 \alpha,$$

giving the same result as before.

ILLUSTRATIVE EXAMPLE 141

METHOD III. *Geometrical Method.* The center of gravity can also be found by regarding the given volume as the sums and differences of simpler volumes of which the center of gravity is already known.

The volume is obtained by taking the complete sphere $OPsQ$ and subtracting from it the cone $OPrQ$ and the spherical segment $PrQs$. The center of gravity of the sphere and cone are known, — that of the segment $PrQs$ is most easily found by regarding it as the difference between the sector $CPsQ$ and the cone $CPrQ$. Thus we regard the original figure as made up of

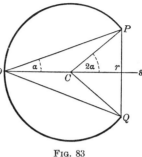

FIG. 83

(sphere $OPsQ$) − (cone $OPrQ$) − (sector $CPsQ$) + (cone $CPrQ$).

The volumes of these, and the distances of their centers of gravity from O measured along OC, are as follows:

FIGURE	VOLUME	DISTANCE OF C.G. FROM O
+ sphere	$\tfrac{4}{3}\pi a^3$	a
− cone $OPrQ$	$-\tfrac{1}{3}(2a\cos^2\alpha)(\pi a^2\sin^2 2\alpha)$	$\tfrac{3}{4}(2a\cos^2\alpha)$
− sector $CPsQ$	$-\tfrac{2}{3}\pi a^3(1-\cos 2\alpha)$	$a+\tfrac{3}{8}a(1+\cos 2\alpha)$
+ cone $CPrQ$	$\tfrac{1}{3}(a\cos 2\alpha)(\pi a^2\sin^2 2\alpha)$	$a+\tfrac{3}{4}a\cos 2\alpha$

In this table the negative sign denotes that a figure is to be removed, so that its volume must be reckoned as of negative sign.

Denoting the distance of any center of gravity from O by x, and using the formula

$$\bar{x} = \frac{\sum mx}{\sum m}$$

of § 86, we obtain as the distance of the center of gravity of the whole figure from O

$$\bar{x} = \frac{\tfrac{4}{3}\pi a^4 - \tfrac{1}{4}(2a\cos^2\alpha)^2(\pi a^2\sin^2 2\alpha) - \tfrac{2}{3}\pi a^4(1-\cos 2\alpha)\{1+\tfrac{3}{8}(1+\cos 2\alpha)\}}{\tfrac{4}{3}\pi a^3 - \tfrac{1}{3}(2a\cos^2\alpha)(\pi a^2\sin^2 2\alpha) - \tfrac{2}{3}\pi a^3(1-\cos 2\alpha) + \tfrac{1}{3}(a\cos 2\alpha)(\pi a^2\sin^2 2\alpha)}$$

$$\frac{+\tfrac{1}{3}(a\cos 2\alpha)(\pi a^2\sin^2 2\alpha)a(1+\tfrac{3}{4}a\cos 2\alpha)}{\tfrac{4}{3}\pi a^3 - \tfrac{1}{3}(2a\cos^2\alpha)(\pi a^2\sin^2 2\alpha) - \tfrac{2}{3}\pi a^3(1-\cos 2\alpha) + \tfrac{1}{3}(a\cos 2\alpha)(\pi a^2\sin^2 2\alpha)},$$

which, after reduction, gives
$$\bar{x} = a\cos^2\alpha,$$
the same result as before.

CENTER OF GRAVITY

GENERAL EXAMPLES

1. A plane quadrilateral $ABCD$ is bisected by the diagonal AC, and this diagonal is divided in the ratio $a:b$ by the diagonal BD. Prove that the center of gravity of the quadrilateral lies in AC and divides it into two parts in the ratio $2a+b:2b+a$.

2. A uniform wire is bent into the form of a circular arc and the two bounding radii, and the center of gravity of the whole is found to be at the center. Show that the angle subtended by the arc at the center is $\tan^{-1}(-\tfrac{4}{3})$.

3. The three feet of a circular table are vertically below the rim and form an equilateral triangle. Prove that a weight less than that of the complete table cannot upset it.

4. A triangular table is supported by three legs at the middle points of its sides, and a weight W is placed on it in any position. It is found that the table will just be upset if a weight P is placed at one angular corner. The corresponding weights needed to upset it at the other corners are Q, R. Prove that $P+Q+R$ is independent of the position of the weight W.

5. Weights are nailed to the three corners of a triangular lamina, each proportional to the length of the opposite side of the triangle, and of combined weight equal to the original weight of the lamina. Show that the center of gravity of the triangle is at the center of the nine-point circle.

6. A uniform triangular lamina of weight W and sides a, b, c is suspended from a fixed point by strings of lengths l_1, l_2, l_3 attached to its angular points. Show that the tensions of the strings are
$$Wkl_1, \quad Wkl_2, \quad Wkl_3,$$
where $\qquad k = [3(l_1^2 + l_2^2 + l_3^2) - a^2 - b^2 - c^2]^{-\frac{1}{2}}.$

7. Explain how a clock hand on a smooth pivot can be made to show the time by means of watchwork, carrying a weight round, concealed in the clock hand.

8. A spindle-shaped solid of uniform material is bounded by two right circular cones of altitudes 6 and 2 inches with a common circular base of radius 1 inch. It is suspended by a string attached to a point on the rim of the circular base. Find the inclination of the axis of the spindle to the vertical when it is hanging freely.

9. A pack of cards is laid on a table, and each projects beyond the one below it in the direction of the length of the pack to such a distance that each card is on the point of tumbling, independently of those below it. Prove that the distances between the extremities of successive cards will form a harmonic progression.

EXAMPLES 143

10. Prove that the center of gravity of any portion PQ of a uniform heavy string hanging freely is vertically above the intersection of the tangents at P, Q.

11. A hemispherical shell has inner and outer radii a, b. Show that the distance of its center of gravity from its geometrical center is
$$\frac{3}{8}\frac{(a+b)(a^2+b^2)}{a^2+ab+b^2}.$$

12. An anchor ring is cut in two equal parts by a plane through its center which passes through its axis. Find the center of gravity of either half.

13. Prove that the pull exerted by a man in a tug of war is $\frac{a}{b}$ of his weight, where a is the horizontal projection of a line joining his heels to his center of gravity, and b is the height of the rope above the ground.

14. Prove that a horse weighing W pounds can exert a horizontal pull of Wa/h pounds at a height h above the ground by advancing his center of gravity a distance a in front of its position when he is standing upright on his legs.

15. A rod of varying density and material is supported by a man's two forefingers, across which it rests in a horizontal position. The man moves his fingers toward one another, keeping them in the same horizontal plane, and allowing the rod to slip over one or both of his fingers. Show that when his fingers touch, the center of gravity of the rod will be between the points of contact of his fingers with the rod.

16. A semicircular disk rests in a vertical plane with its curved edge on a rough horizontal and an equally rough vertical plane, the coefficient of friction being μ. Show that the greatest angle that the bounding diameter can make with the vertical is
$$\sin^{-1}\frac{\mu+\mu^2}{1+\mu^2}\frac{\pi}{2}.$$

17. A hemisphere of radius a and weight W is placed with its curved surface on a smooth table, and a string of length $l\,(l<a)$ is attached to a point on its rim and to a point on the table. Prove that the tension of the string is
$$\frac{3}{8}W\frac{a-l}{\sqrt{2\,al-l^2}}.$$

18. A triangular lamina of weight W is supported by three vertical strings attached to its angular points so that the plane of the triangle is horizontal; a particle of weight W is placed at the orthocenter of the triangle. Prove that the tensions of the strings are given by
$$\frac{T_1}{1+3\cot B\cot C}=\frac{T_2}{1+3\cot C\cot A}=\frac{T_3}{1+3\cot A\cot B}=\frac{W}{2}.$$

19. Find the center of gravity of a lamina bounded by a parabola and a line perpendicular to its axis.

20. Find the center of gravity of the volume cut from a solid paraboloid by a plane perpendicular to its axis.

21. Find the center of gravity of the area inclosed by two radii of an ellipse.

22. Find the center of gravity of the volume cut off from a solid ellipsoid by a plane through the center.

23. Find the center of gravity of half of an ellipsoidal shell, this being bounded by two similar concentric and coaxial ellipsoids, and a plane through the center.

24. A right circular cone whose base is of radius r is divided into two equal parts by a plane through the axis. Prove that the distance of the center of gravity of either half from the axis is $\dfrac{r}{\pi}$.

25. Find the center of gravity of a lamina bounded by the semicubical parabola $x^3 = ay^2$, the axis of x, and the ordinate $x = a$.

26. Find the center of gravity of a single loop of the curve
$$r = a \sin 3\theta.$$

27. Find the center of gravity of an octant of a sphere.

28. A cylindrical hole of radius a is drilled through a hemisphere of radius b so that the radius perpendicular to the base of the hemisphere is also the central line of the hole. Find the center of gravity of the figure.

29. Find the center of gravity of the area inclosed between the two circles
$$x^2 + y^2 = a^2; \ x^2 + y^2 = 2\,ab.$$

30. Find the center of gravity of a lens made of homogeneous glass, having spherical surfaces of radii r, s, and of which the thickness is t at the center and zero at the edge.

CHAPTER VII

WORK

109. Measurement of work. There are various kinds of work, but in mechanics we are concerned only with the work done in moving bodies which are acted on by forces. Such work is described as mechanical work. We say that mechanical work is done whenever a body is moved in opposition to the forces acting on it, as, for instance, in raising a weight, in dragging a heavy body over a rough surface, or in stretching an elastic string. In the first case work is performed against the force of gravity, in the second case against the frictional force exerted on the moving body by the rough surface, and in the third case against the tension of the string.

Obviously in estimating the amount of work done, two factors have to be taken into account, namely the amount of the force acting on the body and the distance through which the body is moved in opposition to this force. The amount of work will clearly be directly proportional to the force, — in raising a weight of 200 pounds through a given distance we do twice as much work as in raising a weight of 100 pounds through the same distance. It will also be proportional to the distance moved, — in raising a weight through two feet we do twice as much work as in raising the same weight through one foot. Thus the amount of work done varies as the product of the force and the distance.

The amount of work done in raising a weight of one pound through a height of one foot is called *one foot pound*.

From what has been said, it is clear that the work done in raising a weight of w pounds through a height of h feet is wh foot pounds. Also, the work done in moving a body a distance of s feet in opposition to a force of F pounds weight is Fs foot pounds. Thus we may say that *the work done in moving a body through any distance against a uniform force is the product of the distance and the force.*

Suppose, for instance, that it is found that the force required to drag a railway train along a level track is equal to the weight of 10,000 pounds, then the work done in hauling this train a distance of 100 miles

$$= 100 \times 5280 \times 10{,}000 \text{ foot pounds.}$$

110. Rate of performing work. Work frequently has to be done within a given time, so that it is often necessary to measure the rate at which work is being done. The rate of doing work in which 33,000 foot pounds are done per minute is called *one horse power* (1 H. P.).

This unit was introduced by Watt, and was supposed to measure the rate of working of an ordinary horse. It is found, however, that very few horses are capable of working continuously at one horse power for any length of time.

As an example of the calculation of horse power, let us find the horse power required of an engine to haul a train at 30 miles an hour, the frictional resistance being equal to the weight of 10,000 pounds. A velocity of 30 miles an hour = 44 feet per second, so that the work done per second = 44 × 10,000 foot pounds. Since one horse power = 550 foot pounds per second, we see that the horse power required

$$= \frac{44 \times 10{,}000}{550} = 800 \text{ horse power.}$$

This gives the horse power required to haul the train at a steady speed of 30 miles per hour. We shall find that if the speed is not constant the horse power will be different, part of the work being used up in producing the acceleration of the motion. For the present, however, we confine our attention to motion with uniform velocity.

Absolute Unit of Work

111. We have already seen that besides the practical unit of force, which is the weight of a unit mass, there is also a second unit of force, known as the absolute unit, which is defined as being a force capable of producing unit acceleration in unit mass. As the practical unit produces acceleration g in unit mass, where g is

MEASUREMENT AND UNITS 147

the acceleration due to gravity, it follows that the practical unit is g times the absolute unit.

In practical British units, the unit force is the pound weight. In absolute units, the corresponding unit is known as the *poundal;* it is the force which will produce unit acceleration in a mass of one pound.

The practical unit of work, as we have said, is the work done in raising a mass of one pound through one foot, i.e. in moving through one foot the point of application of one pound weight. There is also an absolute unit of work, namely the work done in moving through one foot the point of application of one poundal. This unit is called the *foot poundal*. Since one pound weight is equal to g poundals, we obviously have the relation

$$1 \text{ foot pound} = g \text{ foot poundals.}$$

EXAMPLES

1. At what speed can a horse of 1 horse power draw a cart weighing 1 ton, friction being supposed to cause a horizontal force equal to one fortieth of the weight of the cart?

2. A body resisted by a force of P poundals is moved against this resistance with a velocity v. What horse power is required?

3. At what rate can a steam roller of 7 horse power and weight 1 ton roll a path, the resistance due to friction being equal to the weight of the roller?

4. A snail weighing $\frac{1}{4}$ ounce climbs a wall 6 feet in height in 4 hours. At what horse power is he working?

5. A load of bricks weighing 5 tons has to be raised to the top of a house 50 feet in height by 10 laborers, each of whom works at an average rate of $\frac{1}{12}$ horse power. How long ought the job to take?

6. The piston of an engine has an area of a square feet and a stroke of l feet, and the engine makes p revolutions per minute. If the pressure per unit area acting on the piston is p pounds weight per square foot, prove that the horse power at which the engine is working is

$$\frac{p \, l \, a \, n}{33,000}.$$

7. A locomotive has a circular piston of diameter 17 inches, and stroke 26 inches. It makes 250 revolutions per minute, the pressure being 225 pounds weight per square inch. Find its horse power.

WORK

8. If 200 horse power is required to drive a steamer 150 feet long at a speed of 9 knots, prove that 25,600 horse power will be required to drive a similar steamer, 600 feet long, similarly immersed, at 18 knots, assuming that the resistance is proportional to the wetted surface and the square of the velocity through the water. Prove also that the cost of coal per ton of cargo will be the same in the two steamers.

9. Fifty horse power is transmitted from one shaft to another by means of a belt moving over two wheels on the shafts with a linear velocity of 250 feet per minute. Find the difference of tensions on the two sides of the belt.

10. A locomotive consumes $1\frac{1}{4}$ pounds of coal per horse-power-hour. How much coal is required to haul a train of total weight 1000 tons over 50 miles of level road on which the resistance to friction is 12 pounds weight per ton?

11. A liner of 22,000 horse power makes a run of 3300 miles in six days. Find the resistance to the ship's motion.

Work done against a Variable Force

112. If a body is moved in opposition to a force which is not of constant intensity but varies from point to point on the path of the moving body, we can no longer use the formula Fs for the amount of work performed.

To calculate the amount of work done, we divide up the whole range over which motion takes place into an infinite number of infinitesimally small ranges, each of these ranges being so small that the force opposing the motion may be regarded as of constant magnitude during the motion through any one of them.

If ds is any small range at a distance s from the starting point, and if F is the intensity of the force opposing the motion while the body moves through the small range ds, then the work done in moving through this range is $F ds$. The total work done, the sum of the amounts of work done in all the ranges, is, accordingly,

$$\int F\,ds.$$

Work done in stretching an Elastic String

113. As an example of the use of this formula, let us find the work done in stretching an elastic string. Let the natural length of the string be l, and let λ denote its modulus of elasticity.

WORK OF STRETCHING A STRING

When the length of the string is x, its tension T, by the formula of § 39, is given by
$$T = \frac{x-l}{l}\lambda.$$

In stretching the string through a further distance dx, — i.e. from length x to length $x + dx$, — the work done
$$= T\,dx$$
$$= \frac{\lambda}{l}(x - l)\,dx.$$

By integration, we find that the work done in stretching a string from length a to length b
$$= \int_{x=a}^{x=b} \frac{\lambda}{l}(x-l)\,dx$$
$$= \frac{\lambda}{2\,l}((b-l)^2 - (a-l)^2)$$
$$= \frac{\lambda}{2\,l}(b + a - 2\,l)(b - a).$$

The distance stretched is $b - a$, while $\dfrac{\lambda}{2\,l}(b + a - 2\,l)$ is the tension when half of the stretching has been completed, i.e. when $x = \tfrac{1}{2}(a + b)$.

Thus we have found that

The work done in stretching an elastic string from any length a, greater than the natural length of the string, to a length b, is equal to the tension at length $\tfrac{1}{2}(a + b)$ multiplied by $(b - a)$.

Obviously, if the tension is measured in pounds weight and the extension $(b - a)$ in feet, the product will give the amount of work measured in foot pounds. If the tension is measured in poundals, and $(b - a)$ in feet, the product will give the amount of work in foot poundals.

150 WORK

Work represented by an Area

114. Let PQ represent the path described by a moving body, and let us draw ordinates at each point in PQ to represent, on any scale we please, the force opposing the motion of the body at that point. Let s, r be two adjacent points, and let ss', rr' be the ordinates at these points.

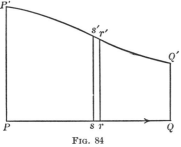

Fig. 84

Then the area of the small strip $ss'rr'$ may, in the limit, be supposed equal to sr multiplied by ss'. On the scale on which we are representing forces, this product will represent the distance sr multiplied by the force opposing the motion of the body from s to r. In other words, the small area $ss'rr'$ will represent the work done in moving the body from s to r.

By addition of such small areas, we find that the complete area $PP'QQ'$ represents the work done in moving from P to Q.

115. This method gives a simple way of investigating the work done in stretching an elastic string, already calculated in § 113. Let OP be the natural length. For the sake of definiteness suppose that the end O is held fast, and that as the string is stretched the point P moves along the line OP. Let it be required to find the work done in stretching the string from a length OA to a length OB.

Let Q be any point of the line $OPAB$, and let QQ' be drawn to represent the tension when the length of the string is OQ.

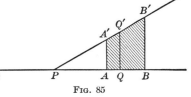

Fig. 85

For different positions of Q, the ordinate QQ' will be of different heights. Since, by Hooke's law, the tension is proportional to the extension, the height of the ordinate QQ' (representing the tension) will always be in the same ratio to PQ (the extension). Thus Q' is always on a certain straight line through P. If AA', BB' are the ordinates which represent the tensions at A, B, this line will, of course, pass through the points A', B'. The work done in stretching the string through the range AB is now, in accordance with § 114, represented by the area $AA'B'B$, the area which is shaded in fig. 85.

GRAPHICAL REPRESENTATION OF WORK 151

The area of this figure is clearly equal to AB multiplied by the ordinate at the middle point of AB. This ordinate represents the tension of the string when its length is equal to $\frac{1}{2}(OA+OB)$, so that we again obtain the result of § 113, namely

(work done) = (range of stretching, AB)
× (tension at halfway stage of stretching).

116. The indicator diagram. The graphical representation of work explained in § 114 is made use of in practical engineering. Suppose that OO' is the distance traveled by a piston inside a cylinder. When the piston is in any position P, let the pressure acting on the piston be measured, and let a line PP' be drawn at right angles to OO' to represent it on any assigned scale. As the piston moves along the range OO' and then back along the range $O'O$, the point P' will describe a closed curve $AP'BP''A$, which is called the *indicator diagram* of the motion of the piston.

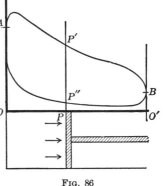

Fig. 86

The work done by the steam *on* the piston in its forward motion is, as we have seen, represented by the area $AP'BO'POA$ inclosed between the curve $AP'B$ and the axis OO'. This work is expended in moving the piston forward in opposition to the thrust in the piston rod. Similarly the work done by the steam *on* the piston in its backward motion is represented by the curve $BO'POAP''B$ inclosed between $BP''A$ and the axis OO', this area being taken negatively, since the piston is now moving in opposition to the pressure at work on it.

Thus the whole work done on the piston is represented by the difference of these two areas, and this is easily seen to be the area $AP'BP''A$ of the indicator diagram itself. Hence, to find the rate at which an engine is performing work, it is only necessary to measure the area of its indicator diagram and the number of revolutions per unit time.

WORK DONE AGAINST FORCE OBLIQUE TO DIRECTION OF MOTION

117. So far we have only considered cases in which the force acts in a direction exactly opposite to that in which the particle moves. We may, however, have to calculate the work when the motion makes any angle with the direction of the force.

When a body is moved at right angles to the force acting on it, the work done will clearly be nil; e.g. in moving a weight about on a horizontal surface no work is done against gravity.

We can now find the amount of work done when a body is moved in a direction making any angle with the force acting on it. Let a body be moved from P to Q, a small distance ds of its path, while acted on by a force R, of which the line of action makes an angle ϕ with QP. Resolve R into two components, $R\cos\phi$ along QP and $R\sin\phi$ perpendicular to QP. The work done against the force R is the same as the work which would be done if these two forces $R\cos\phi$, $R\sin\phi$ were acting on the body simultaneously. The work done against the former force would be $R\cos\phi\, ds$; that against the latter would be nil. Thus the whole amount of work done is $R\, ds \cos\phi$.

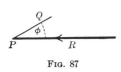

Fig. 87

118. Let R have components X, Y, Z, and let the element of path PQ have direction cosines l, m, n. The direction cosines of the line of action of R are

$$\frac{X}{R}, \ \frac{Y}{R}, \ \frac{Z}{R},$$

and since this makes an angle $\pi - \phi$ with PQ, we must have

$$\cos(\pi - \phi) = l\frac{X}{R} + m\frac{Y}{R} + n\frac{Z}{R}.$$

Hence
$$R\, ds \cos\phi = - ds\, (lX + mY + nZ)$$
$$= - (X\, dx + Y\, dy + Z\, dz),$$

where dx, dy, dz are the projections of ds on the axes. This gives an analytical expression for the work done in a small displacement. By integration, we can find the work done in any motion.

WORK OF RAISING BODIES AGAINST GRAVITY 153

119. Work of raising a system of bodies against gravity. If a particle of mass m is moved a distance ds along a path making an angle ϕ with the vertical (upwards), the work done is $mg \cos\phi \, ds$. Since the distance through which the particle is raised is $ds \cos\phi$, we may say that the work done is equal to the weight of the body (mg) multiplied by the distance through which the particle is raised.

By taking the particle along any path, and adding together the amounts of work done on the successive elements of the path, we find that the total work done against gravity is equal to the weight of the particle multiplied by the total vertical distance through which the body has been raised.

120. Let us suppose that we move a number of particles of masses $m_1, m_2, \cdots$. Let their heights above the ground before the motion be $h_1, h_2, \cdots$, and let their heights at the end of the motion be $h_1', h_2', \cdots$. The work done against gravity on the first particle is $m_1 g (h_1' - h_1)$; by addition of such quantities, the total work done against gravity

$$= m_1 g (h_1' - h_1) + m_2 g (h_2' - h_2) + \cdots$$
$$= g \left(\sum m_1 h_1' - \sum m_1 h_1 \right). \tag{35}$$

Now let M be the total mass of the particles, and let H, H' denote the heights of the center of gravity of all the particles above the ground before and after the motion respectively. Then, by the formula of § 86, we have

$$H = \frac{\sum m_1 h_1}{\sum m_1} = \frac{\sum m_1 h_1}{M},$$

so that
$$\sum m_1 h_1 = MH,$$

and, similarly,
$$\sum m_1 h_1' = MH'.$$

Thus the total work, as given by expression (35), becomes

$$g(MH' - MH) = Mg(H' - H).$$

WORK

Thus the total work done against gravity is equal to the total weight of the particles multiplied by the vertical height through which the center of gravity of the particles has been raised.

WORK PERFORMED AGAINST A COUPLE

121. THEOREM. *If a rigid body acted on by a system of forces be given any small rotation through an angle ϵ about any axis, the work done is $G\epsilon$, where G is the moment about this axis of the forces opposing the motion.*

Let the axis of rotation be supposed to be a line perpendicular to the plane of the paper, meeting it in the point L. Let a typical force be a force F acting on the particle A of the body.

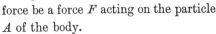

FIG. 88

As the result of the rotation, let A move to a position A', so that the angle ALA' is equal to ϵ, the angle through which the body has been turned.

Then, during the rotation, the point of application of the force F moves from A to A', and, therefore, the work done

$$= F \cdot AA' \cdot \cos \phi,$$

where ϕ is the angle between F and AA',

$$= AA' \times \text{component of } F \text{ along } AA'$$
$$= \epsilon \times LA \times \text{component of } F \text{ along } AA'$$
$$= \epsilon \times \text{moment of } F \text{ about the axis of rotation}.$$

If the rigid body is acted on by a number of forces applied to its different particles, we find, on summation, that the total work done

$$= \epsilon \times \text{sum of the moments of all these forces about the axis of rotation}$$
$$= G\epsilon, \text{ where } G \text{ is the moment about the axis of rotation of all the forces}.$$

VIRTUAL WORK

EXAMPLES

1. A man who weighs 140 pounds walks up a mountain path at a slope of 30 degrees to the horizon at the rate of 1 mile per hour. Find his rate of working in raising his own weight in horse power.

2. At what horse power is an engine working which hauls a train of 1000 tons up an incline of 1 in 200 at 12 miles an hour, the resistance due to friction being $\frac{1}{200}$ of the weight of the train?

3. An automobile weighing 1 ton can run up a hill of 1 in 60 at 8 miles an hour. Taking the resistance due to friction as $\frac{1}{50}$ of the weight of the car, find at what rate it could run down the same hill, assuming the horse power developed by the engine to remain the same.

4. A cargo of stone weighing 18 tons is unloaded from a barge on to a quay 30 feet above the barge by cranes worked by an engine. If the unloading takes three hours, find the average horse power at which the engine has been working.

5. Assuming that a man in walking raises his center of gravity through a vertical height of one inch at every step, find at what horse power a man is working in walking at 4 miles an hour, his stride being 33 inches, and his weight 168 pounds.

6. A cyclist and his machine weigh 200 pounds, and he rides up an incline of 1 in 80 at 15 miles an hour. His bicycle is geared to 72 inches, and the length of his cranks is 7 inches. Find the average vertical pressure of his foot on the pedal, assuming this pressure to exist only during the *downward* motion of the pedal.

7. A single-screw ship has engines of 5000 horse power, and, when working at full power, the engines make 75 revolutions per minute. Find the couple transmitted by the shaft.

8. When one body rolls on another, there is found to be a couple opposing the motion, equal to that produced by the normal reaction at the end of an arm of length l, where l is called the coefficient of rolling friction.

If a railroad truck runs on wheel of radius a, show that the resistance to its motion produced by rolling friction is l/a times its weight.

The Principle of Virtual Work

122. By a small displacement is meant for the present a motion in which each particle of a system is displaced from its original position through a distance which is so small that it may be treated as an infinitesimal quantity of which the square may be neglected. If the system is under the action of forces, work will be done in performing any small displacement. Since the displacement is supposed to be a small quantity, the work performed will also be a small quantity.

156 WORK

If any particle is in equilibrium, the resultant force acting on it vanishes, so that the work done in any small displacement of the particle vanishes to a higher order than the displacement. If a rigid body, or system of rigid bodies or particles, is in equilibrium, and any small displacement is given to it, the work done on each particle is nil, so that the aggregate work done is nil.

123. The forces acting on the particles of the system may, as in § 50, be divided into two classes:

(a) forces applied to the bodies from outside;

(b) pairs of actions and reactions acting between the particles of the bodies, or between two bodies in contact.

In calculating the work done in a small displacement, we must take account of the work done against all the forces of both classes, but shall find that a great number of the terms arising from the forces of the second class cut one another out.

124. Let us first consider the pair of forces which constitute the action and reaction between two particles P, Q of a rigid body. Let the amount of each force be R, its direction being QP or PQ according as it acts on P or Q. Let the effect of a small displacement be to move P, Q to P', Q' respectively, and let $P'p$, $Q'q$ be perpendiculars drawn from P', Q' to PQ. The work done against the force R acting on P is $R \times Pp$, while that done against the force R acting on Q is $-R \times Qq$. Thus the total work performed

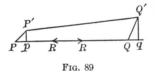

Fig. 89

$$= R(Pp - Qq)$$
$$= R(PQ - pq)$$
$$= R(PQ - \text{projection of } P'Q' \text{ on } PQ).$$

Since the body is rigid, the length $P'Q'$ is equal to the length PQ, and since the displacement is, by hypothesis, small, the angle between $P'Q'$ and PQ is small. Thus the projection of $P'Q'$ on PQ

$= P'Q'$, except for small quantities of order higher than the first,
$= PQ$,

so that the work performed vanishes.

VIRTUAL WORK 157

125. Again, the work performed against the pair of forces which constitute action and reaction between two smooth surfaces can be seen to vanish.

First consider the case in which one body is held at rest while the second is made to slide over its surface. In such a displacement the work performed, if any, is performed against the reaction which acts on the moving body. Since the force acts along the normal, while its point of application necessarily moves in the tangent plane, — i.e. at right angles to the normal, — we see that the work done is nil.

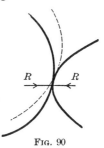

FIG. 90

The most general motion possible for the two surfaces is compounded of a motion of the kind just described and a motion in which the two surfaces move as a rigid body. The work done in the first part of the displacement has just been seen to be zero, the work done in the second part of the displacement vanishes by § 124; hence the total work vanishes, proving the result required.

126. The results just proved are not true if the contact between the surfaces is rough. The work done in such a case depends on the magnitude of the frictional forces, and as it is generally as difficult to determine the amount of these forces as to solve the whole problem, the method of virtual work is not of any value in such cases.

127. We have now seen that a large number of forces may be left out of account altogether in calculating the work done in a small displacement, and the principle of virtual work, which states that when a system is in equilibrium the work done in any small displacement is zero, requires us only to calculate the work performed against external forces, and not that performed against the internal actions and reactions of rigid bodies.

128. Systems of pulleys. An important application of the principle of virtual work is the following: Let us suppose that we have any arrangement of pulleys and inextensible ropes, the ropes having two free ends, — to one of which the weight to be raised is attached, and to the other of which the power is applied. Let these two free ends of rope be called the weight end and the power end respectively, and let us suppose that the arrangement is such that, in order to move the weight end through 1 inch, the power end must be moved through n inches. Let a weight W be attached to the weight end, and let us suppose that it is found that a force P must be applied to the power end to maintain equilibrium.

We now have forces P and F in equilibrium. To find the relation between them, let us give the system a small displacement. Let us move the weight W a distance ds, then, if the rope is not to be stretched, we must suppose the power P moved through a distance $n\,ds$. The work done by external force consists solely of the work performed on the power end of the rope, namely $P\,n\,ds$, and the work performed in moving the weight against gravity, namely $W\,ds$. These are of opposite signs,— if we raise the weight, $W\,ds$ must be taken positively and $P\,n\,ds$ negatively, and *vice versa*. If the system was initially in equilibrium, the total work performed by external forces in this small displacement must vanish, so that the equation of equilibrium is seen to be

$$W\,ds - P\,n\,ds = 0,$$

so that
$$P = \frac{W}{n},$$

giving the relation between power and weight.

This investigation assumes that friction, etc., may be neglected, and also neglects the weight of the moving ropes and pulleys.

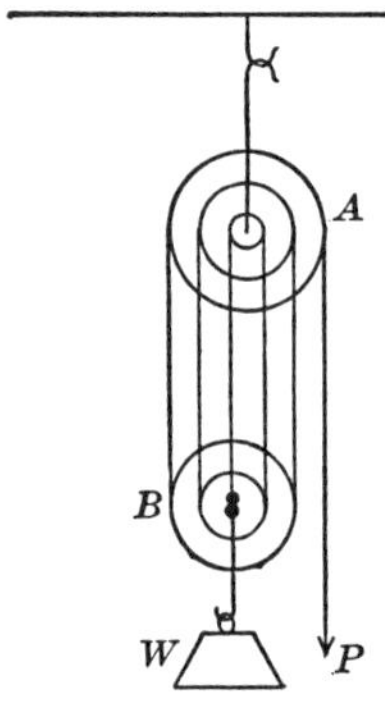

As an instance of a system of pulleys, let us consider the arrangement shown in fig. 91.

FIG. 91

There are two blocks of pulleys, A and B. The former is fixed, while the latter is free to move, and has the weight W suspended from it. The rope, starting from the power end, passes first round a pulley of block A, then round one of block B, then round one of block A, and so on any number of times, until finally its end is fastened to block B. To find the relation between P and W, we need only find the number n. Let us suppose that in addition to the free power end of the rope the number of vertical ropes is s. Then, if we pull the power end until the weight end is raised

VIRTUAL WORK

1 inch, we shall shorten each of these s ropes by 1 inch, and so lengthen the power end by s inches. Thus $n = s$, so that, in this case, $P = \dfrac{W}{s}$.

For instance, with two pulleys in the lower block and three in the upper block the value of n will be 5, so that each pound of power will support 5 pounds of weight, — a man pulling with a vertical pull of 100 pounds could support a weight of 500 pounds, and as soon as his pull exceeds 100 pounds, he will raise the weight of 500 pounds.

ILLUSTRATIVE EXAMPLES

1. As a first example of the principle of virtual work, let us suppose that we have an endless elastic string of natural length a, modulus λ, placed over a sphere of radius b, and allowed to stretch under gravity. We might, of course, find the amount of stretching in the equilibrium position by resolving forces, but we can get it more readily by the method of virtual work. Let us suppose that, when in equilibrium, the string lies on a small circle of angular radius θ. Let a small displacement be given, this consisting of each element of the string being displaced down the surface of the sphere, so that the string forms a new circle of angular radius $\theta + d\theta$. The length of the string when forming a circle of angle θ was $2\pi b \sin\theta$; the increase in this when θ is changed to $\theta + d\theta$ is $d\theta \dfrac{\partial}{\partial \theta}(2\pi b \sin\theta)$ or $2\pi b \cos\theta\, d\theta$. The work done in stretching the string by this amount is $T \cdot 2\pi b \cos\theta\, d\theta$, where T is the tension. Work

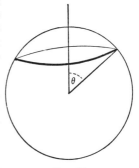

Fig. 92

is also done against (or, in this particular case, with) the force of gravity. The height of the center of gravity of the string when forming a circle of angle θ is $b\cos\theta$; on increasing θ to $\theta + d\theta$, this increases by $-b\sin\theta\, d\theta$, so that the work done against gravity is $-wb\sin\theta\, d\theta$. We have now calculated all the work performed in the small displacement; by the principle of virtual work, the total amount of this work must be nil, so that

$$-wb\sin\theta\, d\theta + T \cdot 2\pi b \cos\theta\, d\theta = 0.$$

Thus,
$$T = \frac{w}{2\pi}\tan\theta,$$

and the length of the string corresponding to tension T is, as we have seen,

$$a\left(1 + \frac{T}{\lambda}\right).$$

Hence
$$a\left(1 + \frac{w}{2\pi\lambda}\tan\theta\right) = 2\pi b \sin\theta,$$

an equation giving θ.

160   WORK

2. *Gearing of a bicycle*. As a second example, let us apply the principle of virtual work to the mechanism of a bicycle. Let the length of the crank be a, and let the bicycle be geared to b inches, so that each revolution of the pedals causes the machine to move as far forward as it would in one revolution of a wheel of b inches diameter. Let us find what pressure must be exerted on the pedal by a rider in order that the machine may move forward against an opposing frictional force of w pounds weight.

Let us give the machine a small displacement, the cranks being supposed to turn through an infinitesimal angle ϵ, and the wheels and machine moving forward accordingly. Since the gearing is to b inches, the distance moved by the machine as a whole will be $\tfrac{1}{2} b\epsilon$ inches, while the distance moved by the pedal, taking the machine itself as frame of reference, will be $a\epsilon$. Let W pounds weight be the force exerted on the pedal when the machine is just on the point of motion, so that the machine is in equilibrium under this force acting on the pedal, and the backward pull of w pounds due to friction. The equation of virtual work is

$$W \cdot a\epsilon - w \cdot \tfrac{1}{2} b\epsilon = 0,$$

so that the required force is
$$W = \frac{b}{2\,a} w.$$

Thus the force is directly proportional to the gearing of the machine, but inversely proportional to the length of the cranks.

3. *Four rods of equal weight w and length a are freely jointed so as to form a rhombus $ABCD$. The framework stands on a horizontal table so that CA is vertical, and the whole is prevented from collapsing by a weightless inextensible string of length l which connects the points B, D. It is required to find the tension in this string.*

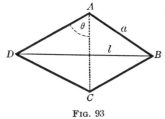

Fig. 93

To find the tension by the principle of virtual work, we must of course find a small displacement such that work is done in opposition to the tension, or otherwise the tension would not enter into the equations at all. Since the string is inextensible, it is not possible in actual fact to stretch it and so perform work against its tension. We can however imagine it to be stretched in spite of its actual inextensibility, or, what comes to the same thing, we can imagine it replaced by an extensible string of the same length and having the same tension. It is now easy to arrange a displacement of the kind required.

Let us imagine that the framework is displaced in such a way that A moves vertically downwards towards C, while C remains at rest. Let the displacement be such that the angle DAC is increased from θ to $\theta + d\theta$. The length l of the string which corresponds to the angle θ is given by

$$l = 2\,a \sin \theta,$$

from which, by differentiation, we obtain

$$dl = 2\,a \cos \theta\, d\theta,$$

ILLUSTRATIVE EXAMPLES

giving the relation between the increments dl, $d\theta$ in l and θ. The work done against the tension of the string (T) in this displacement is $T\,dl$. The height of the center of gravity of the whole figure above C is initially $\frac{1}{2}AC$, or $a\cos\theta$, so that, as in § 120, the work done against gravity is

$$4\,w\,d\,(a\cos\theta).$$

Thus the total work performed by external forces in the displacement is

$$4\,w\,d\,(a\cos\theta) + T\,dl,$$

or, on substituting the values of dl and $d(a\cos\theta)$,

$$-4\,wa\sin\theta\,d\theta + T\cdot 2\,a\cos\theta\,d\theta.$$

For equilibrium this must vanish. We must therefore have

$$T = 2\,w\tan\theta,$$

giving the required tension.

4. *A rod of length l and weight w is suspended by its two ends from two points at the same height and distant l apart, by two strings each of length a. Find the couple required to hold the rod in a position in which it makes an angle θ with its equilibrium position.*

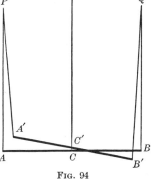

Fig. 94

In equilibrium the strings are vertical, the two ends A, B of the rod lying exactly underneath the two points of suspension P, Q.

As the rod is turned from its equilibrium position, we can imagine its middle point to rise gradually along the vertical line through the original position of this middle point. When the rod has been turned through any angle θ, let the height through which this point has risen be x.

Then the projection of the length PA' on a vertical line will be $a - x$, while its projection on a horizontal plane, being equal to the horizontal projection of AA', will clearly be $l\sin\dfrac{\theta}{2}$.

Thus, expressing that the length of the displaced string PA' remains equal to its original value a, we have

$$a^2 = (a - x)^2 + l^2\sin^2\frac{\theta}{2}. \tag{a}$$

To find the couple required to hold the rod at an angle θ, let us suppose that the rod is held in equilibrium in this position by a couple G, and that a small displacement occurs in which θ is increased to $\theta + d\theta$. The work done against the couple is equal, by § 121, to $-G\,d\theta$, the negative sign being taken, since the couple aids, instead of opposing, the motion. The work done against gravity is equal to $w\,dx$. Thus the equation of equilibrium is

$$-G\,d\theta + w\,dx = 0.$$

To obtain the relation between $d\theta$ and dx we differentiate equation (a), obtaining

$$-2(a-x)\,dx + l^2 \sin\frac{\theta}{2}\cos\frac{\theta}{2}\,d\theta = 0.$$

Thus
$$G = w\frac{dx}{d\theta}$$

$$= \frac{wl^2 \sin\dfrac{\theta}{2}\cos\dfrac{\theta}{2}}{2(a-x)}$$

$$= \frac{wl^2 \sin\theta}{4\sqrt{a^2 - l^2 \sin^2\dfrac{\theta}{2}}},$$

giving the couple required.

EXAMPLES

1. A square $ABCD$ is formed by joining four equal rods by freely moving hinges. The points A, C are joined by an elastic string of natural length equal to a diagonal of the square, and of modulus λ. What forces must be applied to the points B, D to stretch the string to double its length?

2. Three spheres each of radius a and weight w are tied to a point P by strings of natural length l and modulus λ, and hang freely, touching each other. Find the depth of their centers below P.

3. The mechanism by which a Japanese umbrella is opened is such that each rib turns through an angle of 5° for every inch that the sliding piece is moved up the stick. If there are 18 ribs, each of weight $\frac{1}{2}$ ounce, and having their centers of gravity 10 inches from their pivots, find with what force the sliding piece must be pushed up the stick to open the umbrella, when the stick is held vertically, and the ribs are inclined at an angle of 30° to it.

4. The hands of a clock are balanced with counterpoises, so as to be in equilibrium in any position. When the time indicated by the clock is 5.10, a bird of weight w suspends itself from a point on the minute hand which is six feet from the pivot. How large a vertical thrust must be applied to the hour hand, also at a point six feet from the pivot, to restore equilibrium?

5. A clock is wound by raising a weight of 20 pounds through a distance of 3 feet, this enabling the clock to run for 30 hours. The pendulum and escapement are removed, so that the hands will "race" unless held fast. How large a couple must be applied to the minute hand to prevent this occurring?

6. The coupling between two English railway carriages consists of a rod with a right-handed and a left-handed screw cut at its opposite ends and turning in nuts attached to the carriages. If the pitch of each screw is one inch, and the rod is turned by a force of 56 pounds acting at best advantage at the end of a lever 15 inches long, find the force by which the carriages are drawn together.

Potential Energy

129. It will have been noticed that we are concerned with two different kinds of work. The first is typified by the work done in raising a weight in opposition to gravity, the second by the work done against friction in hauling a train along a level road. The essential difference between the two kinds is that work of the first kind can be recovered from the system of bodies by making these bodies themselves perform mechanical work, whereas work of the second kind, when once expended, can never be regained. In raising a weight we may be said to be storing up work rather than spending it, for the weight can at any time be made to yield back all the work devoted to raising it. If we raise a weight w through a distance h, the work done on the weight is wh; on letting the weight descend to its original position, the work done for us by the weight is wh, so that the total work performed on the weight is nil.

On the other hand, in hauling a mass a distance s against a frictional force F, the work performed is Fs. To bring the mass back to its original position, we have to expend an additional amount of work Fs, so that the total work performed is $2 Fs$. This brings out the essential difference between the two types of work and between the two systems of forces against which the work is performed.

130. DEFINITION. *When the forces acting on a system of bodies are of such a nature that the algebraic total work done in performing any series of displacements which bring the system back to its original configuration is nil, the system of forces is said to be a conservative system.*

The algebraic work being nil, the work done on the system in taking it to any configuration is equal and opposite in sign to the work done on the system in allowing it to resume its former configuration, so that all the work spent can be regained. The work is accordingly stored up, or conserved.

164   WORK

A small amount of reflection will show that a system of forces is conservative if the only forces which come into play are some or all of the following:

(a) gravity;
(b) reactions in which the contact is perfectly smooth;
(c) tensions of strings, extensible or inextensible.

On the other hand, if any one or more forces of the following types come into play (so that work is performed against them), the system of forces is non-conservative:

(a) reactions in which the contact is rough;
(b) resistance of the air.

131. THEOREM. *The work done on a system of bodies acted on by conservative forces, in moving from one configuration P to a second configuration Q, is independent of the series of configurations through which the system moves in passing from P to Q.*

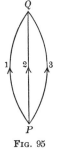

FIG. 95

To prove this, let us denote the work done in passing from P to Q through one series of configurations by W_1, that done in passing through any second series by W_2, and that done in returning from Q to P by any third series of configurations by W_3. If we pass from P to Q by the first series and back from Q to P by the third, the total work done is nil, so that

$$W_1 + W_3 = 0.$$

So, also, if we pass from P to Q by series 2 and back from Q to P by series 3,

$$W_2 + W_3 = 0.$$

Thus $W_1 = W_2$, which proves the theorem.

132. DEFINITION. *Taking any configuration P as standard, the work done in moving a system of bodies from the configuration P to the configuration Q is spoken of as the potential energy of configuration Q.*

POTENTIAL ENERGY   165

The potential energy, accordingly, measures the work which has been stored up in placing the system in configuration Q.

THEOREM. *The work done in moving a system from a configuration (1) to a second configuration (2) against conservative forces is $W_2 - W_1$, where W_1, W_2 are respectively the potential energies in configurations (1) and (2).*

For if P is the standard configuration, the work from P to (1) is W_1; the work from P to (1) plus that from (1) to (2) is W_2, so that the work from (1) to (2) is $W_2 - W_1$.

133. THEOREM. *If a system of bodies is in a configuration of potential energy W, and if x, y, z are the coördinates of any particle, the resultant force acting on the particle has components*

$$-\frac{\partial W}{\partial x}, \quad -\frac{\partial W}{\partial y}, \quad -\frac{\partial W}{\partial z}.$$

To prove this, let us imagine that we give the system a small displacement, which consists in moving the single particle at x, y, z a distance dx parallel to the axis of x. If X, Y, Z are the components of the force acting on it, the work we do in the displacement is, as in § 118, equal to $-X\,dx$. This work is also equal to the increase in the potential energy, namely $\dfrac{\partial W}{\partial x}\,dx$, so that we have

$$-X\,dx = \frac{\partial W}{\partial x}\,dx.$$

Thus $X = -\dfrac{\partial W}{\partial x}$, and similarly we may prove that

$$Y = -\frac{\partial W}{\partial y}, \quad Z = -\frac{\partial W}{\partial z}.$$

134. THEOREM. *If a system of bodies is in a configuration of potential energy W, and if θ is an angle giving the orientation of a rigid body of the system about any line, the moment about this line of the forces acting on the rigid body (reckoned positive if tending to rotate it in the direction of θ increasing) is*

$$-\frac{\partial W}{\partial \theta}.$$

For, let us give the system a small displacement, which consists in turning the body in question through a further angle $d\theta$ about the selected line, so that θ becomes changed into $\theta + d\theta$. The increase in potential energy is $\dfrac{\partial W}{\partial \theta} d\theta$, while the work performed is, by the theorem of § 121, equal to $-G\, d\theta$, where G is the moment about the axis of all the forces acting on the rigid body.

Thus
$$\frac{\partial W}{\partial \theta} d\theta = -G\, d\theta,$$
so that
$$G = -\frac{\partial W}{\partial \theta},$$
the result required.

135. THEOREM. *In a position of equilibrium of a system of bodies, the potential energy W is either a maximum or a minimum.*

The potential energy is a function of all the coördinates of all the particles of which the system of bodies is composed, say

$$x_1, y_1, z_1;\ \ x_2, y_2, z_2;\ \text{etc.}$$

If the position is one of equilibrium, each particle is in equilibrium, so that the components of the forces acting on each particle vanish separately by § 33. By § 133 the condition for this is

$$\frac{\partial W}{\partial x_1} = 0,\ \ \frac{\partial W}{\partial y_1} = 0,\ \ \frac{\partial W}{\partial z_1} = 0,$$
$$\frac{\partial W}{\partial x_2} = 0,\ \text{etc.}$$

But these are exactly the conditions that W shall be a maximum or a minimum.

136. The converse of this theorem is also true.

THEOREM. *If the potential energy of a system of bodies is either a maximum or a minimum in any configuration, then the configuration is one of equilibrium.*

For, with the notation of the previous section, if W is a maximum or a minimum, it follows that

$$\frac{\partial W}{\partial x_1} = 0,\ \ \frac{\partial W}{\partial y_1} = 0,\ \ \frac{\partial W}{\partial z_1} = 0.$$

Since $-\dfrac{\partial W}{\partial x_1}$, $-\dfrac{\partial W}{\partial y_1}$, $-\dfrac{\partial W}{\partial z_1}$ are the components of the force acting on particle (1), these equations indicate that particle (1) is in equilibrium. Similarly, it follows that the other particles are in equilibrium, giving the result.

137. An important special case of these theorems arises when the only forces which do any work in a displacement are the weights of the bodies of which the system is composed. If M is the mass of the whole system, and if h is the height of its center of gravity above any standard horizontal plane, the potential energy is, by § 120, Mgh, and this is a maximum or a minimum when h is a maximum or a minimum. Thus we have the theorem:

In a system of bodies in which the only forces which perform work in a displacement are those of gravity, the configurations of equilibrium are those in which the height of the center of gravity is a maximum or a minimum.

EXAMPLES

1. Two uniform rods, each of length l, are freely jointed at their extremities and placed over a smooth cylinder of radius a of which the axis is horizontal. Find the angle which the rods make with the horizontal when in equilibrium.

2. An elliptic disk is weighted so that its center of gravity is halfway between its center and one extremity of its major axis. Show that if its eccentricity is greater than $\dfrac{1}{\sqrt{2}}$ there will be four positions of equilibrium in which the disk stands vertical on a horizontal plane, but otherwise only two.

3. A horizontal rod of weight W has its center pierced by a fixed vertical screw on which it turns, one revolution raising or lowering it by $\tfrac{1}{2}$ inch. If there is no friction, find the couple required to hold it at rest.

4. A plug of weight W is made in the shape of a pyramid of square cross section. It is placed with its axis vertical in a square hole of side c, the depth of its vertex in this position being d below the plane of contact. Find the couple required to hold it turned through an angle θ and still having its axis vertical.

5. A smooth parabolic wire is placed with its axis vertical. Two beads are strung on it, and are connected by a string which passes through a smooth ring at the focus. Show that there are an infinite number of positions of equilibrium.

6. A smooth bowl in the shape of an ellipsoid of semi-axes a, b, c has one axis vertical. Find the couple required to hold a rod of length l in a horizontal position in the bowl, making an angle θ with a position of equilibrium.

168 WORK

KINETIC ENERGY

138. Suppose that a moving particle is acted on by a force of which the direction is opposite to that of the motion of the particle. The effect of the force, according to the second law of motion, is to produce a retardation in the velocity of the particle. The velocity of the particle decreases so long as the force acts, so that if the force continues to act for a sufficient time, the particle must ultimately be reduced to rest.

Consider, for example, a hammer striking a nail. The reaction between the hammer and nail is a force in the direction opposite to that of the motion of the hammer, and this ultimately brings the hammer to rest. Again, when a particle is projected vertically upwards, its weight after a time reduces it to rest, after which of course it falls back to the ground.

By the time that the moving body has been reduced to rest the point of application of the force, which has moved with the moving body, has moved through a certain distance. Thus a certain amount of work has been done by the moving body. We are thus led to the conception of the *motion* of a body possessing a capacity for doing work.

For instance, in the previous examples, the motion of the hammer has driven the nail into position, and the motion of the particle projected into the air has raised it to a certain height above the earth's surface.

139. Let us suppose that a particle moving with velocity v is opposed by a force P (in absolute units) acting in the direction opposite to that of the motion of the particle. Let the particle describe a distance ds in opposition to this force in time dt, and let its velocity change from v to $v - dv$ in this time. The particle then has a retardation $\dfrac{dv}{dt}$ in the direction of its motion, or, what is the same thing, an acceleration $\dfrac{dv}{dt}$ in the direction in which P is acting, so that by the second law of motion

$$P = m\frac{dv}{dt}.$$

KINETIC ENERGY

The work done by the particle in moving the distance ds in opposition to the force P is

$$P\,ds = m\frac{dv}{dt}\,ds,$$

or, since $\dfrac{ds}{dt}$ is the same as the velocity v of the particle,

$$P\,ds = mv\,dv.$$

Integrating, we find that the whole work done by the particle before being reduced to rest is

$$\int P\,ds = \tfrac{1}{2}mv^2. \tag{36}$$

Since P has to be measured in absolute units (cf. § 22), it follows (§ 111) that the work $\tfrac{1}{2}mv^2$ will also be measured in absolute units.

Thus whatever the magnitude of the force opposing the motion of a particle, the work performed by the particle before being reduced to rest is the same, namely $\tfrac{1}{2}mv^2$ absolute units of work.

The quantity $\tfrac{1}{2}mv^2$ (measured in absolute units) is called the kinetic energy of a moving particle. It is equal to the amount of work which can be performed by the particle before being reduced to rest.

Suppose, for instance, that the resistance offered by a nail to being driven into a board is equal to the weight of 5000 pounds, i.e. that it would require a weight of 5000 pounds to *press* it into the board. Suppose that it is driven into the board by being struck with a hammer, of which the head weighs 10 pounds, and hits the nail with a velocity of 50 feet per second. Let s be the distance the nail is driven in at each stroke measured in feet, then the work done by the hammer at each stroke is that of moving a force of 5000 pounds weight — or $5000 \times g$ poundals — through a distance of s feet. It is therefore equal to $5000\,gs$ foot poundals. The kinetic energy of the hammer in striking the nail is

$$\tfrac{1}{2}mv^2 = \tfrac{1}{2}\cdot 10 \cdot 50^2 = 12{,}500$$

in absolute foot-pound-second units. Thus from the relation (36) we have the equation

$$5000\,gs = 12{,}500,$$

in which, since the units are foot-pound-second units, we may take $g = 32$, and so obtain

$$s = \tfrac{25}{320} \text{ feet} = \tfrac{15}{16} \text{ inches}.$$

140. Theorem. *During the motion of a particle under any system of forces, the increase in kinetic energy is equal to the total work done on the particle by external agencies.*

Let us consider motion of a particle from one position P to a second position Q, and let the velocities of the particle at these two points be v_P, v_Q respectively.

Let us examine the motion over any element ds of the path, and let the velocities at the beginning and end of this element be v and $v + dv$. Let P be the force, or component of force along ds, which acts on the particle while it describes the element ds of its path. If dt is the time taken to describe this element of path, the acceleration is $\dfrac{dv}{dt}$, and since the force acting in the direction of motion is P, we have, by the second law of motion,

$$P = m\frac{dv}{dt}.$$

Hence, as in § 139, $\quad P\,ds = m\dfrac{dv}{dt}ds$

$$= m\frac{ds}{dt}dv$$

$$= mv\,dv.$$

Integrating over the whole path from P to Q, we obtain

$$\int_P^Q P\,ds = m\int_P^Q v\,dv$$

$$= \tfrac{1}{2}m\,v_Q^2 - \tfrac{1}{2}m\,v_P^2 \qquad (37)$$

$$= \text{increase in kinetic energy}.$$

The left-hand side of this equation represents the work done on the particle, proving the result required.

141. The work performed on the particle by external forces may be regarded also as equal to *minus* the work performed by the particle on external agencies. For if P is the force acting on the particle along ds, it follows from the equality of action and reaction that the force acting on the external agencies from the particle is $-P$, so that the total work performed by the particle is $-P\,ds$. Thus the theorem can be stated in the following alternative form:

During the motion of a particle under any system of forces, the decrease in kinetic energy is equal to the total work done by the particle against external agencies.

142. If the system of forces acting on the particle is a conservative system, the value of $-\int_P^Q P\,ds$, the total work performed by the particle on external agencies, is equal, by § 132, to $W_Q - W_P$. Thus equation (37) becomes

$$W_Q - W_P + \tfrac{1}{2} m (v_Q^2 - v_P^2) = 0,$$

or again $\qquad W_Q + \tfrac{1}{2} m v_Q^2 = W_P + \tfrac{1}{2} m v_P^2, \qquad (38)$

so that the sum of the potential and kinetic energies is the same at Q as at P, proving the theorem.

The sum of the potential and kinetic energies is called the *total energy* of the particle.

Conservation of Energy

143. The kinetic energy of a system of bodies is obviously equal to the sum of the kinetic energies of the separate particles. The potential energy of the system, as has been seen, is the sum of the potential energies of its particles.

Thus the total energy of a system is equal to the sum of the total energies of the separate particles. Since the total energy of each particle remains constant, it follows that the total energy of the system remains constant.

The fact that the total energy remains constant is spoken of as the *Conservation of Energy*. An equation expressing that the total energy at one instant is equal to that at any other instant is spoken of as an *equation of energy*.

144. As an illustration, let us consider the firing of a stone from a catapult.

Work is performed in the first place in stretching the elastic of the catapult, and the work is stored as potential energy of the stretched elastic. As soon as the catapult is released, the stone is acted on by the tension of

172   WORK

the elastic; the stone moves under the accelerating influence of this tension, and the tension of the elastic slackens. While this is in progress the stone is acquiring kinetic energy, while the stretched elastic is losing potential energy. By the theorem just proved, the kinetic energy gained by the stone must be just equal to the potential energy lost by the elastic.

When the stone escapes from the catapult, most of the potential energy of the elastic will have disappeared, having been transformed into the kinetic energy of the stone. After this a further transformation of energy may take place while the stone is in motion. If the stone moves upwards, its potential energy will increase, so that there must be a corresponding decrease in its kinetic energy — its speed must slacken. On the other hand, if the stone moves downwards, the potential energy will decrease, so that its kinetic energy will increase — it will gain in velocity.

145. A very important deduction from the principle of the conservation of energy is the following:

THEOREM. *If a particle slide along any smooth curve, being acted on by no forces except gravity and the reaction with the curve, and if u, v be the velocities at two points P, Q of its path, then*

$$v^2 = u^2 + 2gh, \qquad (39)$$

where h is the vertical distance of Q below P, — i.e. is the vertical projection of the path PQ described by the particle.

Let h_P, h_Q denote the heights of P and Q above any horizontal plane — for instance, the earth's surface. Then when the particle is at P its kinetic energy is $\tfrac{1}{2}mu^2$, and its potential energy is mgh_P. Thus its total energy is

$$\tfrac{1}{2}mu^2 + mgh_P.$$

Similarly at Q its total energy is

$$\tfrac{1}{2}mv^2 + mgh_Q.$$

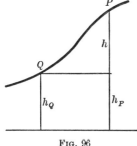

FIG. 96

Since the system of forces acting is a conservative system, the total energy remains unaltered. Thus

$$\tfrac{1}{2}mu^2 + mgh_P = \tfrac{1}{2}mv^2 + mgh_Q,$$

so that
$$v^2 - u^2 = 2g(h_P - h_Q) = 2gh,$$

proving the theorem.

CONSERVATION OF ENERGY

146. The theorem of § 145 is clearly true when the particle is ascending, in which case h is negative — or if the particle ascends during part of its path and descends during the remainder. Moreover, the particle may move under *any* conservative system of forces, provided only that the whole potential energy arises from the weight of the particle, and the theorem remains true.

It is true, for instance, of a particle tied to an inextensible string, or of a particle moving freely in a vacuum.

To illustrate the use of the theorem, let us suppose that a bicyclist, riding with a velocity of 15 miles an hour, comes to the top of a hill of height 60 feet, down which he coasts. Let us find his velocity at the bottom, on the supposition that friction, air resistance, etc., may be neglected.

Taking the top and bottom of the hill to be the points P, Q respectively of the theorem just proved, we have, from the data of the problem,

$$h = 60 \text{ feet},$$
$$u = 15 \text{ miles per hour} = 22 \text{ feet per second}.$$

Thus, using foot-second units, we have

$$v^2 = u^2 + 2\,gh = 22^2 + 2\cdot 32\cdot 60 = 4324,$$

so that $\quad v = 66$ feet per second, approximately,

$$= 45 \text{ miles per hour}.$$

Thus the velocity of the bicycle, if unchecked by friction or air resistance, would be one of about 45 miles per hour.

EXAMPLES

1. An automobile running 40 miles an hour comes to the foot of a steep hill, and at the same instant the engine is shut off. To what height up the hill will the automobile go before coming to rest (neglect friction, etc.)?

2. A laborer has to send bricks to a bricklayer at a height of 10 feet. He throws them up so that they reach the bricklayer with a velocity of 10 feet per second. What proportion of his work could he save if he threw them so as only just to reach the bricklayer?

3. A gun carriage of mass 3 tons recoils on a horizontal plane with a velocity of 10 feet per second. Find the steady pressure that must be applied to it to reduce it to rest in a distance of 3 feet.

4. A ship of 2000 tons moving at 30 feet a minute is brought to rest by a hawser in a distance of 2 feet. Find in tons what pull the hawser has to sustain.

5. A bicycle and rider weigh 200 pounds, and, when riding along a level road at 25 miles an hour, the rider suddenly applies a brake which presses on the tire with a force equal to the weight of 50 pounds. If the coefficient of friction between the brake and the tire is $\frac{1}{3}$, find how far the machine will go before coming to rest.

6. In the last question, how far will the machine go if, instead of the road being level, it is down an incline of 1 in 20?

7. A bullet fired with a velocity of 1000 feet per second penetrates a block of wood to a depth of twelve inches. Prove that if it were fired through a board of the same wood, two inches thick, its velocity on emergence would be about 913 feet per second. (Assume the resistance of the wood to the bullet to be constant.)

8. Two equal weights P and P are supported by a string passing over two small smooth pulleys A and B in the same horizontal line, and a weight $W = \dfrac{2}{\sqrt{3}} P$ is attached to the middle point of the string between A and B. Prove that W will continue to descend until WAB forms an equilateral triangle, and examine what will happen after this.

9. A string of natural length l and modulus λ is suspended between two points A, B in the same horizontal line and at a distance h apart, and has a weight W attached to its middle point. The weight W is held at rest midway between the points A, B, and is suddenly set free. Find how far it will fall before being brought to rest by the strings.

10. A heavy particle hangs by a string of natural length l, which it stretches to a length l', the other end of the string being fixed. The particle is pulled down to a length $2l'$ below the point of support, and is then set free. How high will it rise?

11. Determine the horse power which could be obtained from the kinetic energy of a river at a place where the width is 100 feet, the mean depth 20 feet, and the mean velocity $4\frac{1}{2}$ miles per hour. (A cubic foot of water weighs 62.5 pounds.)

12. The river of the last question ends in a waterfall of which the bottom is 50 feet below the river bed. Find the horse power which could be obtained from the water.

13. A locomotive burns $1\frac{1}{4}$ pounds of coal per horse-power-hour. How much coal must be burned, beyond that consumed in overcoming gravity, friction, etc., in giving to a train of 300 tons a velocity of 55 miles an hour?

Stable and Unstable Equilibrium

147. Let us consider a system at rest in a position of equilibrium, and capable of moving from this position by only one path, over which it may, of course, move in either direction. As an illustration of a system of this kind we may take a locomotive standing

STABILITY AND INSTABILITY 175

on a pair of rails, a door turning about a hinge, or a bead sliding on a wire. The system is supposed to be acted on by any number of conservative forces, but to be in a position of equilibrium under these forces.

Let P denote the position of equilibrium, and let W_P be the potential energy when the system is in configuration P. Let x denote any coördinate which measures how far the configuration of the system has moved from P — for instance, returning to our former illustrations, x might denote the distance the locomotive had moved along the track, the angle through which the door had turned about its hinges, or the distance the bead had moved along the wire. The value of x will of course be considered positive if the system moves in one direction, and negative if it moves in the other.

As the system moves away from its equilibrium configuration P, the value of x will change. The value of W, the potential energy, will also change, and as it depends only on the value of x if the forces are conservative, we may say that W is a function of x.

By a well-known theorem, we can expand W in powers of x in the form

$$W = W_P + x\left(\frac{\partial W}{\partial x}\right)_P + \tfrac{1}{2}x^2\left(\frac{\partial^2 W}{\partial x^2}\right)_P + \cdots, \qquad (40)$$

in which the subscript P denotes (as it has already been supposed to denote in the case of W_P) that the quantity is to be evaluated in the configuration P. Since the configuration P is supposed to be one of equilibrium, we have by the theorem of § 135,

$$\left(\frac{\partial W}{\partial x}\right)_P = 0,$$

so that equation (40) becomes

$$W = W_P + \tfrac{1}{2}x^2\left(\frac{\partial^2 W}{\partial x^2}\right)_P + \cdots. \qquad (41)$$

For configurations near to P, x is small, so that the term $\tfrac{1}{2}x^2\left(\frac{\partial^2 W}{\partial x^2}\right)_P$ in equation (41), although itself small, is yet very large compared with the terms in x^3, x^4, etc., which follow it.

Thus, for configurations near to P, we may neglect these latter terms altogether, and write the equation in the form

$$W - W_P = \tfrac{1}{2} x^2 \left(\frac{\partial^2 W}{\partial x^2}\right)_P. \tag{42}$$

The value of $\left(\dfrac{\partial^2 W}{\partial x^2}\right)_P$ may be either positive or negative.

If it is positive, then $W - W_P$ is positive whatever the value of x, so that the potential energy W in every configuration near to P is greater than that in configuration P. In other words, W is a *minimum* at P.

So also if $\left(\dfrac{\partial^2 W}{\partial x^2}\right)_P$ is negative, $W - W_P$ is negative for all small values of x, and we find that W is a *maximum* at P.

148. Suppose now that the system is placed at rest in some configuration near to P. This configuration is not one of equilibrium, so that the system cannot remain at rest. To determine the direction in which it begins to move, we need only notice that as the system moves it acquires kinetic energy, and as this must, by § 143, be acquired at the expense of its potential energy, we see that the system will begin to move in such a direction that its potential energy will be diminished.

A glance at equation (42) will show whether this direction is towards or away from P. We see that if $\left(\dfrac{\partial^2 W}{\partial x^2}\right)_P$ is positive, the value of x^2 must decrease, so that the motion will be towards P, whatever the value of x. Similarly, if $\left(\dfrac{\partial^2 W}{\partial x^2}\right)_P$ is negative, the value of x^2 must increase, so that the motion will be always away from P.

We have now seen that if the system is placed in a configuration adjacent to P, the question of whether the motion which ensues is towards or away from P does not depend on the configuration in which the system is placed, but depends on the sign of $\left(\dfrac{\partial^2 W}{\partial x^2}\right)_P$.

STABILITY AND INSTABILITY 177

We have seen that if P is a configuration of equilibrium, and if the system is slightly displaced from P to a neighboring configuration, then

(a) if $\left(\dfrac{\partial^2 W}{\partial x^2}\right)_P$ is positive, the system, when set free, will return to its original position of equilibrium;

(b) if $\left(\dfrac{\partial^2 W}{\partial x^2}\right)_P$ is negative, the system when set free will move farther away from its original position of equilibrium.

Equilibrium of the first kind is called *stable* equilibrium; equilibrium of the second kind is called *unstable* equilibrium.

We can summarize the results as follows:

Sign of $\left(\dfrac{\partial^2 W}{\partial x^2}\right)_P$	Potential Energy W	Equilibrium
+	minimum	stable
−	maximum	unstable

149. THEOREM. *Positions of stable and unstable equilibrium occur alternately.*

We can assume that we are dealing only with finite forces, so that the function W will always be finite: it can never pass through the values $W = \pm \infty$. It must be continuous, for, by hypothesis, the work done in placing the system in any configuration must have a definite value, so that the potential energy can have only one value for a given configuration. Also the differential coefficients of the potential energy must be finite, for these measure the forces (§ 133) which can have only finite values in any given configuration.

Thus if the graph of the function W is drawn, we see that it must consist of portions in which W is alternately increasing and decreasing. On passing from a portion in which W increases to one in which it decreases, we pass through a point at which W is a maximum, while in passing from a region in which W decreases to one in which it increases, we pass through a minimum. Thus maximum and minimum values of W must occur alternately, or, what is the same thing, configurations of stable and unstable equilibrium must occur alternately.

150. Examples of these two kinds of equilibrium can be found in the illustrations already employed.

1. Locomotive moving on a pair of rails. Let h be the height of the center of gravity in any position, let x denote distances measured horizontally along the track, and let M be the mass of the locomotive. The potential energy is then Mgh. The condition for equilibrium in the configuration $x = 0$ is

$$\frac{d}{dx}(Mgh) = 0,$$

or $\frac{dh}{dx} = 0$, expressing that the value of h must be either a maximum or a minimum. The table on page 177 shows that if h is a minimum, — i.e. if the center of gravity is at its lowest point, — the equilibrium will be stable.

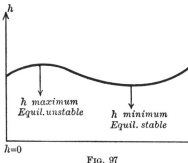

Fig. 97

Thus, if the locomotive is moved slightly from this position, it will roll back to it again. If h is a maximum, — i.e. if the center of gravity is at its highest point, — the equilibrium will be unstable. The locomotive is now at the summit of a hill, and if displaced to either side of the summit, will continue rolling down the hill.

NOTE. If the moving parts of the engine are not "balanced" properly, the center of gravity may not always be at the same height above the rails, so that the maxima and minima of h do not necessarily occur at points where the height of the track is a maximum or a minimum. For instance, a position of equilibrium might occur where the track was not level, or again a position of stable equilibrium might occur at a point at which the track was at its highest point, the height of the center of gravity above the rails being of course a minimum at this point. Thus if the engine were displaced to a point slightly lower on the track, and set free, it would return of itself to the highest point. The principle here is the same as that of mechanical toys which, on being placed at rest at the foot of an inclined plane, start to roll up the plane as soon as set free.

We notice that positions of stable and unstable equilibrium must occur alternately, as already proved in § 149.

2. Door turning on hinges. Here again the potential energy is Mgh, where h is the height of the center of gravity of the door above any standard

STABILITY AND INSTABILITY 179

level. As the door turns on its hinges, its center of gravity describes a circle about the line of hinges. If this line is perfectly vertical, the circle described by the center of gravity lies entirely in a horizontal plane, so that every position is one of equilibrium, and the question of stability or instability does not arise. If, however, the line of hinges is not perfectly vertical, the circle will lie in an inclined plane. The points at which the height above the standard horizontal plane is a maximum or minimum are two in number:

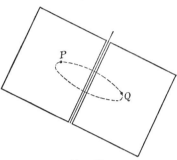

FIG. 98

P, the highest point of the circle, at which equilibrium is unstable;

Q, the lowest point of the circle, at which equilibrium is stable.

3. Bead sliding on wire. To obtain a definite problem, let us suppose that the bead P slides on an elliptic wire placed so that its major axis AA'

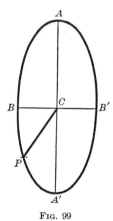

FIG. 99

is vertical, and let it be acted on by its weight, and also by the tension of a stretched elastic string of which the other end is tied to the center of the ellipse. Let a, b be the semi-axes of the ellipse, and let l, λ be the natural length and modulus of the string, l being greater than a, so that the string is always stretched. Let w be the weight of the bead.

The first step is to calculate the potential energy in any configuration. Let the configuration be specified by the eccentric angle ϕ of the point on the ellipse occupied by the bead. The height of the bead above the center of the ellipse is then $a \cos \phi$, so that that part of the potential energy which arises from gravitational forces is $wa \cos \phi$. The length of the string r is given by

$$r^2 = a^2 \cos^2 \phi + b^2 \sin^2 \phi, \qquad (a)$$

and the work done in stretching the string from length l to length r is (§ 113)

$$\frac{\lambda}{2\,l}(r-l)^2.$$

This may be taken to be the part of the potential energy which arises from the stretching of the string. Thus the total potential energy will be

$$W = wa \cos \phi + \frac{\lambda}{2\,l}(r-l)^2. \qquad (b)$$

180   WORK

The positions of equilibrium are now given by $\dfrac{dW}{d\phi} = 0$, or

$$wa \sin\phi - \frac{\lambda}{l}(r-l)\frac{dr}{d\phi} = 0,$$

or, substituting for r from equation (a),

$$wa \sin\phi + \frac{\lambda}{l}(a^2 - b^2)\sin\phi \cos\phi - \frac{\lambda(a^2 - b^2)\sin\phi \cos\phi}{\sqrt{a^2 \cos^2\phi + b^2 \sin^2\phi}} = 0.$$

Rationalizing, we find that roots are given by $\sin\phi = 0$, and also by

$$\left[wa + \frac{\lambda}{l}(a^2 - b^2)\cos\phi \right]^2 (a^2 \cos^2\phi + b^2 \sin^2\phi) - \lambda^2 (a^2 - b^2)^2 \cos^2\phi = 0,$$

which reduces to

$$\left[wa + \frac{\lambda}{l}(a^2 - b^2)\cos\phi \right]^2 \left[(a^2 - b^2)\cos^2\phi + b^2 \right] - \lambda^2 (a^2 - b^2)^2 \cos^2\phi = 0, \quad (c)$$

an equation of the fourth degree in $\cos\phi$.

The roots of $\sin\phi = 0$ are $\phi = 0, \pi$, so that there are always two positions of equilibrium at A, A', the ends of the major axis. Equation (c), being of the fourth degree, may have 0, 2, or 4 real roots in $\cos\phi$. The equation as it stands has been obtained by squaring both sides of the equation to be satisfied, and in doing this we have doubled the number of roots of the true equation. Thus the true equation will only be satisfied by 0, 1, or 2 real roots in $\cos\phi$. In other words, between A and A', on either side of the wire, there can be at most *two* positions of equilibrium.

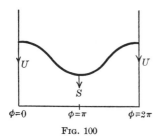

Fig. 100

It would be a tedious piece of work to find the actual values of the roots for $\cos\phi$, and then determine the signs of the values of $\dfrac{d^2W}{d\phi^2}$ corresponding to these roots. The question is, however, very much simplified by using the general theory of stable and unstable configurations.

If we put $\lambda = 0$ in expression (b), we obtain as the potential energy in the case in which λ is vanishingly small in comparison with w,

$$W = wa \cos\phi,$$

of which the graph is shown in fig. 100. Here there are only two positions of equilibrium, namely $\phi = 0$ and $\phi = \pi$, the former being unstable (U) and the latter stable (S).

STABILITY AND INSTABILITY 181

Again, if we put $w = 0$ in expression (b), we obtain as the potential energy in the case in which λ is infinitely great in comparison with w,

$$W = \frac{\lambda}{2l}(r - l)^2,$$

and the graph of W in this case is shown in fig. 101. There are four positions of equilibrium,

$$\phi = 0, \quad \frac{\pi}{2}, \quad \pi, \quad \frac{3\pi}{2},$$

which are respectively unstable, stable, unstable, and stable.

The general case in which λ stands in a finite ratio to w is intermediate between the two extreme cases which have been considered. The graph for W in the general case can be obtained by compounding the two graphs already drawn. To obtain the ordinate corresponding to any value of ϕ, we multiply the corresponding ordinates in the graphs already obtained by the appropriate constants, and add. The two ordinates give the two terms of expression (b) separately: their sum gives the total value of W as required.

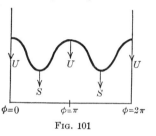

FIG. 101

From this geometrical construction it is clear that $\phi = 0$ remains a configuration of unstable equilibrium. The configuration $\phi = \pi$ is also a configuration of equilibrium, but may be either stable or unstable. Between these two configurations there may be one other configuration of equilibrium, as in fig. 101; or there may be none, as in fig. 100. Since, by § 149, stable and unstable configurations occur alternately, it is clear that if the configuration $\phi = \pi$ is stable, there can be no other configuration of equilibrium between this and $\phi = 0$, while if $\phi = \pi$ is unstable, there must be one configuration of equilibrium between $\phi = \pi$ and $\phi = 0$, and this must be stable.

The stability or instability of the configuration $\phi = \pi$ accordingly determines the nature of the solution for a given value of λ. This stability or instability is in turn determined by the sign of $\dfrac{\partial^2 W}{\partial \phi^2}$ at $\phi = \pi$. To determine this, let us write $\pi - \phi = \theta$ near to $\phi = \pi$, and neglect terms smaller than θ^2. We have, to this approximation,

$$r^2 = a^2 \cos^2\phi + b^2 \sin^2\phi$$
$$= a^2 - (a^2 - b^2)\sin^2\theta$$
$$= a^2 - (a^2 - b^2)\theta^2,$$

so that by equation (b),

$$W = wa \cos\phi + \frac{\lambda}{2l}(r - l)^2$$
$$= -wa(1 - \tfrac{1}{2}\theta^2) + \frac{\lambda}{2l}\left[a\left(1 - \frac{1}{2}\frac{a^2 - b^2}{a^2}\theta^2\right) - l\right]^2.$$

Thus
$$\frac{\partial^2 W}{\partial \theta^2} = wa - \frac{\lambda(a^2 - b^2)(a - l)}{al}.$$

It follows that equilibrium at $\phi = \pi$ is stable or unstable according as

$$\lambda < \text{ or } > \frac{wa^2 l}{(a^2 - b^2)(a - l)}.$$

To sum up, there are two cases:

I. $\lambda < \dfrac{wa^2 l}{(a^2 - b^2)(a - l)}$. The only positions of equilibrium are $\phi = 0$ and $\phi = \pi$, which are respectively unstable and stable.

II. $\lambda > \dfrac{wa^2 l}{(a^2 - b^2)(a - l)}$. There are positions of equilibrium $\phi = 0$ and $\phi = \pi$, both unstable, and also an intermediate position which is stable. This last position is determined by equation (c).

Critical and Neutral Equilibrium

151. If the value of $\dfrac{\partial^2 W}{\partial x^2}$ at a position of equilibrium is zero, the equilibrium is called *critical*. So far, we have not discovered what happens when a system is slightly displaced from a position of critical equilibrium.

In general, the value of W in the neighborhood of any position of equilibrium can be expanded in the form (cf. equation (41))

$$W = W_P + \tfrac{1}{2} x^2 \left(\frac{\partial^2 W}{\partial x^2}\right)_P + \tfrac{1}{6} x^3 \left(\frac{\partial^3 W}{\partial x^3}\right)_P + \tfrac{1}{24} x^4 \left(\frac{\partial^4 W}{\partial x^4}\right)_P + \cdots . \quad (43)$$

If $\dfrac{\partial^2 W}{\partial x^2}$ vanishes at P, the most important term in the value of $W - W_P$ is that in x^3, so that we have approximately

$$W - W_P = \tfrac{1}{6} x^3 \left(\frac{\partial^3 W}{\partial x^3}\right)_P.$$

Here $W - W_P$ changes sign on passing through $x = 0$, the configuration of equilibrium, so that the graph of W is as shown in

STABILITY AND INSTABILITY

fig. 102, having a horizontal tangent and point of inflection at P. On one side the potential energy is less than at P, on the other side it is greater.

Let Q, Q' be two adjacent configurations on these two sides of P. If the system is placed at Q, it must move so that its potential energy decreases, and therefore moves away from P. If it is placed at Q', for the same reason it must move at first towards P, but it will move beyond P and will then continue to move away from P, — for it cannot come to rest until its potential energy is again equal to that at Q', and this cannot happen in the neighborhood of P. Thus

Fig. 102

if the system starts from *any* configuration in the neighborhood of P, it will ultimately be moving away from P. In other words, the equilibrium is *unstable*.

Thus if $\dfrac{\partial^2 W}{\partial x^2} = 0$ at P, the equilibrium is, in general, unstable. An exception has to be made when $\dfrac{\partial^3 W}{\partial x^3} = 0$; for then we have

$$W - W_P = \tfrac{1}{24} x^4 \left(\frac{\partial^4 W}{\partial x^4}\right)_P.$$

This case may be treated as in § 148, and we find that the equilibrium is stable or unstable according as $\left(\dfrac{\partial^4 W}{\partial x^4}\right)_P$ is positive or negative.

152. Higher degrees of singularity may be treated in the same way, and we easily obtain the following general rules:

If the first differential coefficient which does not vanish is of odd order, the equilibrium is unstable.

If the first differential coefficient which does not vanish is of even order, the equilibrium is stable or unstable according as this differential coefficient is positive or negative.

It is possible for all the differential coefficients to vanish, in which case the problem is best treated by other methods.

For instance, if the potential energy is of the form

$$W = x^2 e^{-\frac{1}{x^2}},$$

it will be found that all the differential coefficients vanish in the configuration given by $x = 0$. On drawing a graph of the function W it appears that the equilibrium is stable.

It may be that all the differential coefficients of W vanish because W is a constant throughout the whole of a range surrounding the configuration under consideration. If this is so, the system may be displaced, and there will be no force tending to move it from its new configuration — every configuration is one of equilibrium. Equilibrium of this kind is called *neutral equilibrium*.

A case of neutral equilibrium has already occurred in Ex. 2, p. 179, — a door free to swing about a vertical line of hinges. A second case is that of a sphere rolling on a horizontal plane.

Systems possessing Several Degrees of Freedom

153. So far we have considered only systems which are limited to moving through a single series of configurations — systems with only a single degree of freedom. The determination of the stability or instability of a system having more than one degree of freedom is a more complex problem.

If the potential energy is absolutely a minimum in a position of equilibrium, so that every possible motion involves an increase of potential energy, then the equilibrium is stable. This obviously can be proved by the same argument as has served when there is only one degree of freedom.

If the potential energy is not an absolute minimum, — that is to say, if displacements are possible in which the potential energy decreases while moving away from the position of equilibrium, — then the configuration is one of unstable equilibrium. This will be proved later. It cannot be proved by the methods used in this chapter, and so we defer the question until later (Chapter XII).

EXAMPLES 185

GENERAL EXAMPLES

1. Prove that the horse power of an engine which overcomes a resistance of R pounds at a speed of S miles an hour is
$$RS \div 375.$$

2. A train weighing, with the locomotive, 500 tons is kept moving at the uniform rate of 30 miles an hour on the level, the resistance of air, friction, etc., being 40 pounds per ton. Find the horse power of the engine.

By how much must this horse power be increased if the rate is to be maintained while water is taken up from a trough between the rails to the amount of 20 pounds per foot passed over, the height to which the water is raised above the trough being 10 feet, and the kinetic energy imparted to the water in the trough, as well as that of the motion of the water taken up, relatively to the tank, being neglected?

3. The sides of a conical hill are of such a shape that a given mass will just rest on them without slipping. A man wishes to move this mass from a point at the base of the hill to a second point diametrically opposite to the first. Show that the work of dragging it over the hill is less than the work of dragging it round the base of the hill, in the ratio $2 : \pi$.

4. Show that the work a man does in dragging a weight up a hill from a given point A to the summit B depends only on the positions of A and B, and is independent of the shape of the hill, provided he keeps always in the vertical plane through A and B.

5. A catapult is made by tying the two ends of a piece of elastic, natural length a, modulus of elasticity λ, to the two prongs of a forked piece of wood, distant l apart, l being greater than a. A stone of mass m is placed at the middle point of the catapult, and is drawn back until the string is stretched to double its natural length. If it is then set free, find the velocity with which it will leave the catapult.

6. If, in the last question, the stone is projected vertically upwards from the catapult, find the height to which it will rise before coming to rest.

7. A necklace of mass m is made of beads threaded on a light string of modulus λ. It is held in a horizontal plane, with the string unstretched, resting on the surface of a smooth right circular cone of semivertical angle α, of which the axis is vertical. If the necklace is let go, how far will it slip down the cone before coming to rest?

8. A fly wheel is of radius 2 feet 6 inches, and the weight of the spokes, etc., may be neglected in comparison with that of the rim. It is rotating at the rate of 250 revolutions per minute about a fixed axle, which is 3 inches in diameter, the coefficient of friction between the wheel and axle being $\frac{1}{20}$. If it is left to itself, find how many revolutions it will make before stopping.

9. A spider hangs from the ceiling by a thread of modulus of elasticity equal to its weight. Show that it can climb to the ceiling with an expenditure of work equal to only three quarters of what would be required if the thread were inelastic.

10. A fine thread having two masses each equal to P suspended at its ends is hung over two smooth pegs in the same horizontal line, distant $2a$ apart. A mass Q is then attached to the middle point of the portion of the string between the pegs and allowed to descend under gravity. Show that its velocity after falling a depth x will be
$$\sqrt{\left\{\frac{2\,g\,(x^2 + a^2)\,(Qx + 2\,Pa - 2\,P\sqrt{x^2 + a^2})}{Q(x^2 + a^2) + 2\,Px^2}\right\}}.$$

11. Assuming that the attraction of the earth on a body outside the earth falls off inversely as the square of the distance of the body from the earth's center, find with what velocity a shot would have to be fired vertically upwards from the earth's surface so as never to return to the earth at all.

12. A steam hammer of weight 30 tons is pressed down partly by its weight and partly by the pressure of steam in a vertical cylinder acting on a piston which moves with the hammer. The area of the piston is 4 square feet, and the steam pressure is 225 pounds to the inch. If the hammer is raised a height of 2 feet above its block, and set free, find the velocity with which it will strike the block.

13. The ends of a uniform rod of length l are connected by a string of length a which is placed over a smooth peg. Show that the rod can only hang in a horizontal or vertical position, and examine the stability or instability of these positions.

14. Two equal uniform rods are rigidly jointed in the shape of the letter L, and placed astride a smooth circular cylinder of radius a. Find the smallest length of the rods consistent with stability of equilibrium, the rods being constrained to remain in a vertical plane perpendicular to the axis of the cylinder.

15. A cube of stone of edge a rests symmetrically and with its base horizontal on a rough circular log of diameter b. Show that the equilibrium is stable or unstable according as $b >$ or $< a$.

16. A rocking stone rests on a fixed stone, the contact being rough, and the common normal at the point of contact being vertical. If ρ, ρ' be the radii of curvature of the surfaces of the two stones at the point of contact, and if h be the height of the center of gravity of the movable stone, show that the equilibrium of the rocking stone will be stable or unstable according as
$$\frac{1}{h} > \text{ or } < \frac{1}{\rho} + \frac{1}{\rho'}.$$

EXAMPLES

17. A ladder of length h and weight w stands in a vertical position on a rough floor, an elastic string being tied to its topmost point and to a point in the ceiling at a height b above the floor, its tension being T. Show that the equilibrium is stable or unstable according as

$$T > \text{ or } < \frac{w(b-h)}{2b}.$$

18. If the tension in question 17 is equal to $w(b-h)/2b$, determine whether the equilibrium is stable or unstable.

19. If in question 14 the rods are of the critical length which separates stability from instability, show that the equilibrium is neutral, so that the rods can, within certain limits, rest in any position in the plane perpendicular to the axis of the cylinder.

20. Show that the rods in the last question are in stable equilibrium as regards displacements in which the plane of the rods rotates about a vertical axis, and find the couple required to hold the rods in a position in which the plane makes any given angle θ with the axis of the cylinder.

21. Find the smallest length of the rods in the last question, in order that the equilibrium may be stable for all possible displacements.

22. The radii of curvature at the blunt and pointed ends of a hard-boiled egg are a and b respectively, and the egg can just be made to balance on its blunt end when stood on a rough horizontal surface. Show that it can be made to balance on its pointed end if stood inside a hemispherical basin of radius less than

$$\frac{a(c-b)}{a+b-c},$$

where c is the longest axis of the egg. If the radius of the basin is just equal to $a(c-b)/(a+b-c)$, would the equilibrium be stable or unstable?

23. A, B, C are three equidistant smooth pegs in the same horizontal line, and a heavy uniform string has its ends tied to A, C, and is looped over B. Show that there may or may not be a position of equilibrium in which the two catenaries AB, BC are unequal, and that if there is such a position it will be stable.

Show also that the position of equilibrium in which the middle point of the string is at B is unstable or stable according as an unsymmetrical position of equilibrium does or does not exist.

CHAPTER VIII

MOTION OF A PARTICLE UNDER CONSTANT FORCES

154. The simplest case of motion of a single particle occurs when the particle is acted upon only by constant forces and moves in a straight line.

If P is the component force in the direction of the motion of the particle, there will, by the second law of motion, be an acceleration f given by
$$P = mf,$$
where m is the mass of the particle. Since the forces are, by hypothesis, constant, the acceleration f is also constant.

Let the particle start with a velocity u, and move with a constant acceleration f. In time t the increase in velocity is ft, so that, after any time t, the whole velocity is $u + ft$. Denoting this velocity by v, we have
$$v = u + ft. \tag{44}$$

By definition, v is equal to $\dfrac{ds}{dt}$, where s is the space described from the beginning of the motion. We accordingly have
$$\frac{ds}{dt} = u + ft,$$
an equation giving the rate of increase of s at any instant. Integrating, we obtain
$$s = ut + \tfrac{1}{2}ft^2, \tag{45}$$
no constant of integration being needed, because the distance described at time $t = 0$ has to be 0, from the definition of s.

By equation (44), $u = v - ft$, so that equation (45) can be written
$$s = vt - \tfrac{1}{2}ft^2. \tag{46}$$

MOTION UNDER CONSTANT FORCES

This gives the distance described in time t, when we know the velocity v with which the particle arrives at the end of its journey.

Combining equations (45) and (46), we have

$$s = \tfrac{1}{2}(u+v)t, \qquad (47)$$

showing that the space described is the arithmetic mean of ut and vt: the former is the space that would be described if the particle maintained its original velocity u through the whole time t; the latter is that which would be described if the particle had its final velocity throughout the whole time.

Combining equation (47) with equation (44), which can be written in the form

$$ft = (v-u),$$

we obtain, on eliminating t,

$$2fs = v^2 - u^2, \qquad (48)$$

an equation connecting the space described with the initial and final velocities.

This last equation may also be deduced from the equation of energy. Since the work done on the particle is equal to the change in its kinetic energy, we have

$$Ps = \tfrac{1}{2}mv^2 - \tfrac{1}{2}mu^2,$$

and since $P = mf$, equation (48) follows at once.

Body falling under Gravity

155. The simplest application of these equations is to the motion of a body which is allowed to fall freely under the influence of gravity, so that the acceleration is g.

If the body starts from rest, we put $u = 0$, and measure s vertically downwards. We find from equation (45) that after time t the body has fallen a distance $\tfrac{1}{2}gt^2$, while its velocity is ft. After falling a distance h its velocity is, by equation (48), equal to $\sqrt{2gh}$. This is frequently spoken of as the "velocity due to a height h."

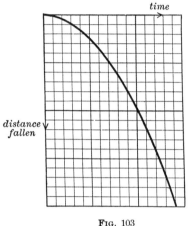

FIG. 103

We notice that the distance fallen varies as the square of the time during which the body has been falling. In fig. 103 the time is measured horizontally, while the distance fallen is measured vertically. The thick curve gives a graphical representation of the distance fallen. Denoting the horizontal distance by x and the vertical by y, we have $x = t, y = \tfrac{1}{2} gt^2$, so that

$$y = \tfrac{1}{2} gx^2.$$

156. This is the equation of a parabola, so that the curve is a parabola. The graph can be obtained experimentally by a method known as Morin's method. A weight P is free to fall vertically in a slot formed in a rod AB, and is arranged so that, as it falls, a pencil attached to it makes a mark on a drum CD which is covered with paper. The drum is made to rotate uniformly. On unrolling the paper from the drum we obtain the graph of fig. 103, — for the horizontal distance is proportional to the time, while the vertical is the distance fallen through. The fact that the curve obtained in this way is accurately a parabola gives experimental confirmation of the fact that motion under gravity is motion with *uniform* acceleration.

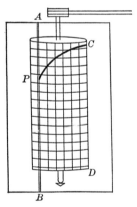

FIG. 104

BODY FALLING UNDER GRAVITY

157. If the body is projected vertically upwards with velocity u, we may measure the distance s vertically upwards, and the acceleration in this direction will be $-g$. Thus we have

$$s = ut - \tfrac{1}{2} gt^2,$$
$$v = u - gt,$$
$$2gs = u^2 - v^2,$$

where s is the distance upwards described after time t, and v is the upward velocity. From the first equation we see that $s = 0$ not only when $t = 0$, but also when $t = \dfrac{2u}{g}$. Thus the particle returns to its original position after time $2u/g$. When $s = 0$ the third equation shows that $u^2 = v^2$. Thus when the particle returns, its velocity is the same as when it started. Clearly this must be so, for the potential energy is the same, and therefore the kinetic energy also is the same.

EXAMPLES

1. If an express train is doubled, the first half being given 5 minutes' start, and attaining its maximum booked speed of 48 miles an hour after moving with constant acceleration for a mile, prove that the two halves will run about 4 miles apart, but the first half will have gone 3 miles before the start of the second half.

2. A train passes another on a parallel track, the former having a velocity of 45 miles an hour and an acceleration of 1 foot per second per second, the latter a velocity of 30 miles an hour and an acceleration of 2 feet per second per second. How soon will the second be abreast of the first again, and how far will the trains have moved in the meantime?

3. A body is dropped from a balloon at a height of 70 feet from the ground. Find its velocity on reaching the ground, if the balloon is (*a*) rising, (*b*) falling, with a velocity of 30 feet a second.

4. A stone is dropped into a well, and the sound of the splash reaches the top after 9 seconds. Find the depth of the well, the velocity of sound being 1100 feet per second.

5. An elevator descends with an acceleration of 5 feet per second per second until its velocity is 20 feet per second, after which its velocity remains uniform. After it has been in motion for 6 seconds, a stone is dropped on to it from the point at which the elevator started. How soon will it strike the elevator?

6. A juggler keeps three balls going with one hand, so that at any instant two are in the air and one in his hand. If each ball rises to a height of 4 feet, show that the time during which a ball stays in his hand is $\tfrac{1}{4}$ second.

7. A body was observed to take t seconds in falling past a hatchway to the bottom of a hold h feet deep. Prove that it fell

$$\frac{1}{2}g\left(\frac{h}{gt}+\frac{1}{2}t\right)^2 \text{ feet,}$$

and struck with velocity

$$\frac{h}{t}+\frac{1}{2}gt \text{ feet per second.}$$

8. A chain 12 feet long hangs from its upper end. If this be released, find the time the chain will take in passing a point 60 feet below the initial position of the highest point.

9. A body whose mass is 5 pounds, moving with a speed of 160 feet per second, suddenly encounters a constant resistance equal to the weight of $\frac{1}{4}$ pound, which lasts until the speed is reduced to 96 feet per second. For what time and through what distance has the resistance acted?

10. Two wagons, coupled together, are pulled along a horizontal track by a steady force, and move over 100 feet in the first ten seconds from rest. The rear wagon is then uncoupled, and it is found that at the end of the next ten seconds the interval between the two wagons is 150 feet. Compare the masses of the two wagons, all resistance being neglected.

11. A balloon of weight W is rising with acceleration f. If a weight w of sand be emptied out of the car, find the increase in the acceleration of the balloon, neglecting the resistance of the air and the buoyancy of the sand.

Motion on an Inclined Plane

158. Suppose we allow a particle to slide down an inclined plane, the contact between the two being supposed perfectly smooth. If m is the mass of the particle, the forces acting on it are its weight mg, and the reaction R normal to the plane. The component down the plane is $mg \sin \alpha$, so that the particle moves with uniform acceleration $g \sin \alpha$.

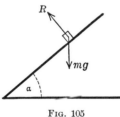

Fig. 105

We can obtain the distance described in the time t from the usual formulæ. If the particle starts from rest, the distance described in time t is $\frac{1}{2} g \sin \alpha \cdot t^2$.

159. Suppose that through a point O we have a great number of smooth wires on which smooth beads are free to slide. Let these

MOTION ON AN INCLINED PLANE 193

wires make all possible angles with the vertical, one of them, OO', being vertical. Let us imagine that the beads are all collected at O and are set free simultaneously. After time t, let the bead which is falling vertically be at P, and let the bead which is falling along a wire inclined at an angle β to the vertical be at Q. This latter bead moves with acceleration $g\cos\beta$. Thus $OP = \tfrac{1}{2}gt^2$, while $OQ = \tfrac{1}{2}g\cos\beta \cdot t^2$. Hence $OQ = OP\cos\beta$, and therefore OQP is a right angle. It follows that Q is on the sphere constructed on OP as diameter, and obviously the same will be true of every other bead. Thus at any instant all the beads

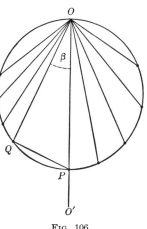

FIG. 106

will be on a sphere of which O is the highest point, and of which the lowest point is at a distance $\tfrac{1}{2}gt^2$ below O. Hence as the motion proceeds the beads will appear to form a sphere which continually swells out in size, the highest point appearing to remain fixed at O, while the lowest point appears to fall freely under gravity.

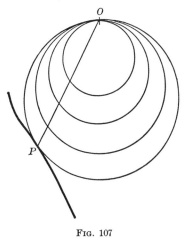

FIG. 107

160. This imaginary experiment indicates a way of solving a practical problem. Suppose we wish to place a smooth plane or wire in such a position that a particle will pass down it from a fixed point O to a given fixed surface in the least time possible. Let us suppose that we fix the apparatus of wires and beads at O, that we set the beads free simultaneously, and watch the increase in size of the sphere which they form. As soon as the sphere reaches such a size that it touches the fixed surface at some point P, one of the beads has arrived at this surface and, moreover, has arrived in shorter time than any of the others. Thus it has found the quickest path from O to the surface. This path is OP, and we can now fix the path without performing the experiment, from the knowledge that a sphere drawn so as to have O as its highest point, and to pass through P, must *touch* the surface at P.

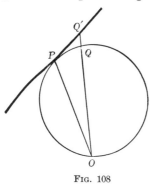

Fig. 108

In the same way, if we wish to find the quickest time from a surface to a fixed point O below it, we have to find a sphere which touches the surface at some point P and has O for its lowest point. Then PO will be the path required. For it is seen at once that the time down all the chords of this sphere which pass through O is the same, so that the time down PO is equal to the time down any other chord QO, and therefore less than the time down the complete path $Q'O$ from the surface to O, of which the chord QO is a part.

ILLUSTRATIVE EXAMPLE

A ship stands some distance from its pier, and it is required to place a chute at some point of the ship's side, so that the time of sliding down the chute on to the pier may be as short as possible.

Clearly the lower end of the chute must just rest on the nearest point O of the pier, and the problem reduces to that of drawing a sphere to have O as its lowest point and to touch the ship's side. Assuming the ship's side to be vertical, the tangents to this circle at the ends of the chute must be horizontal and vertical; whence it is easily seen that the chute must be placed so that it makes an angle of 45° with the vertical.

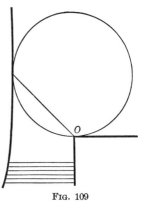

Fig. 109

EXAMPLES

1. A body is projected with a velocity of 20 feet per second up an inclined plane of angle 45°. Find how high up the plane it will go, and how long it will take in going up.

2. Two particles slide down the two faces of a double inclined plane, the angles being α and β. Compare the times they take to reach the bottom, and the velocities they acquire.

3. A body is projected down an inclined plane of length l and height h, from the summit, at the same instant as another is let fall vertically from the same point. Prove that if they strike the base at the same time, the velocity of projection of the first must be
$$\frac{l^2 - h^2}{l} \sqrt{\frac{g}{2h}}.$$

4. Give a construction for finding the line of quickest descent from a fixed point to a circle in the same vertical plane.

5. Particles are sliding down a number of wires which meet in a point, all having started from rest simultaneously at this point. Prove that at any instant their velocities are in the same ratio as the distances they have described.

6. A railway carriage is observed to run with a uniform velocity of 10 miles an hour down an incline of 1 in 250, and on reaching the foot of the incline runs on the level. Find how many yards it will run before coming to rest, assuming the resistance to be constant and the same in each stage of the motion.

7. Prove that if a motor car going at 100 kilometers an hour can be stopped in 200 meters, the brakes can hold the car on an incline of about 1 in 5; and determine the time required to stop the car.

8. A carriage weighing 12 tons becomes uncoupled from a train which is running down an incline of 1 in 250 at a rate of 40 miles per hour. The frictional resistance is 14 pounds weight per ton. Find how far the carriage will go before coming to rest.

9. The pull of the locomotive exceeds the ordinary resistances to the motion of a train by $\frac{1}{50}$ of its whole weight; and when the brakes are full on there is a total resistance of $\frac{1}{15}$ of its whole weight. Find the least time in which the train could travel between two stopping stations on the level 3 miles apart.

10. In the last question, find the time if the track is down a gradient of 1 in 100.

ATWOOD'S MACHINE

161. It is difficult to measure the acceleration produced by gravity from direct observations on a body falling freely, because either the distance fallen must be very great or else the time of falling very small. These difficulties are to some extent obviated in a machine designed by Atwood.

196 MOTION UNDER CONSTANT FORCES

If a string having two equal weights attached to its ends is placed over a smooth vertical pulley, so that the weights hang freely, it is obvious that there will be equilibrium. If the weights are unequal, equilibrium cannot exist. In Atwood's machine the difference between the weights is made small, so that the motion is slow and is therefore easily measured.

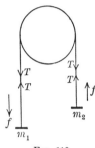

Fig. 110

Let m_1, m_2 be the masses of the weights, of which the former will be supposed to be the greater. When set free, let us suppose that the former descends with an acceleration f. Regarding the string as inextensible, the second mass must ascend with an acceleration f.

The string will be treated as weightless, so that the mass of any element of it may be disregarded. The second law of motion accordingly shows that the resultant force acting on any element must vanish. Thus the forces acting on the string must be in equilibrium (even although the string is not at rest), and it follows, as in § 54, that the tension must be the same at all points, say T.

The forces acting on either mass consist of its weight acting downwards and the tension of the string acting upwards. Thus the resultant downward forces on the two masses are respectively $m_1 g - T$ and $m_2 g - T$. The equations of motion for the two masses are accordingly
$$m_1 g - T = m_1 f,$$
$$m_2 g - T = - m_2 f.$$

If we eliminate T, we obtain
$$f = \frac{m_1 - m_2}{m_1 + m_2} g, \qquad (49)$$

giving the acceleration. On eliminating f, we obtain as the value of the tension
$$T = \frac{2 m_1 m_2}{m_1 + m_2} g. \qquad (50)$$

Clearly, if m_1 is nearly equal to m_2, the acceleration will be small. For instance, if the weights are 100 and 101 grammes, we find that
$$f = \tfrac{1}{201} g = .016 \text{ feet per second per second}.$$

MOVING FRAME OF REFERENCE 197

An acceleration of this smallness could easily be measured; for instance, the heavier mass would only descend eight feet in ten seconds. In practice, the difficulty arises that if the difference of the weights is made too small, the forces acting on the pulley are so evenly balanced that their difference is not sufficient to overcome the friction of the bearings, etc.

Motion referred to a Moving Frame of Reference

162. It has already been seen (§ 25) that the second law of motion remains true when the motion is measured relative to a frame of reference which is not at rest but is moving with a uniform velocity. It is easy to find how the statement of this law must be modified when the frame of reference moves with a known acceleration.

Let α be the acceleration of the frame of reference, let f be the component of the acceleration of a moving particle in the direction of the acceleration α, and let P be the component in this direction of the force acting on the particle. By the second law of motion

$$P = mf', \qquad (51)$$

where f' is the component acceleration referred to a frame of reference at rest. The acceleration f' may, however, be regarded as compounded of the acceleration f of the particle relative to the moving frame of reference, together with the acceleration α of this frame relative to one at rest. Since these accelerations are all in the same direction, we have $f' = f + \alpha$, so that equation (51) becomes

$$P = m(f + \alpha).$$

We can also write this in the form

$$P - m\alpha = mf, \qquad (52)$$

showing that *the motion is the same as if the frame were at rest, provided we imagine the force P diminished by an amount* $m\alpha$.

This result can easily be interpreted physically. Of the force P, a part equal to $m\alpha$ is used up in causing the particle to keep pace with the moving frame of reference. It is only the remaining part, $P - m\alpha$, which is available for producing accelerations relative to the moving frame.

198 MOTION UNDER CONSTANT FORCES

163. Frame moving with vertical acceleration. If the frame of reference moves with an acceleration α vertically downwards, we see that before measuring accelerations relative to this frame, we must suppose the vertical component of force on each particle of mass m to be diminished by $m\alpha$. Whatever forces are acting, there will be amongst them the weights of the particles mg, etc. We can conveniently suppose the diminution $m\alpha$ to be taken from these, so that the weight of a particle, instead of being taken to be mg, will be taken to be $m(g - \alpha)$.

Thus the acceleration of the frame of reference may be allowed for by supposing the acceleration due to gravity to be diminished from g to $g - \alpha$.

For example, if an Atwood's machine is placed in an elevator, then at the instant at which the elevator has an *upward* acceleration α, the acceleration of the masses relative to the machine will be (cf. equation (49))

$$f = \frac{m_1 - m_2}{m_1 + m_2}(g + \alpha),$$

while the tension of the string will be (cf. equation (50))

$$T = \frac{2\, m_1 m_2}{m_1 + m_2}(g + \alpha).$$

164. Effect of earth's rotation on the value of g. As we have seen (§ 25), a frame of reference which is fixed relatively to the earth's surface possesses an acceleration in consequence of the rotation of the earth about its axis.

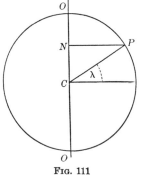

Fig. 111

Let OO' be the earth's axis, and let P be any point on the earth's surface, in latitude λ. Regarding the earth as a sphere of radius a, the point P will describe a circle of radius $a \cos \lambda$ having its center N on the earth's axis. If v is the velocity with which P describes the circle, the acceleration of P is, by § 12, $\dfrac{v^2}{a \cos \lambda}$ towards the center of the circle, — i.e. along PN.

MOVING FRAME OF REFERENCE

Let ω be the angular velocity of the earth, — i.e. let it turn through ω radians per unit time. Then the time of P describing a complete circle is the same as the time required for the earth to perform a complete revolution, namely $\dfrac{2\pi}{\omega}$. This time is also $\dfrac{2\pi a \cos \lambda}{v}$. Hence we have
$$v = a\omega \cos \lambda.$$

The acceleration of the frame of reference is now seen to be
$$\frac{v^2}{a \cos \lambda} = \omega^2 a \cos \lambda$$
along PN. The motion of any particle referred to a frame moving with P may accordingly be calculated as though with reference to a fixed frame, provided the component of force in the direction PN is diminished by $m\omega^2 a \cos \lambda$.

Thus the total force acting may be supposed to consist of the forces which actually do act, combined with a force $m\omega^2 a \cos \lambda$ along NP. Compounding this last force with the earth's attraction, we obtain a force which may be called the *apparent force of gravity* at P. Thus the motion of the frame of reference may be allowed for by using the apparent force of gravity in place of the true attraction of the earth. It is this apparent gravity which is determined experimentally, and which is always meant in speaking of the *weight* of a particle at any point.

To find the apparent weight of a body at the point P, we have to compound its true weight, say mG acting along PC, with a force $m\omega^2 a \cos \lambda$ along NP. Let the latter force be resolved into its components

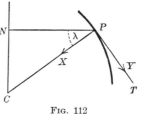

Fig. 112

$$-m\omega^2 a \cos^2 \lambda, \quad m\omega^2 a \cos \lambda \sin \lambda$$
along PC, PT respectively, PT being the tangent at P.

Compounding with the force mG along PC, we find for the components X, Y of the apparent weight along PC, PT respectively,
$$X = m(G - \omega^2 a \cos^2 \lambda), \tag{53}$$
$$Y = m\omega^2 a \cos \lambda \sin \lambda. \tag{54}$$

Squaring and adding, and denoting the apparent weight, as usual, by mg, we obtain

$$m^2g^2 = X^2 + Y^2 = m^2(G^2 - 2\omega^2 a\, G \cos^2\lambda + \omega^4 a^2 \cos^2\lambda). \quad (55)$$

Taking the diameter of the earth to be 7927 miles, and the value of G (the acceleration due to gravity at the North Pole) to be 32.25, we easily find that

$$\frac{\omega^2 a}{g} = \frac{1}{290}, \text{ approximately.}$$

The square of this is so small that to a first approximation it may be neglected, and equation (55) may be written in the form

$$g = G - \omega^2 a \cos^2\lambda.$$

Thus the apparent weight in latitude λ is less than the true weight by $m\omega^2 a \cos^2\lambda$, or about $\frac{1}{290}\cos^2\lambda$ of the whole weight.

The apparent weight does not act along the radius CP. If we suppose it to act at an angle θ with this radius, we obtain, from equations (53) and (54),

$$\tan\theta = \frac{Y}{X} = \frac{\omega^2 a \cos\lambda \sin\lambda}{G - \omega^2 a \cos^2\lambda}$$

$$= \tfrac{1}{290} \cos\lambda \sin\lambda, \text{ approximately,}$$

giving the deviation of the plumb line from the earth's radius at any point.

FRICTIONAL REACTIONS BETWEEN MOVING BODIES

165. It is found experimentally that the relation

$$F = \mu R$$

(in which F, R are the tangential and normal components of the reaction between two bodies) remains very approximately true when the bodies are sliding past one another. The value of μ, the coefficient of friction, is not quite the same as when the bodies are at rest, the latter being always somewhat larger.

Friction between two bodies which are sliding past one another is called *dynamical friction*, that between two bodies at rest being called *statical friction*.

ILLUSTRATIVE EXAMPLES

1. *Two particles of masses m_1, m_2 are placed on two inclined planes of angles α, β, placed back to back, and are connected by a string which passes over a smooth pulley at the top of the planes. If the coefficients of friction between the particles and the planes are μ_1, μ_2, find the resulting motion.*

If motion occurs at all, one particle, say m_1, must move down its plane, while the other, m_2, will move up. Since the string is inextensible, the acceleration of each will be the same, say f in the direction in which motion is taking place.

The forces acting on the first particle are

(a) its weight $m_1 g$ vertically down;
(b) the tension of the string, say T, up the plane;
(c) the reaction with the plane. Let this be resolved into components R, μR normal to and up the plane.

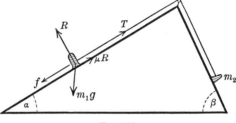

FIG. 113

Since the particle m_1 has no acceleration normal to the plane, the component of the resultant force in this direction must be zero. Resolving in this direction we obtain
$$R - m_1 g \cos\alpha = 0.$$

Resolving down the plane,
$$m_1 g \sin\alpha - \mu R - T = m_1 f,$$
and if we eliminate the unknown reaction R, we obtain
$$m_1 g (\sin\alpha - \mu \cos\alpha) - T = m_1 f. \tag{a}$$

A similar equation can be obtained for the motion of the second particle, namely
$$m_2 g (\sin\beta + \mu \cos\beta) - T = - m_2 f. \tag{b}$$

Solving equations (a) and (b) for f, we obtain
$$f = \frac{m_1 (\sin\alpha - \mu \cos\alpha) - m_2 (\sin\beta + \mu \cos\beta)}{m_1 + m_2} \tag{c}$$
giving the acceleration.

If this value of f comes out negative, we see that the acceleration cannot be in the direction in which motion has been assumed to take place.

If the system starts from rest, motion in the direction assumed is found to be impossible, and we must proceed to examine whether motion in the opposite direction is possible. If this also is found to be impossible, the system will remain at rest.

202 MOTION UNDER CONSTANT FORCES

If, however, the system is known to have been started in motion in the direction assumed, then the acceleration given by equation (c) will be in operation, increasing the velocity if positive, and decreasing it if negative. In the latter case the system will in time be reduced to rest, and we must then examine whether or not it will start into motion in the reverse direction.

2. *To one end of the string of an Atwood's machine a weight of mass m_1 is attached. To the other end a smooth pulley of mass m_2 is attached, over which passes a string with masses m_3, m_4 hanging at its ends. Find the motion.*

Let the mass m_1 be supposed to have an acceleration f, measured downwards. Then m_2 must have an acceleration f upwards. The masses m_3, m_4 will themselves form an Atwood's machine, the whole of which moves upwards with an acceleration f. Thus the tension in the string of this machine, say T_1, is (cf. § 163)

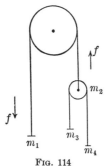

Fig. 114

$$T_1 = \frac{2\, m_3 m_4}{m_3 + m_4}(g + f). \qquad (a)$$

If we denote the tension in the string connecting m_1 and m_2 by T_2, we have as the equation of motion of m_2,

$$T_2 - m_2 g - 2\, T_1 = m_2 f, \qquad (b)$$

while the equation of motion of m_1 is

$$m_1 g - T_2 = m_1 f. \qquad (c)$$

Eliminating T_1 and T_2 from equations (a), (b), and (c), we obtain as the value of the acceleration f,

$$f = \frac{m_1 - m_2 - \dfrac{4\, m_3 m_4}{m_3 + m_4}}{m_1 + m_2 + \dfrac{4\, m_3 m_4}{m_3 + m_4}}.$$

The accelerations of the masses m_3, m_4 relative to m_2 are known, by § 163, to be

$$\pm \frac{m_3 - m_4}{m_3 + m_4}(g + f).$$

3. *At equal intervals on a horizontal circle n small smooth rings are fixed, and an endless string passes through them in order. If the loops of the string between each consecutive pair of rings support n pulleys of masses P, Q, R, $\cdots$ respectively, the portions of string not in contact with the pulleys being vertical, show that the pulley P will descend with acceleration*

$$\frac{\dfrac{1-n}{P} + \dfrac{1}{Q} + \dfrac{1}{R} + \cdots}{\dfrac{1}{P} + \dfrac{1}{Q} + \dfrac{1}{R} + \cdots} g.$$

ILLUSTRATIVE EXAMPLES

The tension of the string must be the same throughout, say T. If the accelerations of the pulleys are $f_P, f_Q, \cdots$, all measured down, we have equations of motion of the type

$$Pg - 2T = Pf_P, \qquad (a)$$

one equation for each pulley. The unknown quantity T enters these equations, as well as the n unknown quantities $f_P, f_Q, \cdots$. Thus there are $n + 1$ unknown quantities, and so far only n equations connecting them. Another equation is therefore required, and this is obtained by noticing that the accelerations $f_P, f_Q, \cdots$ cannot be independent, for the length of the string must remain unaltered.

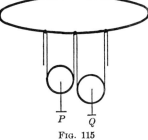

Fig. 115

Let us denote the depths of $P, Q, \cdots$ below the horizontal ring by $s_P, s_Q, \cdots$. Then

$$s_P + s_Q + \cdots$$

must be constant throughout the motion. It follows that

$$f_P + f_Q + \cdots = 0.$$

Substituting the values of $f_P, f_Q, \cdots$ from equation (a), we obtain

$$\left(g - \frac{2T}{P}\right) + \left(g - \frac{2T}{Q}\right) + \cdots = 0,$$

so that

$$2T = \frac{ng}{\frac{1}{P} + \frac{1}{Q} + \cdots}$$

and on substituting this value for $2T$ in equation (a), we obtain the required value of f_P.

EXAMPLES

1. Show that the tension of the string in an Atwood's machine is intermediate between the weights of the two masses. Show also that it is nearer to the smaller than to the larger of these weights.

2. Two weights 16 and 14 ounces respectively are connected by a light inextensible string which passes over a smooth pulley. The weights hang with the strings vertical and the string is clamped so that no motion can take place. If the string is suddenly unclamped find the change in the pressure exerted on the pulley.

3. A string passing across a smooth table at right angles to two opposite edges has attached to it at the ends two masses P, Q which hang vertically. Prove that, if a mass M be attached to the portion of the string which is on the table, the acceleration of the system when left to itself will be

$$\frac{P - Q}{P + Q + M} g.$$

204   MOTION UNDER CONSTANT FORCES

4. Two masses m, m' are tied to the two ends of a string which is slung over a peg, as in an Atwood's machine. The peg is not smooth, the angle of friction between it and the string being ϵ. Find the motion.

5. In question 3, let the coefficient of friction between the table and the weight M be μ, that between the table and string being μ'. Find the motion.

6. A rope hangs over a smooth pulley. Find the uniform acceleration with which a man weighing 10 stone must pull himself up one end of the rope, for the rope to be kept at rest by a weight of 12 stone hanging from the other end.

7. A monkey is tied to one end of the string of an Atwood's machine, the other end having attached to it a weight exactly equal to that of the monkey, and at just the same depth below the pulley. The monkey suddenly starts to climb up his string. Which will rise the faster, the monkey or the weight on the other string?

8. Weights of 10 pounds and 2 pounds, hanging by vertical strings, balance on a wheel and axle. If a mass of 1 pound be added to the smaller weight, find the acceleration with which it will begin to descend, and the tension of each rope. (The inertia of the wheel and axle is to be neglected.)

9. A mass of 5 pounds resting on a smooth plane inclined at 30 degrees to the horizon is connected by a fine thread, which passes over a pulley at the summit of the plane, with a mass of 3 pounds hanging vertically. Compare the pull in the thread when the mass on the plane is held fixed and when it is let go. If the thread is severed or burnt 8 seconds after this mass has been let go, find how far it will rise on the plane before falling back.

10. A light thread passes over two fixed pulleys A and B, and carries between them a movable pulley block C, under which it passes. A mass M is attached to each end of the thread, and a mass m to the movable block. The masses of the pulleys are negligible, and the pulleys are so arranged that all the segments of the thread are vertical. Show that, when the system is let go, the tension in the thread is $mM/(M + \frac{1}{2} m)$ pounds, and find the acceleration with which the mass m falls.

11. An elastic band of mass m, natural length a, and modulus λ is placed round a rough horizontal wheel of circumference b ($> a$). How fast must the wheel be made to rotate for the band to leave the wheel?

12. The elastic band of question 11 is placed on a smooth sphere of circumference b rotating with angular velocity ω. Find the position of rest.

13. If the earth rotates faster and faster until ultimately bodies fly off from its equator, show that by the time this stage is reached the plumb line at any point will be parallel to the earth's axis.

14. A body is placed on a spring balance when in a ship which is sailing along the equator with velocity v. Show that if the balance weighs accurately when the ship is at rest, its reading when the ship is in motion will show an error of $\dfrac{2v\omega}{g}$ times the weight of the body (approximately), where ω is the angular velocity of the earth.

Flight of Projectiles

166. By a projectile here is meant any body which is small enough to be regarded as a particle, and which is projected in such a way that it describes a path under the influence of gravity.

A projectile will, in general, be influenced by the resistance of the air as well as by gravity, but we shall suppose the resistance of the air to be negligible, so that gravity will be the only force which need be taken into account.

To take the simplest case first, let us imagine that the projectile is projected horizontally from the point O (fig. 116), with velocity u. The only force acting is gravity, which has no horizontal component, so that the horizontal velocity remains equal to u throughout the motion. The initial vertical component of velocity is nil, but there is a downward acceleration g. Thus after time t, the horizontal distance described is ut, while the vertical distance fallen is $\frac{1}{2}gt^2$. Denoting the horizontal distance described by x, and the vertical distance fallen by y, we have

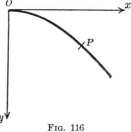

Fig. 116

$$x = ut, \quad y = \tfrac{1}{2} gt^2.$$

The equation of the path described is obtained by eliminating t from these equations, and it is found to be

$$y = \frac{g}{2\,u^2} x^2.$$

This is a parabola, of which the latus rectum is $\dfrac{2\,u^2}{g}$.

Clearly the problem of determining the curve is essentially the same as in § 156. There we have a body falling freely, and tracing its path on a paper *which moves past it* with a uniform horizontal velocity. Here we have a body falling freely, and can imagine it to trace its path on a paper *past which it moves* with a uniform horizontal velocity. The relative motion is the same in the two cases, so that the curves are necessarily the same.

167. At O the velocity of the particle is u, which is the velocity due to a height $\dfrac{u^2}{2g}$. This is equal to a quarter of the latus rectum, and is therefore equal to the depth of O, the vertex of the parabola, below the directrix XM. Thus the total energy of the projectile when at O is equal to that of the same projectile at rest at X, or, of course, at any other point of the directrix, since this is horizontal.

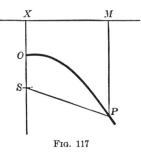

Fig. 117

Since the total energy remains constant, we see that when the particle is at any point P of its path, its kinetic energy is that due to a fall through PM, the distance from P to the directrix.

This is expressed by saying that

The velocity of a projectile at any point is that due to a fall from the directrix.

168. Instead of supposing that the particle is projected horizontally at O, the vertex of the parabola, we can suppose that it has arrived at O in its flight through the air, having been previously projected from some point A. The same reasoning which shows that the part of the path described after passing O is parabolic, will show that the path described before reaching O is parabolic also. Thus the path of a particle projected from any point in any manner is a parabola.

Suppose that a particle is projected from A with velocity v, in a direction which makes an angle α with the horizontal. Let O be the vertex of its path, and let us suppose that when the projectile passes through O, its velocity is u, this velocity being of course horizontal.

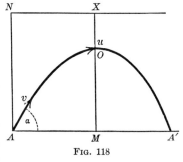

Fig. 118

PROJECTILES 207

There is no horizontal force acting on the particle, so that its horizontal velocity remains unaltered throughout its flight. Thus
$$u = v \cos \alpha.$$

The latus rectum of the parabola is accordingly
$$\frac{2 u^2}{g} = \frac{2 v^2 \cos^2 \alpha}{g}.$$

The velocity v is that due to a fall from the directrix to A, so that if NX is the directrix in fig. 118,
$$AN = \frac{2g}{v^2}.$$

The time of flight from A to O is the time required for gravity to destroy a vertical velocity $v \sin \alpha$; it is therefore $\frac{v \sin \alpha}{g}$. In this time the horizontal distance AM is described with a uniform horizontal velocity u, so that
$$AM = \frac{v \sin \alpha}{g} u = \frac{v^2 \sin \alpha \cos \alpha}{g}.$$

The vertical distance described, OM, is by equation (47) equal to half of the time multiplied by the initial vertical velocity. Thus
$$OM = \frac{1}{2} \frac{v^2 \sin^2 \alpha}{g}.$$

The total range on a horizontal plane, AA', is twice AM, so that
$$AA' = \frac{2 v^2 \sin \alpha \cos \alpha}{g} = \frac{v^2 \sin 2\alpha}{g}.$$

169. If the value of v is fixed (as, for instance, it would be if we were firing a shot with a given charge of powder), while the angle α can be varied, then the range AA' can never exceed $\frac{v^2}{g}$, for the factor $\sin 2\alpha$ can never exceed unity. Thus the greatest range attainable on a horizontal plane with a given velocity of projection

v will be $\dfrac{v^2}{g}$, and to obtain this range we make $\sin 2\alpha = 1$, or $\alpha = 45$ degrees. Thus, to send a projectile as far as possible on a horizontal plane we project it at an angle of 45 degrees.

170. These results can also be obtained analytically. Let us take the point of projection for origin, and the plane in which the flight takes place as plane of xy, the axes of x and y being respectively horizontal and vertical. The x-coördinate of the point reached by the particle after time t is equal to the horizontal distance described in time t with uniform horizontal velocity $v \cos \alpha$. Thus

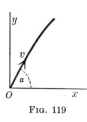

FIG. 119

$$x = v \cos \alpha \cdot t. \tag{56}$$

Similarly the y-coördinate of this point is the distance described in time t, starting with initial velocity $v \sin \alpha$, and with retardation g. Thus

$$y = v \sin \alpha \cdot t - \tfrac{1}{2} g t^2. \tag{57}$$

If we eliminate t between equations (56) and (57), we obtain the equation of the path. It is found to be

$$y = x \tan \alpha - \frac{g x^2}{2 v^2 \cos^2 \alpha}. \tag{58}$$

This can be expressed in the form

$$y - \frac{1}{2} \frac{v^2 \sin^2 \alpha}{g} = - \frac{g}{2 v^2 \cos^2 \alpha} \left(x - \frac{v^2 \sin \alpha \cos \alpha}{g} \right)^2,$$

which is clearly the equation of a parabola, of which the vertex is at the point

$$x = \frac{v^2 \sin \alpha \cos \alpha}{g}, \quad y = \frac{1}{2} \frac{v^2 \sin^2 \alpha}{g}, \tag{59}$$

and of which the latus rectum is of length

$$\frac{2 v^2 \cos^2 \alpha}{g}.$$

PROJECTILES 209

To obtain the range on a horizontal plane, we have to find the point in which the parabola intersects the line $y = 0$. Putting $y = 0$ in equation (58), we obtain at once

$$x = \frac{2 v^2 \cos^2 \alpha}{g} \tan \alpha = \frac{v^2 \sin 2\alpha}{g},$$

agreeing with the value obtained in § 168.

Range on an Inclined Plane

171. Suppose, next, that the projectile is fired so as to strike an inclined plane through O, the point of projection. Let β be the inclination of this plane to the horizon, and let r be the range of the projectile on this plane. Then the coordinate of the point at which the projectile meets the plane must be

$$x = r \cos \beta, \quad y = r \sin \beta.$$

This point is a point on the parabola, so that its coördinates must satisfy equation (58). Substituting these coördinates, we obtain

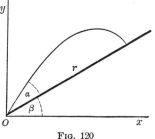

Fig. 120

$$r \sin \beta = r \tan \alpha \cos \beta - \frac{g\, r^2 \cos^2 \beta}{2\, v^2 \cos^2 \alpha},$$

giving as the value of the range r,

$$r = \frac{2 v^2}{g} \frac{\cos \alpha \sin (\alpha - \beta)}{\cos^2 \beta}. \tag{60}$$

Since $\quad 2 \cos \alpha \sin (\alpha - \beta) = \sin (2\alpha - \beta) - \sin \beta, \tag{61}$

it is clear that if α alone is allowed to vary, the range r will be a maximum when $\sin(2\alpha - \beta)$ is a maximum, i.e. when it is equal to unity. To obtain this value, we make

$$\alpha = \frac{\pi}{4} + \frac{\beta}{2}.$$

Thus, to get the maximum range, we project in the direction which bisects the angle between the inclined plane and the vertical.

When projection takes place in this direction, the maximum range R is given by putting $\sin(2\alpha - \beta) = 1$ in the value for r given by equation (60). Thus we have

$$\begin{aligned} R &= \frac{v^2}{g} \frac{2 \cos \alpha \sin (\alpha - \beta)}{\cos^2 \beta} \\ &= \frac{v^2}{g} \frac{\sin (2\alpha - \beta) - \sin \beta}{\cos^2 \beta} \\ &= \frac{v^2}{g} \frac{1 - \sin \beta}{\cos^2 \beta} \\ &= \frac{v^2}{g(1 + \sin \beta)}. \end{aligned} \qquad (62)$$

172. This equation enables us to find the greatest distance which can be reached in any direction by a projectile fired with velocity v. Let us replace β by $\frac{\pi}{2} - \theta$, so that θ is the angle which the direction makes with the vertical. Then the relation between R and θ is

$$R = \frac{v^2}{g(1 + \cos \theta)}. \qquad (63)$$

Regarded as an equation in polar coördinates R, θ, this is clearly the equation of a curve such that we can hit any point inside it with a projectile fired with velocity v, but cannot reach any point outside it. The polar equation of a parabola of latus rectum l, referred to its focus and axis, is known to be

$$R = \frac{\frac{1}{2} l}{1 + \cos \theta}.$$

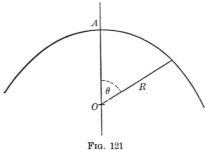

Fig. 121

Comparing this with equation (63), we see that this equation represents a parabola, of which the point of projection is the focus, the axis is vertical, and the semi-latus rectum is v^2/g.

PROJECTILES 211

Envelope of Paths

173. If we imagine all the parabolas drawn, which can be described by projectiles fired from the point O with a given velocity v, we shall obtain a figure similar to fig. 122. The outside curve obviously separates the points which can be reached

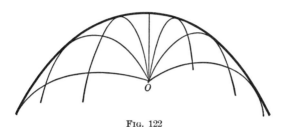

Fig. 122

from those which cannot be reached. Thus this is the parabola of which the equation is given in equation (63). A study of fig. 122 will now show that this curve is the envelope of the system of parabolas which correspond to the different directions of firing.

174. The envelope of the system of parabolas can be found more directly by analytical methods. If we write m for $\tan \alpha$ in equation (58), we obtain the equation of a parabola of the system in the form

$$y = mx - \frac{gx^2}{2v^2}(1 + m^2),$$

and the whole system is obtained by giving different values to m.

The condition for this equation to have equal roots in m is that

$$x^2 = \frac{4gx^2}{2v^2}\left(\frac{gx^2}{2v^2} + y\right),$$

or, in reduced form,

$$x^2 = -\frac{2v^2}{g}\left(y - \frac{v^2}{2g}\right). \tag{64}$$

If x, y satisfy this relation, two parabolas which only differ infinitesimally pass through x, y, and therefore x, y is a point on the envelope. Thus equation (64) is the equation of the envelope, and is easily seen to give the same parabolic envelope as has already been obtained.

175. There is also a very simple geometrical way of determining the envelope of the system of parabolas. We notice first that as the projectiles are all fired from the same point A with the same velocity v, their paths must all have the same directrix NM (fig. 123).

Let any two parabolas of the system intersect in P, and let S, S' be the foci of these parabolas. Let AN, PM be the perpendiculars from A and P to the directrix.

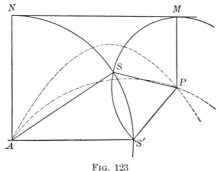

Fig. 123

Then $AS = AS'$, since each is equal to AN, and $PS = PS'$, for each is equal to PM. Thus S, S' are the two points of intersection of two circles of which the centers are A, P.

If the two parabolas are supposed to be adjacent, their foci S, S' are adjacent points, and therefore the two circles touch, and ASP is, in the limit, a straight line. We now have

$$AP = AS + SP$$
$$= AN + PM$$
$$= \text{the perpendicular from } P \text{ on to a fixed horizontal line at a distance } AN \text{ above } MN.$$

The point P, then, satisfies the condition that its distance from this fixed line is equal to its distance from the fixed point A. It therefore is always on a certain parabola of focus A. But also it is always a point on the envelope, this being the locus of the points of intersection of adjacent pairs of the parabolas of the system. Thus the envelope is the parabola just obtained, of which the focus is A, and this is the same parabola as was obtained before.

ILLUSTRATIVE EXAMPLES

1. *A carriage runs along a level road with velocity V, throwing off particles of mud from the rims of its wheels. Find the greatest height to which any of them will rise.*

Let a be the radius of the wheel, then we have seen (p. 9) that any point, such as Q, moves with a velocity $V \cdot QL/a$ in the direction QM at right angles to QL. This will be the velocity of mud projected from Q.

If the angle QLP is θ, the height above the ground at which the mud starts is

$$LN = LP + PN = a(1 + \cos 2\theta),$$

while the vertical component of its velocity is

$$V \cdot (QL/a) \sin \theta = 2V \sin \theta \cos \theta = V \sin 2\theta.$$

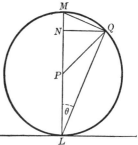

Fig. 124

The mud projected with this vertical velocity will attain a further vertical height

$$\frac{(V \sin 2\theta)^2}{2g},$$

so that the total height attained is

$$a + a \cos 2\theta + \frac{V^2}{2g} \sin^2 2\theta.$$

This may be written as a quadratic function of $\cos 2\theta$ in the form

$$\left(a + \frac{V^2}{2g}\right) - \frac{V^2}{2g} \cos^2 2\theta + a \cos 2\theta$$
$$= \left(a + \frac{V^2}{2g}\right) + \frac{a^2 g}{2V^2} - \frac{V^2}{2g}\left[\cos 2\theta - \frac{ag}{V^2}\right]^2.$$

The maximum value of this expression, as θ varies, occurs when $\cos 2\theta = \frac{ag}{V^2}$, if it is possible for $\cos 2\theta$ to have this value — i.e. if $V^2 > ag$. In this case the maximum height attained is

$$a + \frac{V^2}{2g} + \frac{a^2 g}{2V^2} = \frac{(ag + V^2)^2}{2gV^2}$$

measured from the ground.

If, however, $V^2 < ag$, we cannot make $\left[\cos 2\theta - \frac{ag}{V^2}\right]^2$ vanish. We accordingly make it as small as possible, so that we take $\cos 2\theta = 1$. Thus the mud which carries to the highest point is that which starts at the top point M of the wheel, and obviously this never gets higher than its starting point.

2. *Find what area of a vertical wall can be covered by a fire-hose projecting water with velocity v at a distance h from the wall.*

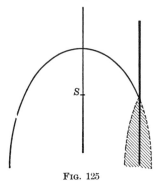
FIG. 125

Let S be the nozzle of the fire-hose, and let us regard it as capable of projecting particles of water in any direction we please with a velocity v. The points which can be reached will, by § 172, be all the points which lie inside a paraboloid of revolution having its axis vertical, S for focus, and latus rectum $\dfrac{2v^2}{g}$. If we take S as origin, and the vertical through S for axis of z, the equation of this paraboloid will be
$$(x^2 + y^2) = \frac{2v^2}{g}\left(\frac{v^2}{2g} - z\right).$$

The curve in which this cuts the vertical wall, of which the equation may be taken to be $y = h$, will be
$$x^2 + h^2 = \frac{2v^2}{g}\left(\frac{v^2}{2g} - z\right),$$
or
$$x^2 = \frac{2v^2}{g}\left(\frac{v^2}{2g} - \frac{h^2 g}{2v^2} - z\right).$$

This is a parabola, of latus rectum $\dfrac{2v^2}{g}$, having its axis vertical, and its vertex at a height
$$\frac{v^2}{2g} - \frac{h^2 g}{2v^2}$$
above S. All the points inside this parabola will be within range of the jet of water. The points on the wall which are outside this parabola will be inaccessible.

EXAMPLES

1. A revolver is fired horizontally from the top of a tower 100 feet high, the bullet leaving the muzzle with a velocity of 600 feet per second. Where will the bullet strike the ground?

2. A rifle bullet, fired horizontally at a height of 10 feet above the surface of a lake, strikes the water at a distance of 500 yards. Find its velocity in feet per second, the resistance of the air being supposed negligible.

3. Prove that the claim for a rifle, that the bullet does not rise more than one inch in a range of 100 yards, implies that the velocity must be greater than 2078 feet per second.

4. Find the greatest range on a horizontal plane of a cricket ball thrown with a velocity of 100 feet per second.

# EXAMPLES 215

5. A shot fired from a gun whose muzzle is close to the ground just clears a man 6 feet high standing 10 yards away, and embeds itself in the ground a quarter of a mile off. Show that the shot rises to a height above the ground which is certainly greater than 22 yards.

6. A projectile has a maximum horizontal range of 256 feet; what is its velocity of projection?

If it be projected with this velocity from a point on the floor of a long corridor 24 feet high, what will be its greatest range, if it is not to strike the ceiling?

7. Prove that the velocity required for a range of 20 miles will be at least 1840 feet per second, with a time of flight of 81.3 seconds.

8. Determine the charge of powder required for the range of 20 miles in the last question, supposing the shot to weigh a ton, and the strength of the powder capable of realizing 100 foot-tons per pound of powder.

9. Show that the range R of a projectile fired from a height h above a level plane with velocity v at an angle α is given by

$$2v^2(h + R\tan\alpha) = gR^2\sec^2\alpha.$$

10. Show that the area of a level plane swept by a gun at a height h above the plane increases proportionally with h, being equal to

$$A + 2h\sqrt{\pi A},$$

where A is the area commanded when the gun is at the level of the plane.

11. A projectile can be fired with a velocity of 1720 feet per second from a fort at a height of 300 feet above a horizontal plane. Find what area of the plane is covered by the gun.

12. A particle is projected so as just to graze the four upper corners of a regular hexagon of side a, placed vertical with one edge resting on a horizontal table Find the highest point in the flight of the particle, and show that the range on the table is $a\sqrt{7}$.

13. A machine-gun is placed on an armored train which runs along a horizontal line of rails with velocity v. The muzzle velocity of shots fired from the gun is V. Find the greatest range

(a) in front of the train;

(b) behind the train.

GENERAL EXAMPLES

1. A train is going at 60 miles an hour, when it comes to a curve having a radius of $\frac{3}{4}$ of a mile. There is a perfectly smooth horizontal shelf in the train, its edge being parallel to the rails and on the side of the shelf away from the center of the curve. A small object stands on the shelf at a distance of 8 inches from the edge. Show that the object will fall off the shelf after the car containing the object has described about 24 yards of the curve, and find what its horizontal velocity will be when it leaves the shelf.

2. A balloon is moving upwards with a speed which is increasing at the rate of 4 feet per second in each second. Find how much the weight of a body of 10 pounds, as tested by a spring balance on it, would differ from its weight under ordinary circumstances.

3. An Atwood's machine is placed, with the string clamped, on one scale of a weighing machine. Show that as soon as the string is unclamped the apparent weight of the machine is diminished by

$$\frac{(m-m')^2}{m+m'}g,$$

where m, m' are the suspended weights.

4. A uniform chain of length l and weight W passes over a smooth peg, hanging vertically on each side. If the chain be running freely, prove that when the length on one side is x, the pressure on the peg is

$$\frac{4x(l-x)}{l^2}W.$$

5. A jet of water falls vertically from a hose to the ground, starting with a velocity which is negligible. Show that the center of gravity of the water which is in the air at any instant is two thirds of the way up from the ground to the hose.

6. A heavy uniform chain of weight w is tied to a string which is pulled up with tension T. Find the tension in the chain at any point.

7. Prove that the shortest time from rest to rest, in which a chain, which can bear a steady load of P tons, can lift or lower a weight of W tons through a vertical distance of h feet is

$$\sqrt{\left(\frac{2h}{g}\frac{P}{P-W}\right)} \text{ seconds.}$$

8. In a system of pulleys with one fixed and one movable block, in which the cord is attached to the axis of the movable block, then passes over the fixed one, then under the movable one, and then over the fixed one, find the weight P which, when attached to the cord, will support a given weight W hung from the movable block. (The blocks are so small that all the straight portions of the cord may be considered parallel.)

Show that, if the weights do not balance, the downward acceleration of W, when let go, will be

$$\frac{W-3P}{W+9P}g,$$

the weight of the cord being neglected, and that of the movable block being included in W.

9. A pulley carrying a total load W is hung in a loop of a cord which passes over two fixed pulleys, and has weights P and Q freely suspended from its ends, each segment of the cord being vertical; show that, when

EXAMPLES 217

the system is let go, W will remain at rest or move with uniform velocity, provided $\dfrac{1}{P} + \dfrac{1}{Q} = \dfrac{4}{W}$ and there is no friction anywhere.

If this relation does not hold, find the acceleration of W.

10. A particle, falling under gravity, describes 100 feet in a certain second. How long will it take to describe the next 100 feet, the resistance of the air being neglected?

If, owing to resistance, it takes .9 second, find the ratio of the resistance (assumed to be constant) to the weight of the particle.

11. Prove that the line of quickest descent from any curve to any other curve in the same vertical plane makes equal angles with the normals to the two curves at the points at which it meets them.

12. Find the position of a point on the circumference of a vertical circle, in order that the time of rectilinear descent from it to the center may be the same as that to the lowest point.

13. Find the line of quickest descent from the focus to a parabola of which the axis is vertical and vertex upwards, and show that its length is equal to the latus rectum.

14. An ellipse is suspended with its major axis vertical. Find the diameter down which a particle can fall in the least time. What is the least value of the eccentricity in order that this diameter may not be the major axis?

15. A bullet is to be fired at a vertical target so as to hit it exactly at right angles. If v is the velocity of the bullet and a the distance of the target from the point of firing, show that the angle of elevation of the shot must be $\frac{1}{2} \sin^{-1}\left(\dfrac{ag}{v^2}\right)$, and show that the point of the target which is hit will be at half the height of the point aimed at.

16. A bullet is fired at a vertical target. Show that the projection of the bullet on the target, as seen by the firer of the shot, appears to move with uniform velocity.

17. A gun fires two shots, one with velocity v at elevation α, and the second with velocity v' at a smaller elevation α', in the same vertical plane. Show that the shots will collide if the interval between firing is

$$\frac{2}{g} \frac{vv' \sin(\alpha - \alpha')}{v \cos\alpha + v' \cos\alpha'}.$$

18. A, B, C are three points in order in a horizontal line, AB being 640 feet; a particle is projected from A with a velocity of 390 feet per second in a direction making an angle $\tan^{-1} \frac{5}{12}$ with AC; at the same instant another particle is projected from B with a velocity of 250 feet per second in a direction making an angle $\tan^{-1} \frac{3}{4}$ with BC; show that these particles will collide, and find when and where.

19. A howitzer gun has a muzzle velocity of 400 feet per second. What distance immediately behind the top of a hill 200 feet high on a level plane is safe, if the gun is distant 1000 yards from the point in the plane vertically below the top of the hill?

20. The sights of a gun are inaccurately marked, the gun carrying always 3 per cent farther than the distance indicated on the sights. A marksman, not aware of the error, aims at a target distant 1000 yards. If the velocity of the shot is 1200 feet per second, show that he will hit the target about one yard too high.

21. The sighting of a rifle is accurate, and to aim at a point on a vertical target distant a feet the rifle ought to be elevated to an angle α. Owing to the unsteadiness of the marksman's hand, the rifle is pointed in directions which lie anywhere within a small angle θ of the true direction. Show that if a succession of shots is fired, the points at which they hit the target will all lie within a small ellipse of semi-axes $a\theta \cos \alpha$ and $a\theta (1 - \tan^2 \alpha)$ respectively.

When $\alpha = \dfrac{\pi}{4}$, the minor axis of this ellipse vanishes, so that the shots ought all to lie in a straight line. Interpret this result.

22. A particle slides down the outer surface of a smooth sphere, starting from rest at the highest point. It leaves the sphere at a point P and describes a parabola in space. Prove that the circle of curvature of the parabola at P will touch the directrix.

23. A particle is projected horizontally from the lowest point of the interior of the surface of a smooth sphere. It leaves the surface of the sphere at P, and after describing a parabola strikes the sphere again at Q. Show that PQ and the tangent at P make equal angles with the vertical.

24. Show that the whole area commanded by a gun planted on a hillside, supposed plane, is an ellipse, whose focus is at the gun, eccentricity the sine of the inclination of the hill, and semi-latus rectum equal to twice the height to which the muzzle velocity of the shot is due.

25. Show that the area of a plane hillside of inclination α, which is commanded by a gun placed on a fort of height H above the hill, is

$$4 \pi h (h + H \cos^2 \alpha) \sec^3 \alpha,$$

where $\sqrt{2\,gh}$ is the muzzle velocity of the shot.

26. A spherical shell of mass m explodes when moving with negligible velocity at a height of h feet above the ground. The shell is divided into very small particles, each of which moves, after the explosion, away from the center of the shell with velocity v, and ultimately falls to the ground. Find the total mass of the fragments which will be found per unit area at any specified distance from the point vertically underneath the shell.

EXAMPLES 219

27. A shell bursts while in the air, all the fragments receiving equal velocities from the explosion. Show that the fragments at any instant lie on a sphere, that the foci of the paths they describe also lie on a sphere, and that the vertices lie on a spheroid.

28. A particle slides down a rough inclined plane AB, starting at rest from A and describing a parabola freely after leaving the plane at B. If F is the focus of the parabola described, show that the angle $AFB = \dfrac{\pi}{2} + \epsilon$, where ϵ is the angle of friction.

29. From a fort a buoy was observed at a depression i below the horizon; a gun was fired at it at an elevation α, but the shot was observed to strike the water at a point whose depression was i'. Show that, in order to strike the buoy, the gun must be fired at an elevation θ, where

$$\frac{\cos\theta \sin(\theta + i)}{\cos\alpha \sin(\alpha + i')} = \frac{\cos^2 i \sin i'}{\cos^2 i' \sin i}.$$

30. Show that the least energy which will project a particle over a wall which is at distance a from the point of projection is

$$\tfrac{1}{2} mga \, \frac{1 + \tan \tfrac{1}{2}\alpha}{1 - \tan \tfrac{1}{2}\alpha},$$

where α is the elevation of the top of the wall at the point of projection.

31. A mill wheel of radius a revolves so that its rim has a velocity V, and drops of water are thrown off from the rim of the wheel. Show that the envelope of their paths is a parabola whose axis is vertical and whose focus is at a distance $\dfrac{a^2 g}{2V^2}$ vertically above the center of the wheel.

CHAPTER IX

MOTION OF SYSTEMS OF PARTICLES

EQUATIONS OF MOTION

176. The present chapter will deal with the motion of systems of particles, taking account of the actions and reactions which may be set up between the different pairs of particles. As a preliminary to this, it will be convenient to recapitulate the results which have been obtained for a single particle, stating these results in a more analytical form than before.

The whole system of forces which act on a particle must, since they act at a point, have a single force as resultant. Let us call this resultant P, and denote its components along three rectangular axes by X, Y, Z.

Also the particle, being regarded as a point, must have a definite acceleration f, and, since f is a vector, this acceleration may be supposed to be compounded of three components f_x, f_y, f_z along the three coördinate axes.

The second law of motion supplies the relation
$$P = mf. \qquad (65)$$

We are, however, told more than this by the second law of motion: we are told that the directions of P and of f are the same. Let λ, μ, ν be the direction cosines of this single direction, then we have
$$X = \lambda P, \quad Y = \mu P, \quad Z = \nu P,$$
and also
$$f_x = \lambda f, \quad f_y = \mu f, \quad f_z = \nu f.$$

From these relations, combined with relation (65), we clearly have
$$\left.\begin{array}{l} X = mf_x \\ Y = mf_y \\ Z = mf_z \end{array}\right\} \qquad (66)$$

EQUATIONS OF MOTION

These are the equations of motion of a particle in analytical form. They simply express the second law of motion in mathematical language.

177. Let x, y, z be the coördinates of the particle at any instant, and let u, v, w be the three components of its velocity. The component u is the velocity, along the axis of x, of the projection of the moving point on the axis of x, and the distance of this point from the origin at any instant is simply x. Thus, by the definition of velocity, we have

$$u = \frac{dx}{dt}, \qquad (67)$$

and similarly, of course,
$$v = \frac{dy}{dt},$$

$$w = \frac{dz}{dt}.$$

The rate at which the x-component of velocity increases is $\frac{du}{dt}$, but it has also been supposed to be f_x, for this is the x-component of the acceleration. Thus we have

$$f_x = \frac{du}{dt},$$

and similarly
$$f_y = \frac{dv}{dt},$$

$$f_z = \frac{dw}{dt}.$$

Using the values just found for u, v, w, these equations become

$$f_x = \frac{d^2x}{dt^2},$$

$$f_y = \frac{d^2y}{dt^2},$$

$$f_z = \frac{d^2z}{dt^2}.$$

Substituting these expressions for the components of acceleration into the equations of motion (66), we obtain these equations in the new form

$$\left. \begin{array}{l} X = m \dfrac{d^2x}{dt^2} \\[4pt] Y = m \dfrac{d^2y}{dt^2} \\[4pt] Z = m \dfrac{d^2z}{dt^2} \end{array} \right\} \quad (68)$$

178. Suppose we have a system of particles, — m_1 at x_1, y_1, z_1; m_2 at x_2, y_2, z_2; etc., — and let the components of force acting on them be X_1, Y_1, Z_1; X_2, Y_2, Z_2; etc.

Then, from the equation just obtained,

$$X_1 = m_1 \frac{d^2 x_1}{dt^2},$$

$$X_2 = m_2 \frac{d^2 x_2}{dt^2}, \text{ etc.,}$$

so that, by addition, $\qquad \sum X = \sum m \dfrac{d^2 x}{dt^2}, \qquad (69)$

where $\sum$ denotes summation over all the particles of the system.

The left-hand member $\sum X$ is the sum of the components along Ox of all the forces acting on all the particles of the system. As in § 50, these forces may be divided into two classes:

(*a*) *external* forces — forces applied to the particles from outside the system;

(*b*) *internal* forces — forces of interaction between pairs of particles of the system.

As in § 50, we find that the contribution to $\sum X$ from the second class of forces is nil. For all these forces fall into pairs, each pair consisting of an action and reaction, of which the components are equal and opposite.

Thus in calculating $\sum X$ we need only take account of *external* forces.

CONSERVATION OF LINEAR MOMENTUM 223

The right-hand term of equation (69), namely $\sum m \dfrac{d^2x}{dt^2}$, can also be modified. Since, by equation (67), we have $u = \dfrac{dx}{dt}$, the value of $\dfrac{d^2x}{dt^2}$ is $\dfrac{du}{dt}$, so that

$$\sum m \frac{d^2x}{dt^2} = \sum m \frac{du}{dt} = \frac{d}{dt}\left(\sum mu\right). \tag{70}$$

By the momentum of a particle, as we have already seen (§ 20), is meant the product of its mass and velocity. The momentum of a particle is therefore a vector of components mu, mv, mw, and mu may be spoken of as the x-component of the momentum. Each particle of the system will have momentum, and the sum of the x-components will be $\left(\sum mu\right)$, the quantity which appears on the right hand of equation (70).

We may now replace equation (69) by

$$\sum X = \frac{d}{dt}\left(\sum mu\right), \tag{71}$$

where $\sum X$ denotes the sum of the x-components of the external forces, and $\sum mu$ is the sum of the x-components of momentum.

CONSERVATION OF LINEAR MOMENTUM

179. When there are no external forces acting, $\sum X = 0$, so that

$$\frac{d}{dt}\left(\sum mu\right) = 0. \tag{72}$$

Similarly we have $\quad \dfrac{d}{dt}\left(\sum mv\right) = 0, \tag{73}$

$$\frac{d}{dt}\left(\sum mw\right) = 0. \tag{74}$$

These equations express that the quantities

$$\sum mu, \quad \sum mv, \quad \sum mw$$

do not vary with the time. That is to say, the components of the total momentum are constant, so that the total momentum, regarded as a vector, is constant. This is known as the principle of the *conservation of momentum*. Stated in words it is as follows:

When any system of particles moves without being acted on by external forces, the total momentum of the system remains constant in magnitude and direction.

Motion of Center of Gravity of System

180. Let us now return to the consideration of the general equations (71),

$$\sum X = \frac{d}{dt}\left(\sum mu\right), \text{ etc.} \qquad (75)$$

Let $\bar{x}, \bar{y}, \bar{z}$ denote the coördinates of the center of gravity of the particles of the system at any instant, and let the components of the velocity of this point be denoted by $\bar{u}, \bar{v}, \bar{w}$. Then we have

$$\bar{u} = \frac{d\bar{x}}{dt}, \text{ etc.} \qquad (76)$$

The value of $\bar{x}$ is, by equation (8),

$$\bar{x} = \frac{\sum mx}{\sum m}.$$

Thus
$$\bar{u} = \frac{d\bar{x}}{dt} = \frac{d}{dt}\left(\frac{\sum mx}{\sum m}\right)$$

$$= \frac{\sum m \frac{dx}{dt}}{\sum m} = \frac{\sum mu}{\sum m},$$

so that if M is the total mass, $\sum m$, of all the particles, we have

$$\sum mu = M\bar{u}.$$

MOTION OF CENTER OF GRAVITY   225

Equation (75) now becomes

$$\sum X = M\frac{d\bar{u}}{dt}, \qquad (77)$$

and similarly we have the equations

$$\sum Y = M\frac{d\bar{v}}{dt}, \qquad (78)$$

$$\sum Z = M\frac{d\bar{w}}{dt}. \qquad (79)$$

Remembering that

$$\frac{d\bar{u}}{dt}, \frac{d\bar{v}}{dt}, \frac{d\bar{w}}{dt}$$

are the components of acceleration of the center of gravity, we see that the motion of the center of gravity is the same as it would be if it were replaced by a particle of mass M, acted upon by a force of components $\sum X, \sum Y, \sum Z$. This force again is simply the force which would be the resultant of all the external forces, if they were all applied to the imaginary particle which we are supposing to move with the center of gravity.

181. In the particular case in which there are no external forces, the center of gravity moves as if it were a particle acted on by no forces, so that its motion will be a motion of uniform velocity in a straight line.

182. The motion of the center of gravity in this particular case, and in the more general case in which external forces act, may accordingly be supposed to be governed by the two following laws:

LAW I. *The center of gravity of every system of particles continues in a state of rest, or of uniform motion in a straight line, except in so far as the action of external forces on the system compels it to change that state.*

LAW II. *When external forces act on the system, the motion of the center of gravity is the same as it would be if all the masses of the particles were concentrated in a single particle which moved with the center of gravity, and all the external forces were applied to this particle.*

226 MOTION OF SYSTEMS OF PARTICLES

These laws may be regarded as the extensions of Newton's Laws I and II to the motion of a system of particles. We can see now why it is often legitimate to apply Newton's second law to the motion of bodies of finite size, as though they were particles (cf. § 26).

The principle of conservation of momentum is often sufficient in itself to supply the solution of a dynamical problem in which only two bodies are in motion.

ILLUSTRATIVE EXAMPLE

A shot of mass m is fired from a gun of mass M, which is free to run back on a pair of horizontal rails. Find the velocity of recoil of the gun, and examine the influence of the recoil on the motion of the shot.

Let us suppose that, before firing, the gun stands pointing at an angle α to the horizon, and let the muzzle velocity of the shot — i.e. the velocity relative to the gun with which the shot emerges — be V.

Let us suppose that the velocity of the shot relative to the earth has components u, v horizontal and vertical, and let the velocity of recoil of the gun be U, measured in the horizontal direction opposite to that in which the gun is pointing.

The system consisting of the gun, powder, and shot is not free from the action of external forces, but these forces, namely the weight of the system and its reaction with the earth, have no horizontal component. Thus the horizontal momentum of the system must remain unaltered by the explosion. This horizontal momentum was zero initially: it is therefore zero when the shot leaves the gun. Thus we have, neglecting the weight of the powder,

$$MU - mu = 0. \quad (a)$$

The velocity of the shot relative to the gun has components

$$u + U, \quad v.$$

This velocity must, however, be a velocity V making an angle α with the horizontal. We therefore have

$$u + U = V \cos \alpha, \quad (b)$$
$$v = V \sin \alpha. \quad (c)$$

From equations (a) and (b) we find

$$\frac{u}{M} = \frac{U}{m} = \frac{V \cos \alpha}{M + m}.$$

Thus the velocity of recoil is

$$U = \frac{m}{M + m} V \cos \alpha.$$

ILLUSTRATIVE EXAMPLE

The components of actual velocity of the shot are

$$u = \frac{M}{M+m} V \cos\alpha,$$
$$v = V \sin\alpha.$$

Thus the actual velocity of the shot is

$$\sqrt{u^2 + v^2} = V\left[1 - \frac{m(2M+m)}{(M+m)^2}\cos^2\alpha\right]^{\frac{1}{2}},$$

while the angle of elevation, θ, is given by

$$\tan\theta = \frac{v}{u} = \frac{M+m}{M}\tan\alpha.$$

EXAMPLES

1. An empty railway truck weighing 8 tons, originally at rest, is run into by a similar truck carrying a load of 24 tons and moving at the rate of a mile an hour, and the two trucks then move on together. Find their common velocity.

2. A gun of mass M fires a shot of mass m horizontally. Show that of the work done by the powder, a fraction $\dfrac{m}{M+m}$ is wasted in producing the recoil of the gun.

3. A particle of mass m slides down a smooth inclined plane of angle α, the plane itself (mass M) being free to slide on a smooth table. Find the acceleration of the particle and the plane.

4. A shell is observed to explode when at the highest point of its path. It is divided into two equal parts of which one is seen to fall vertically. Prove that the other will describe a parabola of which the latus rectum will be four times the latus rectum of the original parabola.

5. A shot of $\frac{1}{2}$ ounce weight strikes a bird of weight 5 pounds while in the air. At the moment of striking, the shot has a horizontal velocity of 1000 feet a second and the bird is flying horizontally in the same direction at a height of 64 feet above the ground, with a velocity of 20 feet a second. Show that the bird will fall at a distance of about 52.2 feet beyond the place where it was struck by the shot.

6. A ship of 5000 tons steaming at 20 knots suddenly runs into a whale whose weight is 12 tons, asleep on the surface of the water. By how much is the ship's speed reduced? (Neglect motion of water.)

7. A mail package weighing 2 hundredweight is thrown out from a train going at 60 miles an hour, with a horizontal velocity relative to the train of 11 feet per second at right angles to the track. It falls into a handcart of weight 3 hundredweight, which is free to move on a level platform, its wheels being set so that its motion will make an angle of 30 degrees with the track of the train. With what velocity will the cart start into motion?

228   MOTION OF SYSTEMS OF PARTICLES

8. A mass of 8 pounds moving north at a speed of 10 feet per second is struck by a mass of 6 pounds moving east at 14 feet per second, and its direction of motion is thereby deflected through 30 degrees, while its speed is increased by 1 foot per second; show that the velocity of the other is diminished by 7.3 feet per second, approximately, and find its new direction of motion.

9. Two ice yachts, each of mass M, stand at rest on perfectly smooth ice, with their keels in the same direction. A man of mass m jumps from the first to the second, and then immediately back again on to the first. Show that the final velocities of the yachts are in the ratio of $M + m : M$.

KINETIC ENERGY

183. We may best begin the study of the kinetic energy of a system of particles by drawing attention to a difficulty which has not so far been encountered in the present book. This difficulty will be best illustrated by a particular example.

Suppose that a ship is moving through the water with a velocity of 20 feet per second, and that a person on deck throws a ball of mass m forward with a velocity of 30 feet per second relative to the ship. If the person were fixed in space, we might say that the work he did was equal to the final kinetic energy of the ball, and was therefore $\frac{1}{2} m (30)^2$, or $450\, m$.

On board ship, however, the ball originally had a velocity of 20 and the thrower increases this velocity to 50. The change in the kinetic energy of the ball is accordingly

$$\tfrac{1}{2} m (50)^2 - \tfrac{1}{2} m (20)^2,$$

or $1050\, m$. If this represents the work done by the thrower, then we are driven to suppose that it would be more than twice as hard to throw the ball on board ship as on land. This would clearly be erroneous.

184. The error lies in this, that the thrower not only imparts a velocity to the ball but also to the ship. If he throws the ball forward he must, from the principle of conservation of momentum, impart a backward velocity to the ship, of momentum equal and opposite to the forward momentum of the ball. The total work performed is equal to the change produced in the total kinetic energy of the ship and the ball.

KINETIC ENERGY

Since, by the third law of motion, no force can act singly, it follows that, in every case of calculation of work from kinetic energy, it will be necessary to consider the kinetic energy of more than one body. For instance, a man throwing a ball on land will not only jerk the ball forward, but will also jerk the whole earth backward, and the energy of both must be taken into account, or we shall get erroneous results.

185. A second difficulty, closely connected with the first, suggests itself at once. Suppose we have a ball thrown with a velocity v along the deck of a ship moving with velocity V. We have seen that we must not suppose the kinetic energy of the ball to be $\frac{1}{2}mv^2$, but is it any more legitimate to suppose it to be $\frac{1}{2}m(v+V)^2$? For the sea in which the ship sails will have a further velocity V' in consequence of the earth's rotation, so that the energy might equally well be taken to be

$$\tfrac{1}{2}m(v+V+V')^2,$$

and so we might go on indefinitely. Knowing of no frame of reference which is absolutely at rest, it would seem to be impossible to find the true value of the kinetic energy. Moreover, it ought to be noticed that the expressions for the kinetic energy referred to different frames of reference differ by more than mere constants. For instance, the difference between the two expressions we have found for kinetic energy relative to the sea and kinetic energy relative to the earth's center is

$$\tfrac{1}{2}m(v+V+V')^2 - \tfrac{1}{2}m(v+V)^2$$
$$= \tfrac{1}{2}mV'^2 + mV'(v+V).$$

This difference not only depends on m and V' but also on v and V. It is not a constant difference, and so does not disappear when we calculate the *increase* in kinetic energy resulting from the action of forces.

The theorems which follow serve the purpose of showing a way through these and similar difficulties.

230   MOTION OF SYSTEMS OF PARTICLES

186. Theorem. *The kinetic energy of any system of moving particles is equal to the kinetic energy of motion relative to the center of gravity of the particles, plus the kinetic energy of a single particle of mass equal to the aggregate mass of the system, moving with the center of gravity.*

Let the particles be m_1 at x_1, y_1, z_1, etc., and let the coördinates be measured with the center of gravity taken as origin. Let the velocities be denoted by u_1, v_1, w_1, etc., and let these also be measured relative to a frame moving with the center of gravity, so that

$$u_1 = \frac{dx_1}{dt}, \text{ etc.}$$

Let the velocity of the center of gravity referred to any frame of reference, moving or fixed (provided only that the directions of the axes do not turn), have components $\bar{u}, \bar{v}, \bar{w}$. Then the velocity of the particle m_1 is compounded of the velocity of the particle, relative to the center of gravity of the system, of components u_1, v_1, w_1, together with the velocity of the center of gravity, of components $\bar{u}, \bar{v}, \bar{w}$. Thus the whole velocity of the particle m_1 has components

$$\bar{u} + u_1, \qquad \bar{v} + v_1, \qquad \bar{w} + w_1.$$

The kinetic energy of the first particle is accordingly

$$\tfrac{1}{2} m_1 [(\bar{u} + u_1)^2 + (\bar{v} + v_1)^2 + (\bar{w} + w_1)^2],$$

so that the kinetic energy of the system is

$$\tfrac{1}{2} \sum m [(\bar{u} + u)^2 + (\bar{v} + v)^2 + (\bar{w} + w)^2],$$

or, on expanding squares,

$$\tfrac{1}{2} \left(\sum m \right) (\bar{u}^2 + \bar{v}^2 + \bar{w}^2)$$
$$+ \bar{u} \sum mu + \bar{v} \sum mv + \bar{w} \sum mw$$
$$+ \tfrac{1}{2} \sum m (u^2 + v^2 + w^2). \qquad (80)$$

Since, when the center of gravity is taken as origin, the coördinates of the particles are x_1, y_1, z_1, etc., we have, by equations (8),

KINETIC ENERGY

$$0 = \frac{\sum mx}{\sum m}, \text{ etc.,}$$

so that $\sum mx = 0$. It follows that $\sum m \frac{dx}{dt} = 0$, or $\sum mu = 0$. Similarly $\sum mv = 0$ and $\sum mw = 0$. The whole of the second line of expression (80) is now seen to disappear, so that we find for the kinetic energy the expression

$$\tfrac{1}{2}\left(\sum m\right)(\bar{u}^2 + \bar{v}^2 + \bar{w}^2) + \tfrac{1}{2}\sum m(u^2 + v^2 + w^2), \qquad (81)$$

which proves the theorem.

187. Next, suppose that the coördinates of the center of gravity at any instant, referred to an imaginary set of fixed axes, are $\bar{x}, \bar{y}, \bar{z}$, the velocity of the center of gravity having, as before, components $\bar{u}, \bar{v}, \bar{w}$.

We have supposed that, relative to the center of gravity, the particle m_1 has coördinates x_1, y_1, z_1, and components of velocity u_1, v_1, w_1. Thus, referred to the imaginary fixed axes, the coördinates of the particle m_1 will be

$$\bar{x} + x_1, \quad \bar{y} + y_1, \quad \bar{z} + z_1,$$

while its components of velocity, as before, are

$$\bar{u} + u_1, \quad \bar{v} + v_1, \quad \bar{w} + w_1.$$

Let the force applied to the particle m_1 have components X_1, Y_1, Z_1. As in § 141, the work done on this particle by the external forces is equal to *minus* the work performed by the particle against these forces. Thus the work done on the particle while it moves over any small element of its path is, by § 118, equal to

$$X_1 d(\bar{x} + x_1) + Y_1 d(\bar{y} + y_1) + Z_1 d(\bar{z} + z_1).$$

The total work done on all the particles in any small displacement is therefore

$$\sum [X_1 d(\bar{x} + x_1) + Y_1 d(\bar{y} + y_1) + Z_1 d(\bar{z} + z_1)],$$

and this can be separated into two parts as follows:

The first part may be taken to be

$$\sum X_1 d\bar{x} + \sum Y_1 d\bar{y} + \sum Z_1 d\bar{z}, \qquad (82)$$

while the second is

$$\sum X_1 dx_1 + \sum Y_1 dy_1 + \sum Z_1 dz_1. \qquad (83)$$

By equation (77), we have

$$\sum X_1 = M \frac{d\bar{u}}{dt},$$

where M is the total mass of the system, expressing that the center of gravity moves as though it were a particle of mass M acted on by a force of components $\sum X_1, \sum Y_1, \sum Z_1$. It is at once clear that expression (82) represents the work done in the motion of this imaginary particle, and this we know must be equal to the increase in its kinetic energy.

The total work done is the sum of expressions (82) and (83). This total work is equal to the increase in the total kinetic energy of the system (by § 140), and this again (by § 186) is equal to the increase in the kinetic energy of motion relative to the center of gravity of the particles, plus the increase in the kinetic energy of the imaginary particle of mass M moving with the center of gravity.

This latter increase, as we have just seen, is represented by expression (82), so that the former must be represented by expression (83).

Thus the increase in kinetic energy relative to the center of gravity is

$$\sum (X_1 dx_1 + Y_1 dy_1 + Z_1 dz_1),$$

and is therefore equal to the work done by the forces, *calculated as though the center of gravity were at rest*.

188. Thus we see that, in the theorem that the increase in kinetic energy is equal to the work done, it is legitimate to calculate both the kinetic energy and the work done by considering motion relative to the center of gravity only; i.e. the system may be treated as though its center of gravity remained at rest.

IMPULSIVE FORCES

As an illustration, consider the problem of firing a shot on board a moving ship. The mass of the shot being small compared with that of the ship, we may suppose the center of gravity of shot and ship to have exactly the motion of the ship. The velocity of the shot relative to this center of gravity may accordingly be taken simply to be that relative to the deck of the ship. The work done by the powder in ejecting the shot from the barrel is the same as though the ship were at rest, so that the velocity of the shot relative to the ship will be the same as though the ship were at rest.

EXAMPLES

1. A cart is moving with velocity V and a man on the cart throws out sand horizontally from the back of the cart at the rate of m pounds per minute, the sand having a velocity v relative to the road. At what rate is the man working?

2. A gun capable of firing a shot vertically upwards to a height h is placed on an armored train running with velocity V. What is the greatest range to which a shot can reach (a) behind the train, (b) in front of the train?

3. In the last question find the nearest point to the track, which is out of range of the gun.

4. A shell of mass M is moving with velocity V. An internal explosion generates an amount E of energy, and thereby breaks the shell into masses of which one is k times as great as the other. Show that if the fragments continue to move in the original line of motion of the shell, their velocities will be

$$V + \sqrt{2\,kE/M}, \qquad V - \sqrt{2\,E/kM}.$$

5. Two men, each of mass M, stand on two inelastic platforms each of mass m, hanging over a smooth pulley. One of the men, leaping from the ground, could raise his center of gravity through a height h. Show that if he leaps with the same energy from the platform, his center of gravity will rise a height

$$h\left(1 - \frac{M}{2\,(M+m)}\right).$$

Impulsive Forces

189. There are many instances in dynamical problems in which the action of a force begins and terminates within so short an interval of time that the action may be regarded as instantaneous. Such forces are called *impulsive forces*. As instances of impulsive forces we may take the forces brought into play by the jerking of an inextensible string, or by the collision between two hard bodies.

The change of momentum produced by the action of an impulsive force is, in general, of finite amount. As the force only acts for an infinitesimal time, the rate of change of momentum must be infinite. By the second law of motion, the rate of change of momentum is equal to the force, so that the force itself, while it lasts, must be infinite. Thus an impulsive force may be regarded as an infinite force acting for an infinitesimal time.

190. At the outset of our study of impulsive forces, it will be well to notice one physical peculiarity of these forces. A perfectly rigid body was defined as one which kept its shape under the action of any forces, no matter how great. At the same time it was mentioned that no perfectly rigid bodies exist in nature. Under the action of infinite, or very great, forces such as occur in impulses, no body may be treated as perfectly rigid.

The consequence of this is that when any impulsive forces are brought into action, relative motion is set up between the different small particles of which continuous bodies are composed. This relative motion possesses energy of a kind which cannot be regained from the system by mechanical processes; in fact, the relative motion of these particles simply represents the heat of the body. Inasmuch as this energy cannot be recovered from the system as mechanical work, we see that the impulsive forces which do work in producing this energy cannot be treated as conservative forces. Thus we see that

The sum of the potential and kinetic energies of a system does not remain constant through the action of impulsive forces.

For clearly part of the total energy is left, after the impulses, in the form of heat.

Consider, for instance, a lead bullet striking a steel target. Suppose that, before striking the target, the bullet is moving horizontally with velocity v at a height h. Its kinetic energy is $\frac{1}{2}mv^2$, its potential energy being mgh. After striking, we may suppose the bullet to have no horizontal velocity, but to fall to the bottom of the target. At the instant at which this fall begins, the kinetic energy is nil, while the potential energy is mgh, as it was before the impact. Thus an amount of energy $\frac{1}{2}mv^2$ has disappeared from the total energy. This has been used up in producing motions of the particles of the bullet and target relative to one another; these show themselves in the form of heat, and also, perhaps, partly in permanent changes of shape,— a dent in the target, or a flattening of the bullet.

IMPULSIVE FORCES

Measure of an Impulse

191. The change of momentum produced by an impulsive force is called the *impulse* of the force. Thus if an impulse I acts on a mass m, changing its velocity (or component of velocity in the direction of the impulse) from u to v, we have

$$I = m(v - u). \tag{84}$$

The force acting at any instant is, by the second law, equal to the rate of change of momentum of the particle (or body) on which it acts. If the force is of constant amount, the whole change of momentum is equal to the product of the force by the time over which it acts. If the force is of variable amount, the change of momentum will be equal to the integral of the force with respect to the time over which it acts. Thus if P is the value of the force acting at any instant of the whole time t, we see that the impulse

$= Pt$, if the force is of constant amount,

$= \int_0^t P\, dt$, if the force is of variable amount.

Work done by an Impulse

192. The work done by an impulse I in changing the velocity of a mass m from u to v will be

$$\tfrac{1}{2} m v^2 - \tfrac{1}{2} m u^2,$$

the increase in the kinetic energy of the mass. Since $I = m(v-u)$, we may write the expression for the work done in the form

$$\tfrac{1}{2} m (v-u)(v+u)$$
$$= I\left(\frac{v+u}{2}\right).$$

Thus *the work done by an impulse is equal to the impulse multiplied by the mean of the initial and final velocities of the mass acted upon.*

236 MOTION OF SYSTEMS OF PARTICLES

If the mass is not moving in the same direction as the line of action of the impulse, the foregoing result will obviously be true if u, v are taken to be the components of the velocities along the line of action.

ILLUSTRATIVE EXAMPLES

1. *A shot of 14 pounds is fired into a target of mass 200 pounds which is suspended by chains so that it is free to start into motion horizontally. If the shot, before impact, was moving with a horizontal velocity of 1000 feet a second, and afterwards remains embedded in the target, find the loss of energy caused by the impact.*

Let V denote the horizontal velocity, measured in feet per second, with which the target and bullet together start into motion after the impact. Then, by the conservation of momentum, equating the momentum before the impact to that after,
$$1000 \times 14 = V \times 214,$$
so that
$$V = \tfrac{14000}{214}.$$

The kinetic energy before impact was $\tfrac{1}{2} \cdot 14 \cdot (1000)^2$; that afterwards is $\tfrac{1}{2} \cdot 214 \cdot V^2$. Thus the loss of energy is

$$\tfrac{1}{2}(14{,}000{,}000 - 214\,V^2) = 6{,}540{,}000 \text{ foot poundals, approximately.}$$

2. *A heavy chain, of length l and mass m per unit length, is held with a length h hanging over the edge of a table, and the remainder coiled up at the extreme edge of the table. If the chain is set free, find the velocity at any stage of the motion.*

Suppose that at any stage of the motion a length x is hanging vertically, so that a length $l - x$ is coiled up on the table. After an infinitesimal time dt let x be supposed to have increased from x to $x + dx$. Then if v is the downward velocity of the chain, we clearly have

$$v = \frac{dx}{dt}.$$

At the beginning of the interval dt, the downward momentum of the chain was that of a mass mx moving with a velocity v. It was accordingly mvx. At the end of the interval, the momentum is that of a mass $m(x + dx)$ moving with a velocity which may be denoted by $(v + dv)$. Thus the gain in momentum is

$$m(x + dx)(v + dv) - mxv,$$

or, neglecting the small quantity of the second order $dv\,dx$, the gain is

$$m(x\,dv + v\,dx).$$

ILLUSTRATIVE EXAMPLES 237

The gain of momentum per unit time is, however, by equation (71), equal to the total force acting, and this is mgx at the beginning of the interval dt and $mg(x + dx)$ at the end. Neglecting the small quantity of the second order $dx\,dt$, we find that the total gain of momentum in the interval dt must be $mg\,x\,dt$.

Thus we have
$$m(x\,dv + v\,dx) = mg\,x\,dt$$
$$= mg\,x\,\frac{dx}{v},$$

or, simplifying,
$$v\,x\,\frac{dv}{dx} + v^2 = gx.$$

To integrate this equation, we multiply by $2x$, and then we obtain
$$v^2 x^2 = \tfrac{2}{3} g x^3 + \text{a constant}.$$

To determine the constant, we note that $v = 0$ when $x = h$, so that the value of the constant must be $-\tfrac{2}{3} g h^3$. Thus we have
$$v^2 = \tfrac{2}{3} g \frac{x^3 - h^3}{x^2},$$

giving the velocity when a length x is off the table. When the last particle of the chain is pulled off, the value of x is l, so that at this instant
$$v^2 = \tfrac{2}{3} g \frac{l^3 - h^3}{l^2}.$$

We notice that this value of v^2 is not the value which would be obtained from the equation of energy. Clearly this equation must not be employed, since impulses are in action all the time, jerking new particles of the chain into motion.

EXAMPLES

1. An empty car of 10 tons weight runs into a similar car loaded with 50 tons of coal, and the two run on together with a velocity of 5 feet per second. What was the velocity of the first car originally, and what was the amount of the impulse between the cars?

2. A stone of weight $\tfrac{1}{4}$ ounce is dropped on to soft ground from a height of 5 feet. Find the impulse exerted before the stone is brought to rest.

3. A mass of 1 ton falls from a height of 16 feet on the end of a vertical pile, and drives it half an inch deeper into the ground. Assuming the driving force of the mass on the pile to be constant while it lasts, find its amount and the duration of its action.

4. A body of mass 10 grammes is moving with a speed of 8 centimeters a second. Suddenly it receives a blow which causes it to double its speed, and to change its direction through half a right angle. Determine the direction of the blow, and the velocity with which the body would have moved off, had it been at rest.

238 MOTION OF SYSTEMS OF PARTICLES

5. The string of an Atwood's machine has masses m_1, m_2 attached to its ends, m_1 being the heavier. After it has been in motion for 1 second m_1 strikes the floor. Find (a) for how long m_2 will continue to ascend, (b) with what velocity m_1 will start into motion again when the string becomes tight.

6. On a certain day one inch of rain fell in 10 hours, the drops falling with a velocity of 20 feet per second. Find the average pressure per square foot on the canvas roof of a tent, supposed horizontal, produced by the impact of the raindrops. (One cubic foot of water weighs $62\frac{1}{2}$ pounds.)

7. The earth, moving in its orbit with velocity V, runs into a swarm of small meteorites, of density one kilogramme to the cubic mile, moving with a velocity v in a direction exactly opposite to that of the earth. Find the rate of decrease of the earth's speed in consequence of its bombardment by the meteorites, and find also the increase in the height of the barometer at different points on the earth's surface, it being assumed that all the meteorites are dissipated into dust before they reach the ground. (The earth's mass is 6×10^{27} grammes, its diameter is 7927 miles.)

8. A uniform chain is coiled in a heap on a horizontal plane, and a man takes hold of one end and raises it uniformly with a velocity v. Show that when his hand is at a height x from the plane, the pressure on his hand is equal to the weight of a length $x + \dfrac{v^2}{2g}$ of the chain.

Elasticity

193. It is a matter of common experience that if we drop a ball of steel on to a hard floor it rebounds to a considerable height, while a ball of wood will rebound to a much smaller height, and a ball of wool, paper, or putty will hardly rebound at all.

When the contact between the surfaces of two bodies is of such a nature that they do not rebound at all after impact, it is said to be *perfectly inelastic*, while if the bodies rebound, the contact is said to be *elastic*. Obviously there are varying degrees of elasticity.

Moment of Greatest Compression

194. Probably the collision of two billiard balls supplies the most familiar instance of an impact with a high degree of elasticity. We shall discuss this impact best by referring the motion of the second ball to a frame of reference moving with the first. Before impact the center of the second ball is approaching that of the first, after impact it is receding. Hence at some instant during the impact, its motion must have changed from one of approaching to one of receding; at this instant the distance between the two centers was a minimum.

Suppose that, before the experiment, we had chalked the two faces of the balls on which the collision takes place. On examining the balls after impact it will be found that the chalk has been disturbed, not only at a single point, but all over a circle of considerable size, — perhaps of diameter half an inch for billiard balls moving with a fair velocity. This shows that at the moment at which the centers of the balls were closest to one another, their distance was less than if they had been placed in contact and at rest, — the balls were compressed.

The instant at which the two centers were nearest is called the *moment of greatest compression.*

In general, for any two surfaces in collision, the instant at which the relative velocity along the common normal vanishes is called the moment of greatest compression. Obviously this is the instant at which the motion of the two surfaces changes from one of approach to one of recession.

195. By the time the moment of greatest compression is reached, the velocities of both bodies will, in general, have been changed, so that forces must have been at work to produce this change. The whole time of action of these forces, the time from the instant at which the bodies first touch to the instant of greatest compression, is so small that these forces may be treated as impulsive. The impulses acting on the two bodies, being action and reaction, must be equal and opposite. If the surfaces are smooth, the direction of these impulses will be along the common normal. If the surfaces are rough, we cannot specify the direction until we know the direction of sliding, if any, of the surfaces over one another. In either case, let us denote the component of the impulse along the common normal by I. The quantity I is called the *impulse of compression.* Clearly it is the forces of which this impulse is composed which reduce the relative normal velocity to zero.

After the moment of greatest compression, a second system of forces must come into play to set up the velocities with which the bodies separate from one another. In fact, at the instant of greatest compression, the compressed parts of the bodies act like a compressed spring, and we can suppose the velocities of separation produced by the action of this imaginary spring. The forces which separate the bodies may again be treated as impulsive, and the component of this impulse along the common normal will be denoted by I'. The impulse I' is called the *impulse of restitution*.

196. When the motion of the bodies before impact is known, we can calculate the velocities at the instant of greatest compression by an application of the principle of conservation of momentum. It is therefore possible to calculate the impulse I, the impulse of compression.

The amount of the impulse I', on the other hand, depends on the nature of the contact between the two bodies; for instance, if the bodies are perfectly inelastic, there is no separation at all after impact, so that $I' = 0$. In general, it is found as a matter of experiment that the impulse I' is connected with the impulse I by the simple law

$$I' = eI,$$

where e is a quantity which depends only on the nature of the contact between the two surfaces, and not on the amount of the impulse I. The quantity e is called the *coefficient of elasticity* for the two bodies.

It is important to understand that this coefficient of elasticity is a quantity entirely different from the coefficients or elastic constants which occur in the theory of elastic solids. Indeed, the term *coefficient of elasticity* is somewhat unfortunate as a description of the quantity e; what is measured is resilience rather than elasticity, and doubtless *coefficient of resilience* would be a better description than *coefficient of elasticity*. The term *coefficient of elasticity* has, however, been generally adopted.

197. The value of e, as we have seen, is zero for perfectly inelastic bodies. For iron impinging on lead, the value of e is about .14, for iron on iron about .66, and for lead on lead about .20. We

PARTICLE IMPINGING ON A FIXED SURFACE 241

notice that resilience depends on the nature of the contact between two bodies, being in this respect similar to the coefficient of friction. The resilience does not arise partly from one body and partly from the other, for if it did the value of e for iron impinging on lead would be intermediate between the values for iron on iron and for lead on lead.

As examples of bodies for which the coefficient of elasticity is large, it is found that the value of e for the impact of two ivory billiard balls is about .81, while for glass impinging on glass it is .94. The most perfect elasticity conceivable is that of two bodies for which $e = 1$, in which case the impulse of restitution is equal to the impulse of compression. The bodies in this case are spoken of as perfectly elastic. The peculiarity of perfectly elastic bodies is that no energy is lost on impact. It is clear that the value of e cannot exceed unity, for if the value of e were greater than unity, the kinetic energy set up by the impulse of restitution would be greater than that absorbed by the impulse of compression, so that the total energy would be increased.

We shall now apply these principles to some important cases of impact.

PARTICLE IMPINGING ON A FIXED SURFACE

Direct Impact

198. Suppose first that the impact is direct — i.e. that, at the instant of collision, the particle is moving along the normal to the surface at the point at which it strikes. Let m be its mass, and v its velocity before impact. At the moment of greatest compression, the particle will be at rest relatively to the plane, so that its momentum is reduced by the impulse of compression from mv to 0. Thus we must have
$$I = mv.$$

If e is the coefficient of elasticity,
$$I' = eI = emv.$$

Thus there is a normal impulse of amount emv, and this generates a velocity ev in the particle. There is no tangential impulse, for there is no sliding of the surfaces past one another. Thus the velocity of rebound is a velocity ev normal to the surface.

Oblique Impact: Smooth Contact

199. If the impact is oblique, let us suppose that the components of velocity along the tangent plane and along the normal before impact are u, v. As before, we find

$$I = mv, \qquad I' = emv,$$

so that the normal velocity after impact, say v', is

$$v' = ev.$$

If the contact is supposed smooth, there can be no force in the tangent plane, so that the momentum in the tangent plane remains unaltered. Thus the velocity in the tangent plane remains equal to u, and the velocity after impact will be one of components u, ev.

Let θ be the angle which the velocity before impact makes with the normal, and let ϕ be the corresponding angle after impact. Then

$$\tan \theta = \frac{u}{v},$$

$$\tan \phi = \frac{u}{ev},$$

so that $\tan \theta = e \tan \phi.$

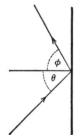

Fig. 126

If the bodies are perfectly elastic, $e = 1$, so that $\theta = \phi$; i.e. the particle rebounds at an angle equal to the angle of incidence. Its reflexion obeys the same law as that of a ray of light.

If the bodies are imperfectly elastic, θ is less than ϕ, so that the path is bent away from the normal.

If the bodies are perfectly inelastic, $e = 0$, so that $\phi = \frac{\pi}{2}$; the particle simply slides along the plane, as of course it obviously must since $I' = 0$.

PARTICLE IMPINGING ON A FIXED SURFACE 243

The kinetic energy before impact is
$$\tfrac{1}{2} m (u^2 + v^2),$$
that after impact is $\quad \tfrac{1}{2} m (u^2 + v'^2).$

Thus there is a loss of kinetic energy of amount
$$\tfrac{1}{2} m (v^2 - v'^2)$$
or
$$\tfrac{1}{2} mv^2 (1 - e^2).$$

This vanishes if $e = 1$, i.e. if the bodies are perfectly elastic. In all other cases there is a loss of energy. We again see that e cannot be greater than unity, or it would be possible to gain energy by causing bodies to impinge on one another.

Oblique Impact: Rough Contact

200. As in the case of a smooth contact, we obtain the relation $v' = ev$ connecting the components of velocity along the normal. The reaction, however, no longer acts entirely along the normal, so that it is not now true that the tangential component of velocity remains unaltered.

Let us consider the case in which the surface of the particle slides in the same direction over the fixed surface during the whole time that the two surfaces are in contact. Then at every instant of the impact there will be a tangential force equal to μ times the normal force, so that the total tangential impulse must be μ times the total normal impulse, and therefore equal to $\mu (I + I')$.

Thus if u' is the tangential velocity after the impact, we have
$$\begin{aligned} m(u - u') &= \mu (I + I') \\ &= \mu (1 + e) I \\ &= \mu (1 + e) mv, \end{aligned}$$
so that $\quad u' = u - (1 + e) \mu v.$

244   MOTION OF SYSTEMS OF PARTICLES

If, as before, we suppose that θ, ϕ are the angles which the path makes with the normal before and after the impact (see fig. 126), we have
$$\tan \theta = \frac{u}{v},$$
$$\tan \phi = \frac{u'}{v'} = \frac{u - (1+e)\mu v}{ev},$$
so that
$$e \tan \phi = \tan \theta - (1+e)\mu.$$

The value of $(1+e)\mu$ is always positive, so that ϕ is less than it would be if the plane were smooth; in other words, the roughness of the plane causes the particle to rebound nearer to the normal.

This equation, however, is only true within certain limits, for we have assumed that there is sliding during the whole time of impact. It may be that at a certain stage of the motion sliding will give place to rolling, and if so the equation we have obtained is no longer valid.

Impact of Two Moving Bodies

201. Suppose now that two bodies A, B of masses m, m' impinge at the point C, the common normal to C being the line CP.

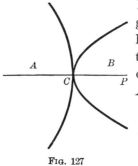

Fig. 127

Let it be supposed that the centers of gravity of the two bodies both lie in the line CP at the moment of impact, and let the components along CP of the velocities of the centers of gravity of the masses A, B respectively be

u, u' before impact,
V, V at the instant of greatest compression,
v, v' after impact.

Then if we denote the impulse of compression by I, and the impulse of restitution by I', we have

$$I = m(u - V) = -m'(u' - V), \tag{85}$$
$$I' = m(V - v) = -m'(V - v'). \tag{86}$$

IMPACT OF TWO MOVING BODIES

From the first line
$$u = V + \frac{I}{m},$$
$$u' = V - \frac{I}{m'},$$
so that
$$u - u' = I\left(\frac{1}{m} - \frac{1}{m'}\right),$$

an equation connecting I with the relative velocity before collision.

Similarly, from equations (86),
$$v - v' = -I'\left(\frac{1}{m} - \frac{1}{m'}\right).$$

The experimental relation $I' = eI$ is now seen to be exactly equivalent to the relation
$$v - v' = -e(u - u'), \tag{87}$$

or, in words: *The normal component of relative velocity of the centers of gravity after collision is equal to e times the relative velocity before collision, and is in the opposite direction.*

This law is known as Newton's experimental law; it expresses the same property of matter as the relation $I' = eI$.

A second relation, connecting velocities before impact with velocities after, is given by the conservation of momentum; we have
$$mv + m'v' = mu + m'u'.$$

Combining this with equation (87), we can determine the velocities v, v' after collision in terms of the velocities u, u' before collision.

Solving, we find that
$$v = \frac{mu + m'u' - em'(u - u')}{m + m'}, \tag{88}$$
$$v' = \frac{mu + m'u' + em(u - u')}{m + m'}, \tag{89}$$

giving the normal velocities.

If the bodies are rough, we find the tangential velocities in the same way as in § 200; while if the bodies are smooth, the velocities in directions perpendicular to CP remain unaltered.

If the center of gravity of the two bodies is at rest, — or, what comes to the same thing, if we measure all velocities relatively to the center of gravity, — we have

$$mu + m'u' = 0,$$

so that
$$v = -e\frac{m'(u-u')}{m+m'},$$

$$v' = e\frac{m(u-u')}{m+m'}.$$

Using the relation $mu = -m'u'$, these become

$$v = -eu,$$
$$v' = -eu',$$

so that the bodies rebound from one another as though they had impinged on a fixed plane of elasticity e.

The kinetic energy, either before or after the collision, is equal to the kinetic energy of a single particle moving with the center of gravity, together with that of the system relative to the center of gravity. The former remains unchanged by collision, so that the loss in the total kinetic energy produced by collision is equal to the loss in the kinetic energy relative to the center of gravity.

If the bodies are smooth, this loss of kinetic energy

$$= \tfrac{1}{2}(mu^2 + m'u'^2 - mv^2 - m'v'^2)$$
$$= \tfrac{1}{2}(mu^2 + m'u'^2)(1 - e^2).$$

Thus the loss of kinetic energy is $(1 - e^2)$ times the original kinetic energy relative to the center of gravity. If the bodies are perfectly elastic, $e = 1$, so that there is no loss of energy; while if $e = 0$, the original energy relative to the center of gravity is all lost.

Impact of Two Smooth Spheres

202. Let us apply the principles just explained to determining the motion, after impact, of two smooth spheres.

At the moment of impact let A, B be the centers of the two spheres, so that the line AB is the common normal to the surfaces at the point of impact C.

IMPACT OF TWO MOVING BODIES 247

As before, let the velocities along AB before impact be u, u', these both being measured in the direction AB, and let the velocities in the same direction after impact be v, v'. Then we have, by the conservation of momentum along AB,

$$mu + m'u' = mv + m'v',$$

and, by Newton's law, $\quad v - v' = -e(u - u').$

From these equations (88) and (89) follow as before.

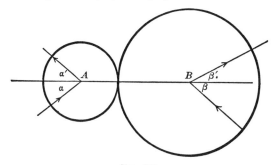

Fig. 128

If the velocities of A before impact make angles α, α' with AB as marked in the figure, the tangential velocities of A before and after impact are

$$u \tan \alpha, \qquad -v \tan \alpha',$$

so that, since the tangential velocities remain unaltered, we must have

$$v \tan \alpha' = -u \tan \alpha;$$

while similarly, from the motion of B,

$$v' \tan \beta' = -u' \tan \beta.$$

Thus equations (88) and (89) become

$$\cot \alpha' = -\frac{mu + m'u' - em'(u - u')}{(m + m')u} \cot \alpha,$$

$$\cot \beta' = -\frac{mu + m'u' + em(u - u')}{(m + m')u'} \cot \beta,$$

giving α', β' in terms of the initial motion.

248   MOTION OF SYSTEMS OF PARTICLES

If, as in the game of billiards, the spheres are of equal mass and the second sphere is originally at rest, we take $m = m'$, $u' = 0$, and obtain

$$\cot \alpha' = -\tfrac{1}{2}(1-e)\cot \alpha, \qquad \beta' = 0.$$

Thus B starts into motion along the line of centers, as it obviously must since the forces which set it in motion act along this line.

Since e is always less than unity and α is necessarily acute, $\cot \alpha'$ must be negative, so that α' will be obtuse. If $e = 1$, then $\alpha' = 90°$. Thus if the spheres were perfectly smooth and perfectly elastic, A would move at right angles to the line of centers after impact; its motion would be the same as if it had impinged on a perfectly smooth and *inelastic* plane.

ILLUSTRATIVE EXAMPLE

A row of similar coins is placed on a rough table, the coins being at equal distances apart and in a straight line. The first coin is projected along this line so as to impinge directly on the second. Find the resulting motion.

Let e be the coefficient of elasticity for an impact between the two coins, and μ the coefficient of friction between the coins and the table. Let m be the mass of each coin, and d the distance between the nearest points of two adjacent coins.

The normal reaction between a coin and the table is mg, so that the frictional force opposing the coin's motion is μmg, and the retardation produced is μg. Thus if a coin is started from its original position with a velocity V, its velocity on reaching the next coin is reduced to u, where

$$V^2 - u^2 = 2\,\mu g d. \tag{a}$$

We now have two coins of equal mass impinging with velocities u, 0. Their velocities after impact, say v, v', are given by the equations

$$v - v' = -eu \text{ (Newton's law)},$$
$$v + v' = u \text{ (conservation of momentum)}.$$

Thus
$$v = \tfrac{1}{2}u(1-e),$$
$$v' = \tfrac{1}{2}u(1+e).$$

After impact the coin originally in motion has a velocity v, and is retarded by a frictional retardation μg. It accordingly comes to rest, if it does not collide again in the meantime, after a distance s given by

$$v^2 = 2\,\mu g s,$$

or
$$s = \frac{v^2}{2g} = \frac{u^2(1-e)^2}{8g}, \tag{b}$$

while the coin which has been started into motion sets off with a velocity

$$v' = \tfrac{1}{2}u(1+e). \tag{c}$$

ILLUSTRATIVE EXAMPLE

The coin before this started into motion with a velocity V given by equation (a). Eliminating u from equations (a) and (c), we find as the relation between successive velocities of starting

$$V^2 = 2\,\mu gd + \frac{4\,v'^2}{(1+e)^2}, \qquad (d)$$

a difference equation with constant coefficients.

If $e = 1$, we notice from equation (b) that $s = 0$, so that each coin remains absolutely at rest after striking the coin next in front of it — it transmits the whole of its momentum to this coin. Also from equation (a),

$$V^2 - v^2 = 2\,\mu gd.$$

After the momentum has been transmitted over the space between n coins, the value of the square of the velocity is reduced by $2\,n\mu gd$. Thus at any point the velocity of a moving coin is that which would be possessed by a coin which had been started with velocity V, and made to move over a distance equal to all the intervals between the coins over which the motion has been transmitted.

If $d = 0$, so that the coins were originally in contact, we have

$$V = \frac{2\,v'}{1+e}.$$

Thus, if there are n coins, the nth coin will start in motion with a velocity

$$V\left(\frac{1+e}{2}\right)^{n-1}.$$

EXAMPLES

1. Hailstones are observed to strike the surface of a frozen lake in a direction making an angle of 30 degrees with the vertical, and to rebound at an angle of 60 degrees. Assuming the contact to be smooth, find the coefficient of elasticity.

2. If the hailstones of the last question rise after impact to a height of 2 feet, find the velocity with which they originally struck the ground.

3. In the last question find the height to which the hailstones will rise in their second rebound from the ice.

4. A ball is dropped on to a horizontal floor and, after rebounding twice, reaches a height equal to half that from which it was dropped. Find the coefficient of elasticity.

5. A bullet strikes a rough target at 45 degrees, and rebounds at the same angle. Show that

$$\mu = \frac{1-e}{1+e}.$$

6. A shot fired from a distance a strikes a target at right angles, and rebounds. Show that it will fall at a distance ae from the target (neglecting the resistance of the air).

7. A sphere of mass m collides with a second sphere of mass m', the contact between them being smooth, and their paths after collision are observed to be at right angles. Prove that $m = em'$.

250   MOTION OF SYSTEMS OF PARTICLES

   8. Two billiard balls stand at rest, and a third ball is made to strike them simultaneously, and is observed to remain at rest after the impact. Show that $e = \frac{2}{3}$.

   9. A particle is projected from a point on a smooth horizontal plane, with velocity V at an elevation α, and after striking the plane rebounds time after time. Show that its total time of flight is $\dfrac{2V \sin \alpha}{g(1-e)}$, and that its total range is $\dfrac{V^2 \sin 2\alpha}{g(1-e)}$.

   10. A player stands at a horizontal distance d from a wall, and throws a ball towards the wall at an inclination α to the horizontal. Show that if it is to return to him after bouncing, he must throw it with a velocity V given by

$$V^2 = \frac{(1+e)gd}{2e \cos \alpha (\sin \alpha - \mu \cos \alpha)},$$

where e, μ are coefficients of elasticity and friction.

   11. In the last question consider the cases of (a) $e = 0$, (b) $\mu = \tan \alpha$, (c) $\mu > \tan \alpha$.

GENERAL EXAMPLES

   1. A particle is placed on the face of a smooth wedge which can slide on a horizontal table; find how the wedge must be moved in order that the particle may neither ascend nor descend. Also find the pressure between the particle and the wedge.

   2. It is required to run trains of 100 tons on a level electric railway, with stations half a mile apart, at an average speed of 12 miles an hour, including half a minute stop at each station. Prove that the electric locomotives must weigh at least an additional 8 tons, taking a coefficient of friction of $\frac{1}{6}$, and supposing the trains fitted with continuous brakes. (Neglect passive resistances.)

   Prove that the railway can be worked by gravity, if the line is curved downward between the stations to a radius of about 46,000 feet; and that the dip between the stations will be about 20 feet, the inclines at the stations about 1 in 33, and the maximum velocity about $23\frac{1}{2}$ miles an hour.

   3. A cylinder of height h and diameter d stands on the floor of a railway car, which suddenly begins to move with acceleration f. Show that the cylinder will only remain at rest relative to the car if f is less than both μg and dg/h.

   4. If a circular hoop is projected, spinning steadily without wobbling, prove that the center describes a parabola, and that the tension of the rim is the weight of a length v^2/g of the rim, where v denotes the rim velocity relative to the center of the hoop.

EXAMPLES

5. A uniform chain 6 feet long, having a mass of 2 pounds per foot, is laid in a straight line along a rough horizontal table, for which the coefficient of friction is $\frac{1}{2}$, a portion hanging over the edge of the table so that slipping is just about to occur. If a slight disturbance sets the chain slipping, find the tension at the edge of the table when x feet have slipped off.

6. Two equal balls A, B, each of mass m, are at a distance a apart. An impulse I acts on A in the direction AB, and a constant force F acts on B in the same direction. Show that A will not overtake B if
$$I^2 < 2\,aFm.$$

7. A bullet weighing one ounce is fired with a velocity of 1200 feet per second at an elevation of 1 degree so as to hit a bird weighing $2\frac{1}{2}$ pounds when the bullet is at the highest point of its path. Supposing the bird to have been at rest when hit and afterwards to fall with the bullet embedded in it, find how far from the point of firing the bird will fall.

8. If a bullet weighing w pounds is fired with velocity v at a body weighing W pounds, advancing with velocity V, prove that the body will retain the velocity
$$\frac{WV - wv}{W + w}, \text{ or } V - \frac{w}{W}(v - u),$$
according as the bullet is embedded, or perforates and retains a velocity u. Calculate the energy liberated, and thence infer the average resistance of the body from the length perforated by the bullet.

9. A pile is being driven in by repeated impacts of a falling weight. How does the extent to which the pile is driven in by each blow depend (a) on the magnitude of the weight, and (b) on the height to which it is raised before being released?

If the weight be 1 ton, and the height from which it falls be 10 feet, and the pile be driven in a tenth of an inch, find the resistance in tons.

10. An inelastic pile, of mass m pounds, is driven vertically into the ground a distance of a feet at each blow of a hammer of mass M pounds, which falls vertically through h feet. Show that the weight which would have to be placed on the top of the pile to drive it slowly into the ground would be
$$M + \frac{M^2 h}{(M + m)a} \text{ pounds.}$$

11. A hammer head of W pounds, moving with a velocity of v feet a second, strikes an inelastic nail of w pounds fixed in a block of M pounds which is free to move. Prove that if the mean resistance of the block to penetration by the nail is a force of R pounds, then the nail will penetrate each blow a distance, in feet,
$$\frac{MW^2}{(M + W + w)(W + w)} \frac{v^2}{2\,gR}.$$

252 MOTION OF SYSTEMS OF PARTICLES

12. In the system of pulleys described in § 128, show that if P is a weight which is not equal to W/n, the acceleration produced in the weight W will be
$$\frac{nP - W}{n^2 P + W} g.$$

13. Two masses m, m' connected by an elastic string are placed on a smooth horizontal table, the masses being at rest and the string unstretched. A blow of impulse P is given to the first mass, in the direction away from the second. Show that when the string is again unstretched, the velocity of the second mass is
$$\frac{2P}{m + m'}.$$

14. Three equal particles are tied at the ends and middle point of an inextensible string, which is placed, fully extended, on a smooth table. The middle particle is jerked into motion in the direction towards and perpendicular to the line joining the other two. Find the loss of energy when the other particles are jerked into motion.

15. A coal train consists of a number of similar trucks hauled by an engine whose weight is just equal to that of three trucks. The train is at rest on a level track, the couplings, which are of equal length, being all equally slack. The engine then begins to move with a constant tractive force, and each truck is jerked into motion as its coupling tightens. Show that the speed of the engine will be greatest just before the tenth jerk occurs.

16. Snow is evenly spread over a roof. If a mass commences to slide, clearing away a path of uniform breadth as it goes, prove that its acceleration is constant, and equal to a third of that of a mass sliding freely down the roof.

17. A heavy, perfectly flexible uniform string hanging vertically with its lowest point at a height h above an inelastic horizontal plane is suddenly allowed to fall on to the plane. Show that the pressure on the table when a length x of the string has fallen on to the table is
$$(3x + 2h) mg.$$

18. Show that if two equal balls impinge directly with velocities $\frac{1+e}{1-e} V$ and V, the former will be reduced to rest.

19. Show that the mass m of a sphere which must be interposed between a sphere of mass M at rest and one of mass M' moving directly on to it with velocity V, in order that the former may acquire the greatest possible velocity from the impact, will be $\sqrt{MM'}$, and that the velocity acquired will be
$$\frac{M' V (1+e)^2}{M + M' + 2m}.$$

EXAMPLES 253

20. Prove that an elastic ball, let fall vertically from a height of h feet on a hard pavement, and rebounding each time vertically with e times the striking velocity, will have described

$$\frac{1+e^2}{1-e^2}h \text{ feet, in } \frac{1+e}{1-e}\sqrt{\frac{2h}{g}} \text{ seconds}$$

before the rebounds cease.

Work this out for $h = 1$, $e = \frac{7}{9}$.

21. A ball is dropped from the top of a tower, height h, and at the same time another ball of equal weight is projected upwards with the velocity $\sqrt{2gh}$ from the base of the tower and collides directly with the falling ball. If the coefficient of restitution be e, prove that the falling ball will, in the rebound, rise to a height short of the top of the tower by $\frac{h}{4}(1-e^2)$.

22. A boy standing on a railway bridge lets a ball fall on the horizontal roof of a car passing under the bridge at 15 miles an hour. If $\mu = \frac{1}{2}$, $e = \frac{4}{7}$ between the roof and the ball, find the least height of the boy's hand above the roof in order that the second rebound of the ball may be from the same point of the roof as the first.

If the boy's hand is at a greater height than this, what will happen?

23. A perfectly elastic particle is projected so as to strike the inside of a surface of revolution of which the axis is a given vertical line. Show that the vertices of all the parabolas described after successive rebounds lie on a surface of which the shape is independent of that of the surface of revolution.

24. Prove that in order to produce the greatest possible deviation in the direction of motion of a smooth billiard ball of diameter a by impact on another equal ball at rest, the former must be projected in a direction making an angle

$$\sin^{-1}\left(\frac{a}{c}\sqrt{\frac{1-e}{3-e}}\right)$$

with the line, of length c, joining the two centers.

25. A pendulum hangs with its bob just in contact with a smooth vertical plane. The bob is drawn aside until it is 5 inches higher than it was, and is then released so as to strike the plane normally; and on the first rebound it rises vertically through 4 inches. What would have been the vertical rise on rebound if the pendulum had been drawn aside through the same angle, but so that the bob strikes at an angle of 60 degrees with the normal?

CHAPTER X

MOTION OF A PARTICLE UNDER A VARIABLE FORCE

203. In almost all the cases of motion of a particle which have so far been considered, the forces acting on the particle have remained constant throughout the whole of the path, so that the acceleration of the particle has been constant. We proceed now to consider the motion of a particle which is acted upon by forces which vary from point to point of the path of the particle.

These problems fall into two classes, according to whether the path described by the particle is or is not given as one of the data of the problem. The former class is the simplest and is considered first. It includes such cases as the motion of a pendulum, in which the "bob" of the pendulum is constrained to describe a circle by the mechanism of suspension of the pendulum, as also that of the motion of a bead on a wire, in which the bead is compelled to describe the path marked out for it by the wire.

Equations of Motion

204. Let s denote the distance described by the particle along its path at any instant t, this distance being measured from any fixed point O on the path. The velocity along the path is then $\dfrac{ds}{dt}$. Calling this v, the acceleration is $\dfrac{dv}{dt}$ or $\dfrac{d^2s}{dt^2}$.

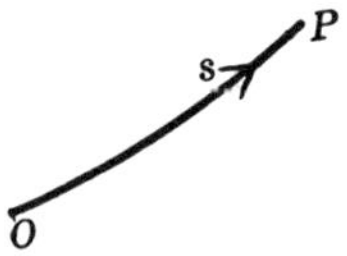

FIG. 129

We can also obtain a value for the acceleration from a knowledge of the forces acting. To find the acceleration, we must resolve all the forces which act on the particle in the direction of the path. If S is the component of force in this direction, the equation of motion of the particle, by the second law of motion, will be

$$S = m\frac{dv}{dt}, \qquad (90)$$

or
$$S = m\frac{d^2s}{dt^2}. \qquad (91)$$

We shall suppose the field of force to be permanent, so that the quantity S may be supposed to depend only on the position occupied by the particle on its path, and not on the instant at which it arrives there. In other words, S is a function of s but not of t. Equation (91) is a differential equation connecting s and t; if we can solve this equation, we shall have a full knowledge of the motion of the particle provided its path is known.

The equation is a differential equation of the second order, but can easily be transformed into one of the first order. For

$$\frac{d^2s}{dt^2} = \frac{dv}{dt} = \frac{dv}{ds}\frac{ds}{dt} = v\frac{dv}{ds},$$

so that the equation can be written

$$S = mv\frac{dv}{ds}.$$

Since S is a function of s, this equation can be integrated with respect to s, so that we obtain

$$C + \int S\,ds = \tfrac{1}{2}mv^2, \qquad (92)$$

where C is a constant of integration.

Since v is equal to $\dfrac{ds}{dt}$, this equation can be written in the form

$$\frac{ds}{dt} = \sqrt{\frac{2}{m}\left(C + \int S\,ds\right)}, \qquad (93)$$

which is an equation of the first degree. If this can be solved, the solution of the problem is complete.

256   MOTION UNDER A VARIABLE FORCE

We notice that the right hand of equation (92) is the kinetic energy of the particle. Also, since the force opposing the motion of the particle along its path is $-S$, its potential energy is $-\int S\,ds$. Thus equation (92) expresses that the sum of the kinetic and potential energies remains constant — it is the *equation of energy* for the motion of the particle. From a knowledge of the total energy at any instant of the particle's motion, we can determine the constant C, and can then proceed to integrate equation (93), if possible.

ILLUSTRATIVE EXAMPLE

Assuming that the value of gravity falls off inversely as the square of the distance from the earth's center, determine the motion of a projectile fired vertically into the air, the diminution of gravity being taken into account.

Let a be the radius of the earth, and g the value of gravity at the surface. Then, at a distance r from the earth's center, the value of gravity will be $\dfrac{ga^2}{r^2}$.

Since the particle moves along a radius drawn through the center of the earth, we may measure all distances from the earth's center, and the distance r from the earth's center may replace the coördinate s of § 204. The value of the force S resolved along the path is $-\dfrac{mga^2}{r^2}$, so that the equation of motion is

$$-\frac{mga^2}{r^2} = m\frac{d^2r}{dt^2}.$$

The equation of energy, as in equation (92), is

$$C - \int \frac{mga^2}{r^2}\,dr = \tfrac{1}{2}mv^2,$$

or
$$C + \frac{mga^2}{r} = \tfrac{1}{2}mv^2. \tag{a}$$

Let us suppose that the particle was projected from the earth's surface with velocity V. Putting $r = a$ in equation (a), the value of v must be V, so that we must have

$$C + mga = \tfrac{1}{2}mV^2, \tag{b}$$

and this equation determines the value of C. Eliminating C from equations (a) and (b), we obtain

$$ga\left(1 - \frac{a}{r}\right) = \tfrac{1}{2}(V^2 - v^2),$$

ILLUSTRATIVE EXAMPLE

giving the velocity at any point in the form

$$v = \sqrt{V^2 - 2ga\left(1 - \frac{a}{r}\right)}. \tag{c}$$

Since $v = \dfrac{dr}{dt}$, this equation becomes

$$\frac{dr}{dt} = \sqrt{V^2 - 2ga\left(1 - \frac{a}{r}\right)}, \tag{d}$$

so that

$$t = \int \frac{dr}{\sqrt{V^2 - 2ga + \dfrac{2ga^2}{r}}}. \tag{e}$$

On performing the integration, we can find the time required to describe any portion of the path. Let us first consider the different types of solution.

We see from equation (c) that v vanishes when

$$r = \frac{2ga^2}{2ga - V^2},$$

so that if $V^2 < 2ga$, there is a positive value of r, intermediate between $+a$ and $+\infty$, for which the velocity vanishes. Thus if $V^2 < 2ga$, the projectile goes to the point at distance $\dfrac{2ga^2}{2ga - V^2}$, and then falls back on to the earth. If $V^2 > 2ga$, we find that there is no positive value of r for which v vanishes, so that the particle goes to infinity: it escapes from the earth altogether.

When $V^2 = 2ga$, the velocity vanishes at infinity; thus the particle just escapes from the earth's attraction, but is left with zero velocity. Its kinetic energy of projection is just used up in overcoming the earth's attraction.

Let us consider first the special case in which $V^2 = 2ga$. We find that equation (e) reduces to

$$t = \int \frac{dr}{\sqrt{\dfrac{2ga^2}{r}}}$$

$$= \frac{2}{3} \frac{r^{\frac{3}{2}}}{\sqrt{2ga^2}} + C',$$

where C' is a new constant of integration.

If we measure time from the instant of projection, we must have $t = 0$ when $r = a$, so that

$$0 = \frac{2}{3}\sqrt{\frac{a}{2g}} + C',$$

and on eliminating C',

$$t = \frac{2}{3a\sqrt{2g}}(r^{\frac{3}{2}} - a^{\frac{3}{2}}). \tag{f}$$

In the case in which $V^2 > 2ga$, we obtain after integration of equation (e),

$$t = -\frac{ga^2}{V^2 - 2ga}\left\{\frac{1}{\sqrt{V^2-2ga}}\log\frac{\sqrt{V^2-2ga+\frac{2ga^2}{r}}-\sqrt{V^2-2ga}}{\sqrt{V^2-2ga+\frac{2ga^2}{r}}+\sqrt{V^2-2ga}}\right.$$
$$\left.+\frac{2b}{r}\sqrt{V^2-2ga+\frac{2ga^2}{r}}\right\} + C', \qquad (g)$$

where C' is a new constant of integration. If the time is to be measured from the instant at which the particle is projected, we must have $t = 0$ when $r = a$, so that

$$0 = -\frac{ga^2}{V^2-2ga}\left\{\frac{1}{\sqrt{V^2-2ga}}\log\frac{V-\sqrt{V^2-2ga}}{V+\sqrt{V^2-2ga}} + \frac{2bV}{a}\right\} + C',$$

and on eliminating C', we can again obtain the time required to describe any portion of the path. The case in which $V^2 < 2ga$ can be treated in a similar manner. This is left as an example for the student.

EXAMPLES

1. In the foregoing illustrative example, suppose that $V^2 < 2ga$, and find
 (a) the greatest height reached;
 (b) the time of flight of the particle.

2. A meteorite falls on to the earth. Assuming it to start from infinity with zero velocity, and to fall directly on to the earth, find the velocity with which it reaches the earth's surface, and the time taken to fall to the earth's surface from a point distant r from the earth's center.

3. A particle falls from distance a into a center of force which attracts according to the law μ/r^2. Show that the average velocity on the first half of the path is to the average velocity on the second half in the ratio

$$\pi - 2 : \pi + 2.$$

4. Find the time of falling to a center of force which attracts according to the law $\mu r^{-\frac{5}{3}}$.

5. A particle moves in a straight line from a distance a to a center of attraction towards which the force is $\frac{\mu}{r^3}$. Show that the time required to reach the center is

$$\frac{a^2}{\sqrt{\mu}}.$$

6. A particle begins to move from a distance a towards a fixed center which repels according to the law μr. If its initial velocity is $\sqrt{\mu a}$, show that it will continually approach the fixed center, but will never reach it.

The Simple Pendulum

205. One of the most important cases of a variable force arises in the motion of a simple pendulum. To obtain a first approximation, we can suppose that the whole weight of the pendulum is concentrated in the bob, which may be treated as a particle, and that this is suspended from a fixed point by a weightless string or rod so that it is constrained to move in a vertical circle.

Let s denote the distance along this circle described by the particle, this distance being measured from the lowest point O. Let the angle PCO between the string and the vertical be denoted by θ, so that $s = a\theta$. The forces acting on the particle consist of its weight and the tension of the string. The latter has no component in the direction in which the particle moves. The former has a component $-mg \sin\theta$. Thus the equation of motion is

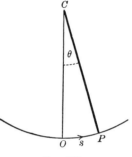

Fig. 130

$$\frac{d^2s}{dt^2} = -g \sin\theta, \qquad (94)$$

where $\theta = s/a$.

206. This equation cannot be solved by elementary mathematics, except in the simple case in which the angle θ is small, — i.e. the case in which the pendulum never swings through more than a small angle from the vertical. Confining our attention to this case, we may replace $\sin\theta$ by θ, and θ by s/a, and write the equation of motion in the form

$$\frac{d^2s}{dt^2} = -\left(\frac{g}{a}\right)s.$$

Thus the acceleration of the bob of the pendulum is proportional to its distance from O, and is towards O.

Writing the equation in the form

$$v\frac{dv}{ds} = -\left(\frac{g}{a}\right)s,$$

and integrating with respect to s, we obtain

$$v^2 = C - \left(\frac{g}{a}\right)s^2. \qquad (95)$$

Clearly the constant C must be positive, and the velocity will vanish as soon as s reaches a value such that

$$C = \left(\frac{g}{a}\right)s^2.$$

Let us denote the two values of s which satisfy this equation by $\pm s_0$, then the motion of the bob is clearly confined within two points at distances s_0 from the point O on opposite sides. Thus we may call s_0 the *amplitude* of the swing.

Replacing C by $\left(\frac{g}{a}\right)s_0^2$, equation (95) becomes

$$v^2 = \frac{g}{a}(s_0^2 - s^2), \qquad (96)$$

so that

$$\frac{ds}{dt} = \sqrt{\frac{g}{a}(s_0^2 - s^2)},$$

and the integral of this equation is

$$t = \int \frac{ds}{\sqrt{\frac{g}{a}(s_0^2 - s^2)}}$$

$$= \sqrt{\frac{a}{g}} \cos^{-1}\left(\frac{s}{s_0}\right) + \epsilon,$$

where ϵ is a constant of integration.

This equation gives

$$\cos^{-1}\left(\frac{s}{s_0}\right) = \sqrt{\frac{g}{a}}(t - \epsilon),$$

so that

$$s = s_0 \cos\left\{\sqrt{\frac{g}{a}}(t - \epsilon)\right\}.$$

THE SIMPLE PENDULUM 261

This equation contains the complete solution of the problem. We notice that the values of s continually repeat at intervals of time t_0 for which

$$\sqrt{\frac{g}{a}}\, t_0 = 2\pi.$$

Thus the motion of the pendulum repeats itself indefinitely. The interval between two instants at which the pendulum is in the same position, namely t_0, given by

$$t_0 = 2\pi \sqrt{\frac{a}{g}},$$

is called the period.

207. Seconds pendulum. To construct a pendulum which is to beat seconds, we choose a so that t_0 shall be equal to two seconds, for a seconds pendulum is one which takes one second to move from left to right, and then one second more to move from right to left. Thus we must have

$$\pi \sqrt{\frac{a}{g}} = 1.$$

In foot-second units we may take $g = 32.19$ for London, and so obtain

$$a = 39.14 \text{ inches},$$

as the length of the seconds pendulum at London.

We notice that the period of a pendulum varies as the square root of its length. Thus, for a pendulum to beat half-seconds, its length would have to be only a quarter of that of the seconds pendulum, and therefore 9.78 inches at London.

Since g varies from point to point on the earth's surface, the length of the seconds pendulum will also vary. If we observe the length of a pendulum and also measure its period with a chronometer, we shall be able to calculate the value of g at the place at which the experiment is performed; in fact, this method affords the easiest and most accurate way of obtaining the value of g at any point of the earth's surface.

262   MOTION UNDER A VARIABLE FORCE

ILLUSTRATIVE EXAMPLE

A pendulum which beats seconds accurately at New York is found to gain 2 seconds a day when taken to Philadelphia. Compare the values of g at Philadelphia and New York.

At Philadelphia the pendulum makes $24 \times (60)^2 + 2$ beats in $24 \times (60)^2$ seconds. Thus the time of a beat is

$$\frac{24 \times (60)^2}{24 \times (60)^2 + 2} \text{ seconds,}$$

and this is equal to $\pi\sqrt{\dfrac{a}{g}}$, where a is the length of the pendulum and g is the value of gravity at Philadelphia. If g_0 denote the value of gravity at New York, we have

$$g_0 = \pi^2 a,$$

$$g = \pi^2 a \left[\frac{24 \times (60)^2 + 2}{24 \times (60)^2}\right]^2$$

$$= \pi^2 a \left(1 + \frac{4}{24 \times (60)^2}\right) \text{ approximately,}$$

so that
$$g = g_0 \div \left(1 + \frac{4}{24 \times (60)^2}\right)$$

$$= g_0 \left(1 - \tfrac{1}{21600}\right) \text{ approximately.}$$

Thus gravity is less at Philadelphia than at New York by about one part in 21,600.

EXAMPLES

1. Calculate the length of a pendulum to beat time to a march of 100 paces a minute.

2. A pendulum which beats seconds in London requires to be shortened by one thousandth of its length if it is to keep time in New York. Compare the values of gravity at London and New York.

3. The value of g at a point on the earth's surface in latitude λ is

$$g = g_0(1 - .00257 \cos 2\lambda),$$

where $g_0 = 32.17$ is the value of g in latitude 45 degrees. Show that the latitude in which a short journey of given length will produce the greatest error in the rate of a pendulum clock is latitude 45 degrees, and find the error per mile in this latitude. (One minute of latitude = 6075 feet.)

4. In a building, at height h feet above the ground, the value of gravity is

$$g_0 - .0001 h,$$

where g_0 is the value of gravity at the foot of the building. In New York, $g_0 = 32.14$. Find the error in the rate of a pendulum clock, produced by taking it from the ground to the top of a building 300 feet high.

SIMPLE HARMONIC MOTION   263

5. The length of a pendulum which makes $2n$ beats per day is changed from l to $l + L$. Show that the pendulum will lose $\dfrac{nL}{l}$ beats per day approximately.

6. A balloon ascends with constant acceleration, and reaches a height of 3600 feet in two minutes. Show that during the ascent a pendulum clock will have lost about one second.

7. A pendulum of length l is adjusted by moving a small part only of the bob of the pendulum, this being of mass equal to one nth of the complete bob. How far must this be moved to correct an error of p seconds a day?

SIMPLE HARMONIC MOTION

208. We have seen that throughout the motion of a pendulum which moves so that its maximum inclination to the vertical is small, the acceleration is proportional to the distance from the middle point of its path, and is directed towards that point. A point which moves in this way is said to move with *simple harmonic motion*. Thus if s is the distance from a fixed point, of a point which moves with simple harmonic motion, we have an equation of the form

$$\frac{d^2s}{dt^2} = -k^2 s,$$

where k is a constant.

Integrating, we obtain, as before in the case of the pendulum (cf. equation (96)),

$$v^2 = k^2(s_0^2 - s^2),$$

and from this again

$$s = s_0 \cos k(t - \epsilon). \tag{97}$$

The constant k is known as the *frequency* of the motion. Thus the frequency of a simple pendulum is $\sqrt{\dfrac{g}{a}}$.

209. A simple geometrical interpretation can be given of simple harmonic motion, and this enables us to obtain a complete knowledge of the motion without any use of the integral calculus or of the theory of differential equations. In fig. 131 let the arm OP rotate about O with uniform angular velocity k, so that P describes a circle of radius a with uniform velocity ka. Let a perpendicular

PN be drawn from P to a fixed diameter AA'. Then we shall find that the point N moves backwards and forwards on the line AA' with simple harmonic motion.

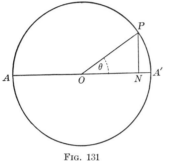

Fig. 131

The acceleration of P is, by § 12, an acceleration $k^2 a$ along PO. This can be regarded as compounded of the acceleration of P relative to N, which must be along NP, and the acceleration of N relative to O, which must be along ON. Thus the acceleration of N is that component of the acceleration of P which is in the direction AA'. This, however, is known to be $k^2 a \sin \theta$, or $k^2 \cdot ON$, along NO. Putting $ON = s$, we have an acceleration $-k^2 s$ in the direction in which s is measured, namely ON. Thus the point N moves with simple harmonic motion.

This geometrical interpretation of simple harmonic motion enables us to obtain expressions for v and s directly. The value of s is ON, or $a \cos \theta$. Let $t = \epsilon$ be an instant at which the point P was passing through the point A' in its motion round the circle, then, at any subsequent instant t, the time since P was at A' will be $t - \epsilon$, so that the angle described by OP will be $\theta = k(t - \epsilon)$. Thus we have
$$s = ON = a \cos k(t - \epsilon). \qquad (98)$$

This is the same result as is contained in equation (97). We notice that the amplitude s_0 of the motion is the same as the radius a of the circle, and that the frequency k is identical with the angular velocity. On differentiating equation (98), we obtain at once for the velocity
$$v = \frac{ds}{dt} = -ka \sin k(t - \epsilon)$$
$$= k \sqrt{a^2 - s^2}.$$

This result can also be obtained by resolving the velocity ka of the moving point P into two components, along and perpendicular

THE CYCLOIDAL PENDULUM   265

to AA'. The former is obviously the velocity of N along AA', and it is at once seen to be of amount $-ka \sin \theta$, or

$$v = -ka \sin k(t-\epsilon)$$
$$= k\sqrt{a^2 - s^2}, \text{ as before.}$$

In this motion, as in the motion of the simple pendulum, the quantity a is called the *amplitude*, while the time $\dfrac{2\pi}{k}$ after which the motion repeats itself is called the *period*.

EXAMPLES

1. A point moves with simple harmonic motion of period 12 seconds, and has an amplitude of 5 feet. Find its maximum velocity, and find its position and velocity one second after an instant at which its velocity is a maximum.

2. A particle moving with simple harmonic motion of period t is observed to have a velocity v when at a distance a from its mean position. Find its amplitude.

3. A particle is free to move along a line AB and is acted on by an attractive force directly proportional to its distance from a point P in AB, and consequently moves with simple harmonic motion. Prove that its average kinetic energy is equal to its average potential energy.

4. A point moving with simple harmonic motion is observed to have velocities of 3 and 4 feet per second when at distances of 4 and 3 feet respectively from its mean position. Find its amplitude and period.

5. A point moves with simple harmonic motion relative to one frame, and the frame itself moves with simple harmonic motion relative to a second frame, the directions of the two motions being parallel, and their periods the same. Show that the motion of the moving point relative to the second frame is simple harmonic motion, of the same direction and period as that of the frame.

6. A weight w is tied to an elastic string of natural length a and modulus λ, and is allowed to hang vertically in equilibrium. The weight is now pulled down vertically through a further distance h. Show that on being set free it will describe simple harmonic motion of amplitude a, provided this does not involve the string ever becoming unstretched. Find the period of the motion.

THE CYCLOIDAL PENDULUM

210. We have seen that the motion of a simple pendulum is simple harmonic motion only so long as the amplitude of the motion is small. It is, however, possible to constrain a particle to move under gravity in such a way that its motion shall be simple harmonic motion no matter how great the amplitude.

266 MOTION UNDER A VARIABLE FORCE

To find the curve in which the particle must be constrained to move, let us go back to equation (94), namely

$$\frac{d^2s}{dt^2} = -g \sin \theta,$$

which is the equation of motion of a particle constrained to move in any curve, provided θ is the angle which the tangent to the curve at a distance s along it makes with the horizontal. For this equation to represent simple harmonic motion, the acceleration $\frac{d^2s}{dt^2}$ must be equal to $-k^2s$. Thus we must have

$$g \sin \theta = k^2 s, \qquad (99)$$

so that $\sin \theta$ must be proportional to s.

211. This relation expresses a property of the cycloid,—i.e. of the curve described in space by a point on the rim of a circle which rolls along a straight line. For, in fig. 132, let P be a point on a cycloid which is formed by a circle rolling along the line EF. When the point on the rim of the moving circle is at P, let A be

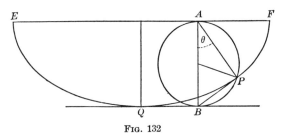

Fig. 132

the point of the circle which is in contact with the line EF, and let AB be the diameter of the circle which passes through A.

At the instant considered, we know that the motion of the point P on the rim of the circle is perpendicular to the line AP (see example 1 on p. 9). Thus since APB is a right angle, the motion must be along BP. Thus BP is the tangent to the cycloid.

If EF is supposed horizontal, the angle θ between the tangent at P and the horizontal is equal to the angle PAB, so that

THE CYCLOIDAL PENDULUM 267

the radius of the circle through P will make an angle 2θ with the vertical.

Suppose that the circle rolls along EF until the tangent to the cycloid at P makes an angle $\theta + d\theta$ with the horizontal. The radius at P must now make an angle $2(\theta + d\theta)$ with the vertical, so that the circle must have rotated through an angle $2\,d\theta$. Since the motion of P may be regarded as one of rotation about A, the small element of path ds described by P will be given by

$$ds = AP \cdot 2\,d\theta.$$

Now $AP = AB \cos \theta = D \cos \theta$, where D is the diameter of the rolling circle. Thus
$$ds = 2D \cos \theta \, d\theta,$$
giving, on integration, $\quad s = 2D \sin \theta.$

No constant of integration is required if we agree to measure s from the point at which $\theta = 0$, i.e. the lowest point of the cycloid.

The property of the cycloid is now proved, and we see that equation (99) is true throughout the motion of a point which describes a cycloid, this being generated by the rolling of a circle of diameter D given by
$$2D = \frac{g}{k^2}.$$

212. If the cycloid is given, the frequency k of the simple harmonic motion will be $k = \sqrt{\dfrac{g}{2D}}$, and the period is $\dfrac{2\pi}{k}$, or

$$2\pi \sqrt{\frac{2D}{g}}.$$

Thus the motion is of the same period as that of a simple pendulum of length $2D$.

213. The importance of cycloidal motion is as follows. It has been seen that the motion of a simple pendulum is only strictly simple harmonic when the amplitude is so small that it may be treated as infinitesimal. For finite amplitudes the motion is not simple harmonic, and consequently the period is different from that of the simple harmonic motion described when the amplitude is very small. Thus the period depends on the amplitude, so that a clock which beats true seconds when the pendulum swings through one angle will gain or lose as soon as the pendulum is made to swing through any different angle. Variations of amplitude must always occur during the motion of any pendulum, and these cause irregularities in the timekeeping of the clock.

We have, however, seen that if a particle describes a cycloid, the period is independent of the amplitude, so that variations of amplitude cannot affect the timekeeping powers of a particle moving in a cycloid.

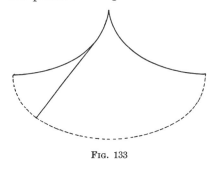

Fig. 133

The simplest way of causing a particle to move in a cycloid is, in practice, to suspend it from a fixed point by a string, in such a way that during its motion the string wraps and unwraps itself about two vertical cheeks. If the curve of these cheeks is rightly chosen, the particle can be made to describe a cycloid, and it can easily be shown that the curves of the cheeks must be portions of two cycloids each equal to the cycloid which is to be described by the particle.

EXAMPLES

1. In cycloidal motion prove that the vertical component of the velocity of the particle is greatest when it has described half of its vertical descent.

2. A particle oscillates in a cycloid under gravity, the amplitude of the motion being b and the period being τ. Show that its velocity at a time t measured from a position of rest is $\dfrac{2\pi b}{\tau} \sin \dfrac{2\pi t}{\tau}$.

3. A particle of mass m slides on a smooth cycloid, starting from the cusp. Show that the pressure at any point is $2\,mg\cos\psi$, where ψ is the inclination to the horizontal of the direction of the particle's motion.

MOTION ABOUT A CENTER OF FORCE 269

Motion of a Particle about a Center of Force

Force Proportional to the Distance

214. Let us suppose that a particle moves under no forces except an attraction to a fixed point O, the force of attraction being directly proportional to its distance from O.

Taking O as origin, let the coördinates of the point P, the position of the particle at any instant, be denoted by x, y, z. Let the force acting on the particle be $\mu \cdot OP$ directed along PO, where μ is a constant. The components of this force along the three coördinate axes are
$$-\mu x, \quad -\mu y, \quad -\mu z.$$

The components of acceleration are, as in § 177,
$$\frac{d^2 x}{dt^2}, \quad \frac{d^2 y}{dt^2}, \quad \frac{d^2 z}{dt^2}.$$

Thus the equations of motion of the particle are

$$m \frac{d^2 x}{dt^2} = -\mu x, \qquad (100)$$

$$m \frac{d^2 y}{dt^2} = -\mu y, \qquad (101)$$

$$m \frac{d^2 z}{dt^2} = -\mu z. \qquad (102)$$

These three equations are all of the same type, namely the type of equation which denotes simple harmonic motion. Thus the foot of the perpendicular from the moving point on to each of the coördinate axes moves with simple harmonic motion.

The solution of equation (100) has already been seen to be
$$x = A \cos p (t - \epsilon),$$
where $p^2 = \mu/m$. This can be written
$$x = A \cos p\epsilon \cos pt + A \sin p\epsilon \sin pt,$$
or again
$$x = C \cos pt + D \sin pt,$$

if we introduce new constants C, D to replace the constants $A \cos p\epsilon$ and $A \sin p\epsilon$. The other two equations have similar solutions, so that we can take the complete solution to be

$$x = C \cos pt + D \sin pt, \qquad (103)$$
$$y = C' \cos pt + D' \sin pt, \qquad (104)$$
$$z = C'' \cos pt + D'' \sin pt. \qquad (105)$$

We can always solve the equations

$$\begin{cases} C + rC' + sC'' = 0, & (106) \\ D + rD' + sD'' = 0, & (107) \end{cases}$$

and so obtain values of r and s for which these relations are true. Let us multiply equations (104) and (105) by these values of r and s, and add corresponding sides to the corresponding sides of equation (103). We obtain

$$(x + ry + sz) = (C + rC' + sC'') \cos pt + (D + rD' + sD'') \sin pt$$
$$= 0, \qquad (108)$$

since equations (106) and (107) are satisfied. The meaning of equation (108) is that for all values of t we have the relation $x + ry + sz = 0$, and, therefore, that throughout its motion the particle remains in the plane of which this is the equation.

The axes of coördinates have been supposed to be chosen arbitrarily. We can always choose the axes so that the plane in which the whole motion takes place is that of xy. The motion is then given by two equations of the form

$$x = C \cos pt + D \sin pt,$$
$$y = C' \cos pt + D' \sin pt.$$

Solving for $\sin pt$ and $\cos pt$ we obtain

$$\sin pt = \frac{C'x - Cy}{C'D - CD'},$$
$$\cos pt = -\frac{D'x - Dy}{C'D - CD'},$$

so that on squaring and adding, we obtain

$$(C'x - Cy)^2 + (D'x - Dy)^2 = (C'D - CD')^2.$$

MOTION ABOUT A CENTER OF FORCE 271

This is the equation of an ellipse.

Thus the most general motion possible for the particle consists in describing the same ellipse over and over again. The period is $2\pi/p$, this being the time required for $\cos pt$ and $\sin pt$ both to repeat their values.

215. The axes of x, y are still undetermined; let us imagine them to be the principal axes of the ellipse.

Then if we suppose the time measured from one of the instants at which the particle is at one of the extremities of the major axis, we shall have equations of the form

$$x = A \cos pt,$$
$$y = B \sin pt.$$

Thus pt is the eccentric angle of the ellipse described by the particle, so that the eccentric angle increases with uniform angular velocity p or $\sqrt{\dfrac{\mu}{m}}$. The motion repeats itself as soon as p increases by 2π. Thus the frequency is p or $\sqrt{\dfrac{\mu}{m}}$, while the period is $2\pi\sqrt{\dfrac{m}{\mu}}$.

216. This motion is realized experimentally in the motion of a pendulum which is not constrained to move in one vertical plane, but of which the deviations from the vertical remain small.

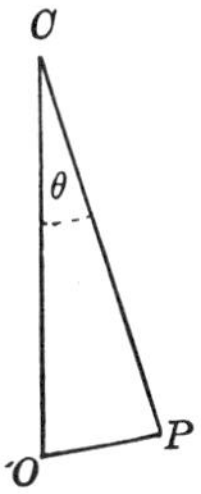

Let the pendulum be of length a, and let its bob be displaced from its equilibrium position O to some near point P, such that the angle PCO may be treated as small. Calling this angle θ, the weight of the bob may be resolved into $mg\cos\theta$ along CP, which is exactly neutralized by the tension of the string, and a force $mg\sin\theta$ along PO. If θ is small, $\sin\theta$ may be put equal to θ, and this in turn to $\dfrac{OP}{a}$. Thus the bob may be supposed to experience a force $\dfrac{mg}{a}OP$ along OP. The motion is therefore of the kind which has been described, the value of μ being $\dfrac{mg}{a}$, and the value of p therefore being $\sqrt{\dfrac{g}{a}}$. Thus we see that a hanging weight

Fig. 134

drawn from its position of equilibrium and projected in any way will always describe an ellipse in the horizontal plane in which it is free to move, having the point immediately below its point of suspension as center.

An arrangement may sometimes be seen at village fairs in England, in which the showman ingeniously takes advantage of this result. A weight is suspended by a string, and a skittle is placed on the floor exactly under the point of suspension of the weight. Passers-by are invited to pay an entrance fee and compete for a prize which is awarded to any one who can project the weight so that on its return it knocks the skittle over. The problem is, of course, as impossible as that of describing an ellipse which shall pass through its own center.

217. Another way in which motion under a force proportional to the direct distance may be realized, is as follows: An elastic string of natural length l has one end fastened to a small particle which is free to move on a smooth horizontal table; the other end, after passing through a small hole in the table, is fastened to a fixed point at a distance l from the hole. If the particle is pulled away from the hole to a point distant r from it, the total length of the string is $l + r$, so that its tension is $\dfrac{r}{l}\lambda$, where λ is its modulus of elasticity. The force acting on the particle, namely the tension of the string, is therefore proportional to the distance of the particle from a fixed point — namely the hole in the table — and its direction is toward the hole. Thus the particle will move in elliptic motion on the table.

EXAMPLES

1. The point P is describing an ellipse under an attractive force to the center, and p is the corresponding point on the auxiliary circle. Show that p moves round the auxiliary circle with uniform velocity.

2. A particle describes an ellipse about a center of force, the attraction being that of the direct distance. Show that the radius vector from the center of the ellipse to the particle sweeps out equal areas in equal times.

3. A particle is describing an ellipse under a force proportional to the distance, when it receives a blow in a direction parallel to the major axis of the ellipse. Show that the minor axis of the new orbit is the same as that of the old, and show how to find the change produced in the major axis.

MOTION ABOUT A CENTER OF FORCE 273

4. A particle is acted on by attractions to a number of centers of force, each being proportional to the distance. Show that it describes an ellipse.

How could a mechanical model be constructed to illustrate this motion?

5. A particle is acted on by a repulsion proportional to its distance from a center of force. Show that it describes a hyperbola.

6. Show that in the last question the radius vector joining the particle to the center of force sweeps out equal areas in equal times.

GENERAL THEORY OF MOTION ABOUT A CENTER OF FORCE

218. Suppose that we have a particle acted on only by a force directed towards a fixed center of force, the magnitude of this force being any function of the distance from the center.

Let O be the center of force, P the position of the particle at any instant, and PP' the direction of the velocity of the particle at this instant. Then the plane OPP' contains the velocity of the particle, which is along PP', and also the acceleration, which is along PO. Hence, after any short interval the veloc-

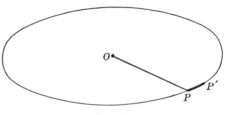

FIG. 135

ity of the particle will still be in the plane OPP'. The particle is still in this plane, say at P', so that the acceleration which is along $P'O$ is also in this plane.

Hence we can show that, after a further small interval, the position, velocity, and acceleration of the particle are all in the plane OPP', and so we can proceed indefinitely.

It follows that the particle will never leave the plane OPP'. We accordingly have the theorem:

The orbit described by a particle about a fixed center of force lies entirely in one plane.

This theorem has already been exemplified in § 214 by the case of the orbit described under an attraction proportional to the distance from the center of force.

Moment of a Velocity

219. The velocity of a point is a vector, and the line of action of this vector may be supposed to be the line through the moving point in the direction of its velocity. We can define the moment of a velocity in just the same way as the moment of a force has been defined. Moreover, all the properties of the moments of a force followed from the fact that forces could be compounded according to the parallelogram law, so that the same properties will be true of the moments of velocities, because velocities also can be compounded according to the parallelogram law.

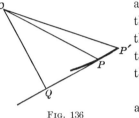

Fig. 136

Let P be a particle describing an orbit about O, and let OQ be a perpendicular from O on to the line through P in the direction of the particle's velocity. Then the moment of the particle's velocity about O is $OQ \times$ (velocity of particle).

After a short interval dt, let the particle be at P'. Its velocity at P' is compounded of its velocity at P together with dt times its acceleration at P. Hence

(moment about O of velocity at P')
= (moment about O of velocity at P)
+ (moment about O of [$dt \times$ acceleration at P]).

The acceleration at P being along PO, the last term of this equation is zero, so that we see that the moments about O of the velocities at P and at P' are equal.

We can extend this step by step as in the former theorem, and obtain finally that

The moment about O of the velocity of a particle describing an orbit about O is constant.

220. We have supposed that the particle moves from P to P' in time dt, so that if v is its velocity at P, then $PP' = v\,dt$. As the particle describes its orbit, the line OP may be regarded as

MOTION ABOUT A CENTER OF FORCE 275

sweeping out an area in the plane of the orbit. The area described in time dt is the small triangle OPP'. We now have

area described in time dt
$= \text{area } OPP'$
$= \tfrac{1}{2} OQ \cdot PP'$
$= \tfrac{1}{2} OQ \cdot v\, dt$
$= \tfrac{1}{2} dt \times \text{moment of velocity about } O.$

Thus the area described per unit time is half the moment of the velocity about O, and this by the theorem of the last section is a constant. Thus we have the theorem:

Equal areas are described in equal times.

Differential Equation of Orbit

221. The theorem just proved, in combination with the theorem of the conservation of energy, enables us to determine the equation of the orbit in which a particle will move. This equation is most conveniently expressed in polar coördinates, the center of force being taken as origin.

If r, θ are the polar coördinates of the particle, the velocity may be regarded as compounded of a velocity $\dfrac{dr}{dt}$ along OP, and a velocity $r\dfrac{d\theta}{dt}$ at right angles to OP.

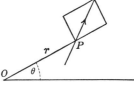

Fig. 137

The velocity is therefore given by

$$v^2 = \left(\frac{dr}{dt}\right)^2 + r^2\left(\frac{d\theta}{dt}\right)^2.$$

The moment of the velocity about O is equal to the moment of the second component, for the moment of the first component vanishes. Thus the moment of the velocity about O is $r \times r\dfrac{d\theta}{dt}$, and since this has a constant value, say h, we have

$$r^2 \frac{d\theta}{dt} = h. \tag{109}$$

276 MOTION UNDER A VARIABLE FORCE

If m is the mass of the particle, and if $f(r)$ is the attraction per unit mass when at a distance r from O, we find that the potential energy of the particle is
$$m\int_{\infty}^{r} f(r)\,dr.$$

The kinetic energy is $\frac{1}{2}mv^2$, or
$$\tfrac{1}{2}m\left[\left(\frac{dr}{dt}\right)^2 + r^2\left(\frac{d\theta}{dt}\right)^2\right].$$

Expressing that the total energy is constant, we have
$$\left(\frac{dr}{dt}\right)^2 + r^2\left(\frac{d\theta}{dt}\right)^2 + 2\int_{\infty}^{r} f(r)\,dr = E, \qquad (110)$$
where E is a constant.

Equations (109) and (110) lead to the differential equation of the orbit. We have, since r and θ are both functions of t,
$$\frac{dr}{dt} = \frac{dr}{d\theta}\frac{d\theta}{dt},$$
so that equation (110) may be expressed in the form
$$\left[\left(\frac{dr}{d\theta}\right)^2 + r^2\right]\left(\frac{d\theta}{dt}\right)^2 + 2\int_{\infty}^{r} f(r)\,dr = E,$$
and on eliminating $\dfrac{d\theta}{dt}$ from this and equation (109), we have
$$\left[\left(\frac{dr}{d\theta}\right)^2 + r^2\right]\frac{h^2}{r^4} + 2\int_{\infty}^{r} f(r)\,dr = E, \qquad (111)$$
the differential equation of the orbit.

LAW OF INVERSE SQUARE

222. Let us now suppose that the attraction follows the law of the inverse square of the distance, so that
$$f(r) = \frac{\mu}{r^2},$$
where μ is a constant. Then
$$\int_{\infty}^{r} f(r)\,dr = -\frac{\mu}{r}, \qquad (112)$$

LAW OF INVERSE SQUARE 277

and equation (111) becomes

$$\left[\left(\frac{dr}{d\theta}\right)^2 + r^2\right]\frac{h^2}{r^4} - \frac{2\mu}{r} = E,$$

whence we obtain
$$d\theta = \frac{h\,dr}{r\sqrt{Er^2 + 2\mu r - h^2}},$$

giving, on integration,
$$\theta = \sin^{-1}\frac{\dfrac{h}{r} - \dfrac{\mu}{h}}{\sqrt{E + \dfrac{\mu^2}{h^2}}} + \epsilon,$$

where ϵ is a constant of integration.

Simplifying, this becomes

$$\frac{h}{r} - \frac{\mu}{h} = \sqrt{E + \frac{\mu^2}{h^2}}\sin(\theta - \epsilon), \qquad (113)$$

and if we compare with the equation

$$\frac{l}{r} - 1 = e\cos\theta,$$

we see that equation (113) represents a conic, having the origin as focus, and being of semi-latus rectum $l = \dfrac{h^2}{\mu}$ and eccentricity $e = \sqrt{1 + \dfrac{Eh^2}{\mu^2}}$. In order that the line $\theta = 0$ may coincide with the major axis of the conic, the value of ϵ must be $-\dfrac{\pi}{2}$.

223. We notice that if

E is positive, then $e > 1$, and the orbit is a hyperbola;
E is zero, then $e = 1$, and the orbit is a parabola;
E is negative, then $e < 1$, and the orbit is an ellipse.

Thus the class of conic described depends solely on the value of E, and not on that of h. And it should be noticed that, if we are given the point of projection of a particle, and also its velocity of projection, the value of E is determined, for by equation (110)

$$E = v^2 - \frac{2\mu}{r}.$$

Thus the class of conic described depends only on the velocity of projection, and not on the direction: the conic is a hyperbola, parabola, or ellipse, according as

$$v^2 > = \text{ or } < \frac{2\mu}{r}.$$

The actual eccentricity depends on both E and h, for if e is the eccentricity, we have

$$e^2 = 1 + \frac{Eh^2}{\mu^2}.$$

224. In order that the particle may describe a circle we must have $e^2 = 0$, and therefore

$$1 + \frac{Eh^2}{\mu^2} = 0.$$

Putting $E = v^2 - \dfrac{2\mu}{r}$ and $h = pv$ (so that p is the perpendicular from the center of force on to the direction of projection), this reduces to

$$\frac{\mu^2}{p^2} - \frac{2\mu}{r} v^2 + v^4 = 0,$$

or,
$$\left(v^2 - \frac{\mu}{r}\right)^2 + \mu^2 \left(\frac{1}{p^2} - \frac{1}{r^2}\right) = 0.$$

Since p is necessarily less than r, neither term in this equation can be negative. Thus, in order that the equation may be satisfied, both terms must vanish, and we must have

$$p = r \text{ and } v^2 = \frac{\mu}{r}.$$

The first equation expresses that the projection must be at right angles to the line joining the particle to the center of force. The second equation, which can be written

$$\frac{\mu}{r^2} = \frac{v^2}{r},$$

shows that the attractive force must just produce the acceleration appropriate to motion in a circle of radius r.

LAW OF INVERSE SQUARE

225. For an elliptic orbit, the periodic time is that required to sweep out an area πab, where a, b are the semi-axes of the ellipse. Since the area is swept out at a rate $\tfrac{1}{2} h$ per unit time, the periodic time T will be

$$T = \frac{\pi ab}{\tfrac{1}{2} h}.$$

The semi latus-rectum l is equal to $\dfrac{b^2}{a}$, and has also been seen to be equal to $\dfrac{h^2}{\mu}$, so that

$$b = \sqrt{al} = h\sqrt{\frac{a}{\mu}},$$

whence
$$T = \frac{2\pi ab}{h} = \frac{2\pi a^{\frac{3}{2}}}{\sqrt{\mu}}. \tag{114}$$

Since this does not depend on the eccentricity, it is clear that the periodic time of any orbit is the same as that in a circle of radius equal to the semi major-axis.

226. The law of force of the inverse square is that of gravitation: the law which we have been investigating is therefore that which governs the motions of the planets in their orbits round the sun, as well as the motions of comets and meteorites. For reasons which cannot be explained here, the conics described by the planets are all of them ellipses of small eccentricity. A wider range is found in the orbits of comets. These bodies generally come from far outside the solar system. To a close approximation many of them may be treated as coming from infinity, and as starting with relatively small velocity. In this case the orbit is approximately parabolic.

Kepler's Laws

227. Long before the theory of the planetary orbits had been worked out mathematically by Newton, three of the principal laws governing the motion of the planets had been discovered empirically by Kepler. Kepler's three laws are as follows:

LAW I. *Every planet describes an ellipse having the sun in one of its foci.*

280   MOTION UNDER A VARIABLE FORCE

Law II. *The areas described by the radii drawn from the planet to the sun are, in the same orbit, proportional to the times of describing them.*

Law III. *The squares of the periodic times of the various orbits are proportional to the cubes of their major axes.*

From the first of these laws Newton proved that the law of force between the planets and the sun must be the law of the inverse square. The third law is seen to express the same fact as equation (114).

Motion of Two Particles about One Another

228. A pair of objects known as a double star is of common occurrence in the sky. This consists of two stars describing orbits about one another, neither star being fixed.

By the theorems proved in Chapter IX, the center of gravity of the two stars must either remain at rest, or else must move with uniform velocity in a straight line, in which case it may, as we have seen, be treated as fixed, provided all motion is measured relative to a frame of reference moving with it.

Let A, B be the positions of the two stars at any instant, and let G be their center of gravity. Let the masses of the stars be m, m', and let a, b denote their distances from G. Then

$$\frac{m}{b} = \frac{m'}{a} = \frac{m+m'}{a+b}. \qquad (115)$$

The complete law of gravitation is expressed by the law

$$F = \gamma \frac{mm'}{r^2},$$

where m, m' are the two masses, r the distance between them, γ a constant whose value can be found by experiment, and F is the force of attraction between the two masses. Thus the force acting on the star B is

$$F = \gamma \frac{mm'}{(a+b)^2},$$

acting along BA. This force can always be regarded as acting from the fixed point G, for its line of action is always BG. Moreover, its magnitude per unit mass of star B is

$$\frac{\gamma m}{(a+b)^2},$$

or, from relations (115), $\dfrac{\gamma m^3}{(m+m')^2 b^2}.$

This is a force $\dfrac{\mu}{r^2}$ acting towards G if we take

$$\mu = \frac{\gamma m^3}{(m+m')^2}.$$

Thus the two stars each describe a conic about the center of gravity of the two. It is possible astronomically to observe the values of the periodic time T and the major axes of the orbits of these conics. From these quantities we can determine the values of μ, so that we know the values of

$$\frac{m^3}{(m+m')^2}, \qquad \frac{m'^3}{(m+m')^2},$$

and these at once lead to the values of m, m'. In this way it has been found possible to determine the masses of some of the stars.

EXAMPLES

(Take the gravitation constant to be $\gamma = 66.6 \times 10^{-8}$ in centimeter-gramme-second units.)

1. Given that the earth attracts as though its mass were concentrated at its center, and that the value of g at the equator, distant 6.378×10^8 centimeters from the earth's center, is 978.1 centimeters per second per second, find the mass of the earth.

2. Taking the masses of the earth and moon as 6.14×10^{27} and 7.94×10^{25} grammes respectively, and assuming their distance apart to be always 3.84×10^{10} centimeters, find the periodic time of the moon.

3. Taking the sun's mass to be 2×10^{33} grammes, and the year to be 365.24 days, find the semi major-axis of the earth's orbit, regarding the sun as a fixed center of force.

4. If the sun's mass is $324{,}000$ times that of the earth, by how much must the result of question 3 be altered when the sun's motion is taken into account?

5. Taking the mass of Jupiter to be $\frac{1}{1080}$ of the mass of the sun, and its greatest distance from the sun to be $498\frac{1}{2}$ million miles, show that, on account of Jupiter's attraction, the sun will describe an ellipse of semi major-axis equal to about 461,000 miles, and find the length of Jupiter's year.

6. The maximum velocity attained by the earth in its orbit is 3,000,000 centimeters per second, and the minimum velocity is 2,920,000 centimeters per second. Find the eccentricity of the earth's orbit.

GENERAL EXAMPLES

1. A particle, attached by a string to a point, has just sufficient energy to make complete revolutions in a vertical circle. Show that the tension of the string is zero and six times the weight of the particle respectively, when the particle is at the highest and lowest points of its path.

2. A particle moves under gravity in a vertical circle, sliding down the convex side of a smooth circular arc. If its velocity is that due to a height h above the center, show that it will fly off the circle when at a height $\frac{2}{3} h$ above the center.

3. If the angle α through which a simple pendulum swings on each side of the vertical is small, but not infinitesimal, show that to a first approximation the time of oscillation is

$$2\pi \sqrt{\frac{l}{g}\left(1 + \frac{1}{16}\alpha^2\right)}.$$

Deduce that a pendulum, which beats seconds accurately when performing infinitesimal oscillations, would lose about 40 seconds a day when attached to a clock which caused it to oscillate to 5 degrees on each side of the vertical.

4. A train is moving uniformly round a curve at 60 miles an hour, and in one of the carriages a seconds pendulum is found to beat 121 times in two minutes. Show that the radius of the curve is about a quarter of a mile.

5. One end of an elastic string, natural length a, modulus λ, is tied to a fixed point on a smooth horizontal table, and the other end is tied to a particle of mass m which rests on the table. If the mass is pulled to a distance $2a$ from the other end of the string, and is then let go, show that it will return to its original position at regular intervals $2(\pi + 2)\sqrt{\frac{am}{\lambda}}$.

6. Two balls weighing W_1 and W_2 pounds are connected by a thread a feet long; and W_1 is held in the hand while W_2 is whirled round. Determine the motion which ensues if W_1 is released from rest when W_2

EXAMPLES

is moving with velocity V at inclination α; and prove that in the air the tension of the thread is

$$\frac{W_1 W_2}{W_1 + W_2} \frac{V^2}{ga} \text{ pounds.}$$

7. Two masses, m_1 and m_2, are connected by a weightless spring of such strength that when m_1 is held fixed, m_2 performs n vibrations a second. Show that if m_2 be held fixed, m_1 will perform $n\sqrt{m_2/m_1}$ vibrations a second, while if both masses are free, they will perform $n\sqrt{\dfrac{m_1 + m_2}{m_1}}$ vibrations per second, the vibrations in every case being in the line of the spring.

8. A particle of mass m, moving in a smooth curved tube of any shape, is in equilibrium under the tensions of two elastic strings in the tube, of natural lengths l, l' and moduli of elasticity λ, λ', of which the other ends are attached to fixed points of the tube. If the particle makes oscillations, large or small, in the tube, show that the time of oscillation is

$$2\pi \sqrt{\frac{m}{\dfrac{\lambda}{l} + \dfrac{\lambda'}{l'}}}.$$

9. A string passes through a small hole in a smooth horizontal table, and has equal particles attached to its ends, one hanging vertically and the other lying on the table at a distance a from the hole. The latter is projected with a velocity $\sqrt{ga}$ perpendicular to the string. Show that the hanging particle will remain at rest, and that if it be slightly disturbed, the time of a small oscillation will be $2\pi\sqrt{\dfrac{2a}{3g}}$.

10. A particle moves in a circular groove, under an attraction $\dfrac{\mu}{r^2}$ to a point P which is in the plane of the circle and distant b from its center. The particle is projected with velocity V from the point of the circle nearest to P. Show that for the particle to perform complete revolutions, the value of V^2 must not be less than $\dfrac{4\mu b}{a^2 - b^2}$.

11. A smooth ellipse, semi-axes a and b, is placed with its major axis vertical, and a particle is projected along the concave side of the arc, with velocity due to a height h above the center. Find the point at which the particle will leave the arc, and show that it will pass through the center of the ellipse if

$$h = \frac{8a^2 + b^2}{6a\sqrt{3}}.$$

12. A particle is constrained to move in a circle of radius a, under an attraction μr per unit mass to a point inside the circle distant c from its center. If the particle be placed at its greatest distance from this point, and started with an infinitesimal velocity, prove that it will pass over the second quadrant of the circle in a time

$$\sqrt{\frac{a}{\mu c}} \log(\sqrt{2}+1).$$

13. A particle describes an ellipse about a center of force in one focus. Show that the velocity at the end of the minor axis is a mean proportional between the velocities at the ends of any diameter.

14. A comet describes a parabola. Show that its velocity perpendicular to the axis of its orbit varies inversely as the radius vector from the sun.

15. A comet of mass m, describing a parabola about the sun, collides with an equal mass m at rest, and the masses move on together. Show that their center of gravity will describe a circle about the sun as center.

16. Assuming that a projectile, after allowing for variations in gravity, describes an ellipse about the earth's center as focus, show that the maximum range on a horizontal plane through the point of projection, for a given velocity v, is

$$\frac{4\,gv^2R^2}{4\,g^2R^2 - v^4},$$

where R is the distance from the earth's center to the point of projection.

17. When the earth is at the end of the major axis of its orbit, a small meteor, of mass one mth of that of the sun, suddenly falls into the sun. Show that the length of the year will be diminished by $\dfrac{2}{m}$ of itself.

18. A planet P moving about the sun S picks up a small meteor, and consequently has its velocity reduced by one nth of its former amount, although unaltered in direction. Treating n as small, show that the eccentricity of the planet's orbit will be reduced by $2\,n\,(e + \cos\theta)$, where θ is the angle between SP and the major axis of the orbit.

Show also that the new major axis will make an angle $\dfrac{2\,n\sin\theta}{e}$ with the old axes.

19. A particle describes an ellipse about the focus. Show that the greatest and least angular velocities occur at the ends of the major axis, and also that if α, β be these angular velocities, the mean angular velocity is

$$\frac{2\,(\alpha\beta)^{\frac{3}{4}}}{\sqrt{\alpha}+\sqrt{\beta}}.$$

EXAMPLES

20. A comet describes a parabola about the sun, its nearest distance from the sun being one third of the radius of the earth's orbit, supposed circular. For how many days will the comet remain within the earth's orbit?

21. If the attraction on a particle varies as the inverse square of the distance from a center of force O, show that there are two directions in which a particle can be projected from a given point P so that its orbit may have a given major axis. If $OP = c$, and if α_1, α_2 are the angles which the two directions of projection make with OP, show that

$$\cot \alpha_1 \cot \alpha_2 = \frac{c}{a} - 1,$$

where a is the semi major-axis.

22. A particle is projected from a point P under a force to a fixed point S at a distance R from P, so as to describe a circle passing through S. The initial velocity is V, and the moment of the velocity about S is h. Show that the particle will describe a semicircle in time

$$\frac{R^2}{4\,h^3}(\pi V^2 R^2 \pm 4\,h\,\sqrt{V^2 R^2 - h^2}).$$

23. A block of mass M, whose upper and lower faces are smooth horizontal planes, is free to move along a groove in a parallel plane, and a particle of mass m is attached to a fixed point in the upper face by an elastic string whose natural length is a and modulus λ. If the system starts from rest with the particle on the upper face, and the string stretched parallel to the groove to $1 + n$ times its natural length, prove that the block will perform oscillations of amplitude

$$\frac{(n+1)\,am}{M+m}$$

and period

$$2\left(\pi + \frac{2}{n}\right)\sqrt{\frac{aMm}{\lambda(M+m)}}.$$

CHAPTER XI

MOTION OF RIGID BODIES

229. The present chapter is devoted to a discussion of the motion of rigid bodies, when the motion is such that the bodies may not be treated as particles.

It has already been proved in § 66 that the most general motion possible for a rigid body is one compounded of a motion of translation and a motion of rotation. As a preliminary to discussing the general motion of a rigid body under the action of forces of any description, we shall examine in greater detail than has so far been done the properties of a motion of rotation.

Angular Velocity

230. We have seen (§ 67) that for every motion of a rigid body in which a point P remains fixed, there is an *axis of rotation*, which is a line passing through P, of which every point remains fixed. If a rigid body is moving continuously we may analyze its motion in the following way. We select a definite particle P of the rigid body, and we refer the motion to a frame of reference having P as origin, and moving so as always to remain parallel to its original position. Relative to this frame, the motion of the body between any two instants is a motion of rotation about P.

Now let the two instants be taken very close to one another, the interval between them being dt. Let us find the axis of rotation of the motion which takes place in the interval dt, and call it PQ. Then PQ is called the axis of rotation at the instant at which the interval dt is taken.

Let us suppose that during the interval dt the rotation of the body about its axis of rotation PQ is found to be a rotation

ANGULAR VELOCITY 287

through an angle $d\theta$. Then the limit, when dt is made to vanish, of the rate $\dfrac{d\theta}{dt}$ is called the angular velocity of the body,—it measures the angle turned through per unit time.

Thus to have a full knowledge of the motion of a rigid body at any instant we must know

(a) the direction and magnitude of the velocity of the point P which has been selected to give a frame of reference;

(b) the direction of the axis of rotation through P;

(c) the magnitude of the angular velocity about the axis of rotation.

231. The angular velocity has associated with it a direction — the axis of rotation — and a magnitude. Thus it may be represented by a line. We shall now prove that it is a vector, i.e. that angular velocities may be compounded according to the parallelogram law.

Let a rigid body have a rotation about P compounded of (a) a rotation of angular velocity ω about an axis PQ, and (b) a rotation of angular velocity ω' about a second axis PQ'. Let the lengths

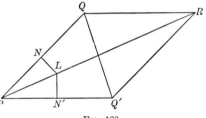

Fig. 138

PQ, PQ' be taken proportional to ω, ω', so that the lines PQ, PQ' will represent the directions and magnitudes of the angular velocities on the same scale.

Let the parallelogram $PQRQ'$ be completed, and let L be any point on the diagonal PR. Let LN, LN' be drawn perpendicular to PQ, PQ' respectively.

In time dt there is, from the first angular velocity, a rotation of the rigid body through an angle $\omega\, dt$ about PQ. The effect of this rotation is to move the particle of the body which originally coincided with L through a distance $LN \cdot \omega\, dt$ at right angles to the plane PLN. Similarly the effect of the rotation about PQ' is to move the same particle through a distance $LN' \cdot \omega' dt$ at right angles to the plane but in the direction opposite to that of the former motion. Thus the total displacement of the particle is

$$LN\omega\,dt - LN'\omega'\,dt. \tag{116}$$

Since L is on the diagonal of the parallelogram, we see that the area of the triangle PLQ is equal to that of the triangle PLQ', so that
$$LN \cdot PQ = LN' \cdot PQ'.$$

Again, since PQ, PQ' are in the ratio of $\omega : \omega'$, this equation may be written in the form
$$LN\omega = LN'\omega',$$

and on comparing with expression (116) we see that the displacement of the particle L vanishes.

Thus the resultant of the two angular velocities is a motion such that the points P and L both remain at rest. It is therefore an angular velocity having PR, the diagonal of the parallelogram, as axis of rotation.

We must next find the magnitude of this angular velocity. Let us denote it by Ω. From Q draw perpendiculars QX, QY to PQ' and PR. The displacement of the particle Q in time dt will be $QY \cdot \Omega\,dt$ at right angles to the plane. This displacement, however, can also be obtained by compounding the displacements produced by the two angular velocities ω, ω'. That produced by the former is nil, since Q is on the axis of rotation; that produced by the latter is $QX\omega'\,dt$. Thus

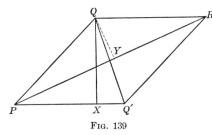

Fig. 139

$$QY \cdot \Omega\,dt = QX \cdot \omega'\,dt. \tag{117}$$

We have $\qquad QY \cdot PR = QX \cdot PQ',$

each being equal to the area of the parallelogram, and on combining this relation with (117), we find

$$\frac{\Omega}{PR} = \frac{\omega'}{PQ'}.$$

ANGULAR VELOCITY 289

Thus if ω' is represented by PQ', then Ω will, on the same scale, be represented by PR.

We have now proved the following:

The resultant of two angular velocities represented by the edges PQ, PQ' of a parallelogram is an angular velocity represented by the diagonal PR of the parallelogram.

Thus angular velocity is a vector, and possesses the properties which have been proved to be true of all vectors.

232. It follows that an angular velocity Ω about an axis of rotation of which the direction cosines are l, m, n may be replaced by three angular velocities $\omega_1, \omega_2, \omega_3$ about the axes of coördinates, such that
$$\omega_1 = l\Omega, \qquad \omega_2 = m\Omega, \qquad \omega_3 = n\Omega. \tag{118}$$

Squaring and adding, we find that
$$\Omega^2 = \omega_1^2 + \omega_2^2 + \omega_3^2. \tag{119}$$

We now see that the motion of a rigid body is given when we know

(a) u, v, w, the components of velocity of the point P;

(b) $\omega_1, \omega_2, \omega_3$, the components of angular velocity.

KINETIC ENERGY OF ROTATION

233. Suppose that at any instant a rigid body is rotating about an axis of rotation PQ with angular velocity Ω.

Let L be any particle of the body, its mass being m, and let LN, the perpendicular distance from L to PQ, be denoted by p. Then the velocity of the particle L is $p\Omega$, and its kinetic energy is $\tfrac{1}{2} m p^2 \Omega^2$.

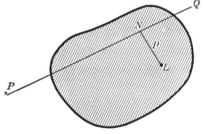

FIG. 140

On summation, the kinetic energy of the whole body is seen to be
$$\tfrac{1}{2} \left(\sum m p^2 \right) \Omega^2.$$

The quantity $\sum mp^2$ is called the *moment of inertia* about the axis PQ.

If we introduce a quantity k, defined by

$$k^2 = \frac{\sum mp^2}{\sum m},$$

so that k^2 is the mean value of p^2 averaged over all the particles of the body, then k is called the *radius of gyration* about the axis PQ.

The kinetic energy can now be written in the form

$$\tfrac{1}{2}\left(\sum mp^2\right)\Omega^2 = \tfrac{1}{2}\sum(m)k^2\Omega^2,$$

so that the energy is the same as if the whole mass were concentrated in a single particle at a distance k from the axis of rotation.

Kinetic Energy of a Rigid Body

234. The point P is at our disposal: let us suppose it to be the center of gravity of the body. Then the most general motion may be compounded of a motion of translation, this being identical with that of the center of gravity, and a motion of rotation about an axis through the center of gravity.

Let V be the velocity of the center of gravity, let Ω be the angular velocity, and let k be the radius of gyration about the axis of rotation through the center of gravity. Let M be the total mass, $\sum m$, of the body.

By the theorem of § 186, the total kinetic energy of the body is the sum of two parts:

(a) the kinetic energy of a single particle of mass M moving with the center of gravity of the body;

(b) the kinetic energy of motion relative to the center of gravity.

The value of part (a) is $\tfrac{1}{2}MV^2$; that of part (b) is $\tfrac{1}{2}Mk^2\Omega^2$. Hence we have for the total kinetic energy

$$\tfrac{1}{2}M(V^2 + k^2\Omega^2). \tag{120}$$

This expression is of extreme importance in itself, but is also of interest because it enables us to prove the following theorem.

235. Theorem. *Let k be the radius of gyration about any axis through the center of gravity, and let k' be the radius of gyration about a parallel axis distant a from the former, then*

$$k'^2 = k^2 + a^2.$$

Let PQ be any axis through the center of gravity G, and let $P'Q'$ be any parallel axis distant a from the former. Suppose that the rigid body has a motion of rotation about $P'Q'$, the angular velocity being Ω.

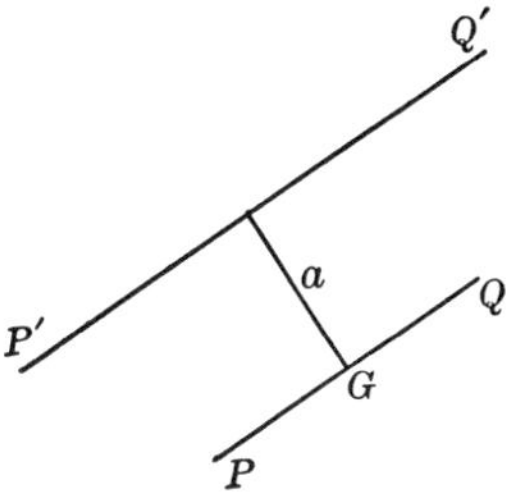

Then the velocity of G is $a\Omega$, and the motion may be regarded as compounded of a motion of translation of velocity $a\Omega$ together with a rotation Ω about the axis PQ. By formula (120), the kinetic energy is

$$\tfrac{1}{2} M(a^2\Omega^2 + k^2\Omega^2).$$

Fig. 141.

It is also $\tfrac{1}{2} Mk'^2\Omega^2$, where k' is the radius of gyration about $P'Q'$. Hence we have

$$\tfrac{1}{2} Mk'^2\Omega^2 = \tfrac{1}{2} M(a^2\Omega^2 + k^2\Omega^2),$$

and the result follows on dividing through by $\tfrac{1}{2} M\Omega^2$.

236. Alternative proof. This theorem may also be proved geometrically.

Let L be any particle of the body, and let the plane of fig. 142 be supposed to be the plane through L at right angles to the two axes of rotation, these axes cutting the plane in the points A, A' respectively. Let $LA = p$, $LA' = p'$, and let LN be drawn perpendicular to AA'. Then $Mk^2 = \sum mp^2$, and also

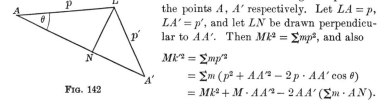

Fig. 142.

$$Mk'^2 = \sum mp'^2$$
$$= \sum m \,(p^2 + AA'^2 - 2p \cdot AA' \cos \theta)$$
$$= Mk^2 + M \cdot AA'^2 - 2AA' (\textstyle\sum m \cdot AN).$$

Now AN is the projection of the line from L to the center of gravity, upon the line AA'. Hence $\sum m \cdot AN = 0$, and we have

$$Mk'^2 = Mk^2 + M \cdot AA'^2,$$

giving the result to be proved, after division by M.

237. From the theorem just proved, it follows that the radius of gyration about any axis can be found as soon as we know that about a parallel axis through the center of gravity, and *vice versa*. We now give some examples of the calculation of radii of gyration.

Calculation of Radii of Gyration

238. Uniform thin rod. Let the rod AB be of length $2a$, and let k be its radius of gyration about an axis through A perpendicular to its length. Let τ be its mass per unit length, and let x be a coördinate which measures distances from A. The element which extends from x to $x + dx$ is of mass $\tau\, dx$, and its perpendicular distance from the axis of rotation is x. Hence

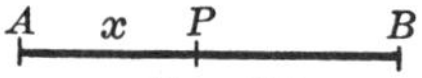

Fig. 143

$$k^2 = \frac{\sum mp^2}{\sum m} = \frac{\int_0^{2a}(\tau\, dx)\, x^2}{\int_0^{2a} \tau\, dx} = \frac{\tfrac{8}{3}\tau a^3}{2\,\tau a} = \tfrac{4}{3} a^2,$$

so that the radius of gyration is $\dfrac{2a}{\sqrt{3}}$.

About the center of gravity, which is distant a from A, the radius of gyration is given by

$$k^2 = \tfrac{4}{3} a^2 - a^2 = \frac{a^2}{3},$$

so that the radius of gyration about the center of gravity is $\dfrac{a}{\sqrt{3}}$.

239. Rectangular lamina. Let us suppose the lamina to be of edges $2a$, $2b$, and let us find its radius of gyration about an axis through its center at right angles to its plane. Let us take axes as in fig. 144, and let σ denote the mass per unit area. Then

$$k^2 = \frac{\sum mp^2}{\sum m} = \frac{\iint (\sigma\, dxdy)(x^2 + y^2)}{4ab \cdot \sigma}.$$

CALCULATION OF RADII OF GYRATION

The integration is over the lamina, and therefore between the limits $x = a$ to $x = -a$ and $y = b$ to $y = -b$. On integrating, we find

$$k^2 = \frac{a^2 + b^2}{3}.$$

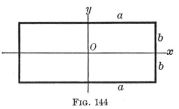

Fig. 144

On taking $b = 0$, the lamina becomes a thin rod, and the result agrees with that obtained in the last section.

240. Homogeneous solid ellipsoid. Let the semi-axes of the ellipsoid be a, b, c, and let us find the radius of gyration about the major axis. Taking the principal axes of the ellipsoid as axes of coördinates, and denoting the density of the ellipsoid by ρ, we have

$$k^2 = \frac{\sum mp^2}{\sum m} = \frac{\iiint (\rho \, dxdydz)(y^2 + z^2)}{\iiint \rho \, dxdydz},$$

where the integration is over the whole volume of the ellipsoid. On performing the integrations, we obtain

$$k^2 = \frac{b^2 + c^2}{5}.$$

EXAMPLES

1. Find the radius of gyration of a rod 12 inches long about a point distant 4 inches from one end.

2. Find the radius of gyration of a circular disk,

(*a*) about an axis through its center perpendicular to its plane;

(*b*) about a diameter.

3. Show that the radius of gyration of a sphere, radius a, about any diameter is $\frac{2}{5} a^2$, and about any tangent line is $\frac{7}{5} a^2$.

4. What is the radius of gyration of a cube about an edge?

5. What is the radius of gyration of a square lamina about a diagonal?

6. Find the radius of gyration of a solid circular cylinder,

(*a*) about an axis;

(*b*) about a generator;

(*c*) about a diameter of one of its ends.

7. Prove that the radius of gyration of a solid conical spindle about its axis is $\sqrt{\frac{3}{10}} \, a$, where a is the radius of its base.

ROUTH'S RULE

241. The following convenient rule, given by Dr. Routh, (*Rigid Dynamics*, § 8), provides an easy way of remembering the values of several radii of gyration. The rule applies to linear, plane, and solid bodies which are

(*a*) rectangular (rod, lamina, or parallelepiped);
(*b*) elliptical or circular (disk or lamina);
(*c*) ellipsoidal, spheroidal, or spherical (solid);

and states that the radius of gyration about an axis of symmetry through the center of gravity is given by

$$k^2 = \frac{\text{sum of squares of perpendicular semi-axes}}{3, 4, \text{ or } 5},$$

where the denominator is 3, 4, or 5 according as the body comes under headings (*a*), (*b*), or (*c*) of the above classification.

ILLUSTRATIVE EXAMPLE

A coin rolls down an inclined plane. Find its velocity after any distance and also its acceleration.

Let the coin be treated as a uniform circular disk, and let a be its radius. When its velocity down the plane is V, its angular velocity will be V/a. The axis of rotation is perpendicular to the plane of the coin. Its semi-axes of symmetry, regarding it as a lamina, will be a, a. The radius of gyration about the axis of rotation through its center is, by Routh's rule,

$$k^2 = \frac{a^2 + a^2}{4} = \tfrac{1}{2} a^2,$$

so that the kinetic energy is

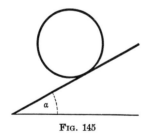

FIG. 145

$$\tfrac{1}{2} M \left[V^2 + \tfrac{1}{2} a^2 \left(\frac{V}{a} \right)^2 \right] = \tfrac{3}{4} M V^2.$$

After rolling a distance s down the plane, the center of gravity of the coin has fallen a distance $s \sin \alpha$, so that from the conservation of energy

$$Mgs \sin \alpha = \tfrac{3}{4} M V^2,$$

and therefore the velocity is given by

$$V^2 = \tfrac{4}{3} sg \sin \alpha.$$

Comparing with the formula (48), $V^2 = 2fs$, for motion under uniform acceleration, we see that the disk rolls down the plane with a uniform acceleration $\tfrac{2}{3} g \sin \alpha$.

MOMENT OF MOMENTUM

EXAMPLES

1. Show that the acceleration of a hoop rolling down a hill of inclination α is $\frac{1}{2} g \sin \alpha$.

2. Find the acceleration of a pair of locomotive wheels running down a gradient of 1 in 50, each wheel consisting of a rim and of spokes of uniform thickness, the weight of the rim being twice that of the spokes and the weight of the axle being half of that of a wheel. (Neglect the thickness of the axle.)

3. Two bicyclists, riding exactly similar machines, coast down a hill, starting with equal velocities at the top. Neglecting the forces of friction and the resistance of the air, show that the heavier rider will reach the bottom first.

4. The pulley of an Atwood's machine is a uniform disk of mass M. When masses m_1, m_2 are attached to the ends of the string, show that the acceleration of m_1 is
$$\frac{m_1 - m_2}{m_1 + m_2 + \frac{1}{2} M} g.$$

5. Two spheres, one a hollow shell and the other a homogeneous solid, roll down hill together, starting simultaneously from rest at the top. Show that their times over any part of the path are in the ratio $5 : \sqrt{21}$.

6. If the masses of the wheels of a carriage are supposed to be all collected at the rim, show that the energy of the carriage when moving with velocity V is $\frac{1}{2} M V^2$, where M is the weight of the complete carriage plus the weights of the wheels.

7. A straight piece of uniform wire is stood vertically on end and allowed to fall over. With what velocity does it strike the ground?

8. A homogeneous solid cigar-shaped spheroid, semi-axes a and b, is stood on its point on a horizontal plane and is allowed to roll over. Find its angular velocity when the end of its minor axis is in contact with the plane, and find the pressure on the plane at this instant.

Moment of Momentum

242. Let x, y, z be the coördinates of any particle of mass m. Let the components of the total resultant force acting on the particle be X, Y, Z. Then the equations of motion are

$$m \frac{d^2 x}{dt^2} = X,$$

$$m \frac{d^2 y}{dt^2} = Y,$$

$$m \frac{d^2 z}{dt^2} = Z.$$

The moment about the axis of x of the force acting on the particle is $yZ - zY$, and from the foregoing equations we have

$$yZ - zY = m\left(y\frac{d^2z}{dt^2} - z\frac{d^2y}{dt^2}\right). \tag{121}$$

The velocity of the particle has components $\dfrac{dx}{dt}, \dfrac{dy}{dt}, \dfrac{dz}{dt}$, so that the moment of this velocity about the axis of z, as defined in § 219, is

$$y\frac{dz}{dt} - z\frac{dy}{dt}.$$

The momentum of the particle is m times its velocity, so that the moment of momentum about the axis of z is m times the moment of the velocity, and therefore

$$m\left(y\frac{dz}{dt} - z\frac{dy}{dt}\right).$$

On differentiating, we have

$$\frac{d}{dt}\left[m\left(y\frac{dz}{dt} - z\frac{dy}{dt}\right)\right]$$
$$= m\left[\left(\frac{dy}{dt}\frac{dz}{dt} + y\frac{d^2z}{dt^2}\right) - \left(\frac{dz}{dt}\frac{dy}{dt} + z\frac{d^2y}{dt^2}\right)\right]$$
$$= m\left(y\frac{d^2z}{dt^2} - z\frac{d^2y}{dt^2}\right)$$
$$= yZ - zY, \tag{122}$$

by equation (121). Thus we have proved that

The rate of change of the moment of momentum of a particle about any axis is equal to the moment, about the same axis, of the forces acting on the particle.

243. Equation (122) is true for each particle of any system of bodies. Let us sum the equation over all particles, then we obtain

$$\frac{d}{dt}\left[\sum m\left(y\frac{dz}{dt} - z\frac{dy}{dt}\right)\right] = \sum(yZ - zY). \tag{123}$$

The right-hand side of this equation is the sum of the moments of the external forces acting on the body or system of bodies, for the internal forces occur in equal and opposite pairs which contribute nothing.

The term $\sum m \left(y \dfrac{dz}{dt} - z \dfrac{dy}{dt} \right)$, which is the sum of the moments of momentum of the separate particles, is called the *moment of momentum of the system*.

Thus equation (123) expresses that

The rate of change of the moment of momentum of any system about any axis is equal to the sum of the moments of the external forces about this axis.

244. Several important consequences of this theorem follow at once.

I. *If a system of bodies is acted on by no external forces, the moment of momentum about every axis remains constant.*

This expresses the principle known as the *conservation of angular momentum*.

The sun affords an instance of a body which may practically be supposed to be acted on by no external forces. It is generally supposed that the sun is gradually shrinking in size; if this is so, we see that its velocity of rotation about its axis must continually increase, in order that its moment of momentum may remain constant.

II. *If all the forces acting on a system are either parallel to a given line, or else intersect this line, then the moment of momentum of the system about this line must remain constant.*

A peg top is acted on only by the reaction at the peg and gravity. The moment of the latter about a vertical line through the peg vanishes, and the moment of the former may be supposed to vanish to a close approximation. Hence the moment of momentum about a vertical through the peg will remain constant, to a close approximation.

III. *If a rigid body is free to rotate about a fixed axis, and if ω is its angular velocity at any instant, then*

$$Mk^2 \frac{d\omega}{dt} = L,$$

where Mk^2 is the moment of inertia about the fixed axis, and L is the sum of the moments about this axis, of all the external forces.

298 MOTION OF RIGID BODIES

To see the truth of this, it is only necessary to notice that a particle of mass m at distance p from the axis has momentum $mp\omega$, so that the moment of momentum of the whole system will be

$$\sum mp^2\omega = Mk^2\omega,$$

and since M and k^2 do not vary with the time, the rate of change of angular momentum will be $Mk^2\dfrac{d\omega}{dt}$.

Oscillation of a Pendulum

245. An important application of the last theorem enables us to find the time of oscillation of a pendulum of any description.

Let O be the pivot about which the pendulum turns, let G be its center of gravity, let $OG = h$, and let the line OG make an angle θ with the vertical at any instant, so that $\omega = \dfrac{d\theta}{dt}$ is the angular velocity of the pendulum about its axis.

Let M be the mass, and k the radius of gyration about its axis, of the whole pendulum. Then the equation of motion is

$$Mk^2\frac{d\omega}{dt} = L,$$

Fig. 146 in which $\omega = \dfrac{d\theta}{dt}$. The value of L is equal to the moment of the weight about the axis through O, and is therefore $Mgh \sin \theta$.

Thus the equation becomes

$$Mk^2\frac{d\omega}{dt} = Mgh \sin \theta,$$

or,

$$\frac{k^2}{h}\frac{d^2\theta}{dt^2} = g \sin \theta.$$

The equation of motion for a simple pendulum of length l is

$$l \frac{d^2\theta}{dt^2} = g \sin \theta,$$

so that we see on comparison that the motion is the same as that of a simple pendulum of length $l = k^2/h$.
For instance, the complete period of small oscillations is

$$2\pi \sqrt{\frac{l}{g}} = 2\pi \sqrt{\frac{k^2}{gh}}.$$

ILLUSTRATIVE EXAMPLE

A ring (e.g. *a dinner napkin ring*) *stands vertically on a table, and a gradually increasing pressure is applied by a finger to one point of the ring in such a way that equilibrium is broken by the point of contact with the table slipping along the table. Find the subsequent motion of the ring.*

We have seen in example 2, p. 109, that it is possible to apply pressure in the manner described.

Let us suppose that when the ring leaves the finger it is observed to be moving with a velocity V forward and a rotation Ω in the direction opposite to that in which it would rotate if it were rolling without sliding. Let v, ω be the values of the velocity and rotation at any instant, measured in the same directions as V and Ω.

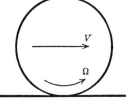

Fig. 147

Let a be the radius of the ring and m its mass. The forces acting on it are

(*a*) its weight mg;

(*b*) the vertical component of the reaction with the table, which is equal to mg since the center of gravity of the ring has no vertical acceleration;

(*c*) the frictional reaction at the lowest point of the ring, which is equal to $mg\mu$ so long as sliding takes place.

By the theorem of § 180 we have

$$m \frac{dv}{dt} = -mg\mu. \qquad (a)$$

We can obtain a second equation from the theorem of § 243. Let us take as axis the axis of the ring at instant t. The moment of inertia at this instant is ma^2. To obtain the moment of momentum we regard the whole motion as compounded of a motion of translation of the center of gravity (velocity v), and a motion of rotation about an axis through the center of gravity (velocity ω). The former contributes nothing to the moment of momentum, so that the whole moment of momentum is

$$ma^2\omega.$$

At the end of a small interval dt the ring will have moved forward a distance $v\,dt$, so that we are now considering the moment of inertia about an axis which is distant $v\,dt$ from the center of gravity of the ring. The moment of inertia after an interval dt is, accordingly, by § 235,

$$m(a^2 + (v\,dt)^2).$$

We may, however, neglect the small quantity of the second order $(dt)^2$ and treat the moment of inertia as though it remained constant and equal to ma^2. The rate of increase of the moment of momentum is, accordingly, $ma^2 \dfrac{d\omega}{dt}$.

The moment of the external forces, measured about the same axis and in the same direction, is

$$-mg\,\mu a,$$

so that we have the equation $\quad ma^2 \dfrac{d\omega}{dt} = -mg\,\mu a,\quad$ (b)

or, simplified, $\quad a \dfrac{d\omega}{dt} = -\mu g,\quad$ (c)

while equation (a) reduces to $\quad \dfrac{dv}{dt} = -\mu g.\quad$ (d)

These relations give the rates of decrease of v and ω so long as sliding is taking place. Sliding clearly ceases as soon as we have $v + \omega a = 0$, for $v + \omega a$ is the forward velocity of the lowest point of the ring. From equations (c) and (d) we have

$$\frac{d}{dt}(v + \omega a) = -2\,\mu g,$$

and initially the value of $v + \omega a$ is $V + \Omega a$. The time required to reduce $v + \omega a$ to zero is, accordingly,

$$\frac{V + \Omega a}{2\,\mu g}.$$

After this interval sliding ceases. The velocity of the ring at this instant is given by

$$v = V - \mu g \cdot \left(\frac{V + \Omega a}{2\,\mu g} \right)$$
$$= \tfrac{1}{2}(V - \Omega a),$$

so that the motion may be either forwards or backwards according as we had initially $V >$ or $< \Omega a$. After sliding has once ceased there is no force tending to start it afresh, so that the ring simply rolls on with uniform velocity v. If $V > \Omega a$, it rolls farther from its point of projection; while if $V < \Omega a$, it will return to the point of projection.

GENERAL THEORY OF MOMENTS OF INERTIA 301

EXAMPLES

1. The line of hinges of a door makes an angle α with the vertical, and the door swings about its position of equilibrium. Show that its motion is the same as that of a certain simple pendulum, and find the length of this pendulum.

2. A target consists of a square plate of metal of edge a and of mass M, hinged about its highest edge, which is horizontal. When at rest it is struck by an inelastic bullet of small mass m moving with velocity v, at a point at depth h below the line of hinges. Find the subsequent motion of the target.

3. A homogeneous sphere is projected without rotation up a rough inclined plane of inclination α and coefficient of friction μ. Show that the time during which the sphere ascends the plane is the same as if the plane were smooth, and that the time during which the sphere slides stands to the time during which it rolls in the ratio $7\mu : 2\tan\alpha$.

4. A sphere of radius a is held at rest at a point on the concave surface of a spherical bowl of radius b. It is suddenly set free and allowed to roll down the surface. Show that the line joining the centers of the two spheres swings in the same way as a simple pendulum of length $\tfrac{7}{5}(b-a)$.

5. A sphere of radius a is held at rest at the highest point of the rough convex surface of a sphere of radius b. It is then set free and allowed to roll down this sphere. Show that the spheres will separate when the line joining their centers makes an angle $\cos^{-1}\tfrac{10}{17}$ with the vertical. Examine the case of $b=0$.

6. A circular hoop, which is free to move on a smooth horizontal plane, has sliding on it a small ring of $1/n$th its mass, the coefficient of friction between the two being μ. Initially the hoop is at rest, and the ring has an angular velocity ω round the hoop. Show that the ring comes to rest relative to the hoop after a time $\dfrac{1+n}{\mu\omega}$.

GENERAL THEORY OF MOMENTS OF INERTIA

Coefficients of Inertia

246. Suppose that a rigid body is rotating about an axis of rotation of which the direction cosines, referred to any three fixed coördinate axes, are l, m, n. Let us take any point O on the axis of rotation for origin, and let L be any particle of mass m_1, distant p from the axis of rotation. Let the coördinates of L be x, y, z, and let $LN(=p)$ be the perpendicular from L on to the axis of rotation.

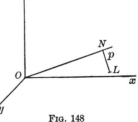

FIG. 148

We have $OL^2 = x^2 + y^2 + z^2$,
$$ON^2 = (lx + my + nz)^2,$$
so that
$$p^2 = OL^2 - ON^2$$
$$= x^2 + y^2 + z^2 - (lx + my + nz)^2$$
$$= x^2(m^2 + n^2) + y^2(n^2 + l^2) + z^2(l^2 + m^2)$$
$$\quad - 2\,mn \cdot yz - 2\,nl \cdot zx - 2\,lm \cdot xy$$
$$= l^2(y^2 + z^2) + m^2(z^2 + x^2) + n^2(x^2 + y^2)$$
$$\quad - 2\,mn \cdot yz - 2\,nl \cdot zx - 2\,lm \cdot xy.$$

Hence the moment of inertia, say I, is given by

$$I = \sum m_1 p^2$$
$$= l^2 \sum m_1(y^2 + z^2) + m^2 \sum m_1(z^2 + x^2) + n^2 \sum m_1(x^2 + y^2)$$
$$\quad - 2\,mn \sum m_1 yz - 2\,nl \sum m_1 zx - 2\,lm \sum m_1 xy$$
$$= l^2 A + m^2 B + n^2 C - 2\,mnD - 2\,nlE - 2\,lmF, \qquad (124)$$

where
$$A = \sum m_1(y^2 + z^2), \text{ etc.,}$$
$$D = \sum m_1 yz, \text{ etc.}$$

The quantities A, B, C are seen to be the *moments of inertia* about the axes of x, y, z respectively. The quantities D, E, F are called *products of inertia*.

By giving different values to l, m, n in equation (124), we can find the moment of inertia about any line through O, as soon as we know the values of the six coefficients A, B, C, D, E, F.

Ellipsoid of Inertia

247. The equation

$$Ax^2 + By^2 + Cz^2 - 2\,Dyz - 2\,Ezx - 2\,Fxy = K,$$

where K is any constant, being of the second degree, represents a conicoid. If r is the radius vector of direction cosines l, m, n, we have

$$r^2(Al^2 + Bm^2 + Cn^2 - 2\,Dmn - 2\,Enl - 2\,Flm) = K,$$

or, from equation (124), $\quad r^2 = \dfrac{K}{I}. \qquad (125)$

GENERAL THEORY OF MOMENTS OF INERTIA 303

Since I is positive for all values of l, m, n, it follows that r^2 is positive for all directions of the radius vector. Thus the conicoid is seen to be an ellipsoid.

This ellipsoid is called the *ellipsoid of inertia* of the point O.

Equation (125) may be written

$$I = \frac{K}{r^2},$$

and now expresses that the moment of inertia about any axis through O is inversely proportional to the square of the parallel radius vector of the ellipsoid of inertia.

Principal Axes of Inertia

248. This physical property of the ellipsoid shows that the ellipsoid itself remains the same, no matter what axes of coördinates are chosen. The ellipsoid has three principal axes, which are mutually at right angles. The directions of these axes are called the principal axes of inertia, at the point O.

If the principal axes of inertia at the point O are taken as axes of coördinates, then the coefficients of yz, zx, xy in the equation of the ellipsoid must disappear. Thus we must have

$$D = E = F = 0.$$

Taking the principal axes of inertia at O as axes of coördinates, equation (124) assumes the form

$$I = l^2 A + m^2 B + n^2 C.$$

The kinetic energy of a rotation of angular velocity Ω is

$$\tfrac{1}{2} I \Omega^2 = \tfrac{1}{2} (l^2 A + m^2 B + n^2 C) \Omega^2$$
$$= \tfrac{1}{2} (A\omega_1^2 + B\omega_2^2 + C\omega_3^2), \qquad (126)$$

where ω_1, ω_2, ω_3 are the components of Ω (see § 232).

GENERAL EQUATIONS OF MOTION OF A RIGID BODY

249. Let O be any point of a rigid body, and let Ox, Oy, Oz be a set of axes moving so that the point O maintains its position in the rigid body, while the axes remain parallel to their original position.

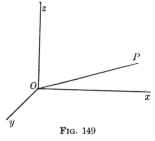

Fig. 149

Let the velocity of O have components u, v, w along these axes. The motion of the rigid body relative to these axes will be a motion of rotation about some axis OP which passes through O. Let us regard this as compounded of rotations ω_x, ω_y, ω_z about the three axes.

Let x, y, z be the coördinates of any point of the rigid body relative to these axes. The velocity of this point relative to the frame supplied by the axes moving with O has components

$$\frac{dx}{dt}, \quad \frac{dy}{dt}, \quad \frac{dz}{dt},$$

while the velocity of this frame in space has components

$$u, \quad v, \quad w.$$

Thus the whole velocity of the point x, y, z has components

$$u + \frac{dx}{dt}, \quad v + \frac{dy}{dt}, \quad w + \frac{dz}{dt}.$$

At any instant let L, M, N denote the sums of the moments of the external forces about the axes of x, y, z respectively, so that, as in § 243,

$$L = \sum(yZ - zY), \text{ etc.}$$

The moment of momentum of a particle of mass m, at x, y, z, about the axis of x is

$$m\left[y\left(w + \frac{dz}{dt}\right) - z\left(v + \frac{dy}{dt}\right)\right].$$

GENERAL EQUATIONS OF MOTION

Hence, by the theorem of § 243,

$$\frac{d}{dt} \sum m \left[y \left(w + \frac{dz}{dt} \right) - z \left(v + \frac{dy}{dt} \right) \right] = L, \qquad (127)$$

and there are similar equations for the other axes.

250. Relative to the moving axes of coördinates the particle m has coördinates x, y, z, so that a rotation ω_x about Ox gives the particle a velocity of components

$$0, \quad -\omega_x z, \quad \omega_x y.$$

Similarly the rotations ω_y, ω_z give velocities respectively of components

$$\omega_y z, \quad 0, \quad -\omega_y x,$$

and

$$-\omega_z y, \quad \omega_z x, \quad 0.$$

Compounding these velocities, we obtain as the components of the resultant velocity, relative to the axes,

$$\frac{dx}{dt} = \omega_y z - \omega_z y,$$

$$\frac{dy}{dt} = \omega_z x - \omega_x z,$$

$$\frac{dz}{dt} = \omega_x y - \omega_y x.$$

Thus $\quad y \dfrac{dz}{dt} - z \dfrac{dy}{dt} = \omega_x (y^2 + z^2) - \omega_y xy - \omega_z xz,$

and on differentiation of this equation with respect to t, we obtain as the value of part of the left-hand member of equation (127)

$$\frac{d}{dt} \sum \left[m \left(y \frac{dz}{dt} - z \frac{dy}{dt} \right) \right]$$

$$= \sum m (y^2 + z^2) \frac{d\omega_x}{dt} - \sum m\, xy \frac{d\omega_y}{dt} - \sum m\, xz \frac{d\omega_z}{dt}$$

$$- \sum m\, yz (\omega_y^2 - \omega_z^2) + \sum m (y^2 - z^2) \omega_y \omega_z$$

$$- \sum m\, zx\, \omega_y \omega_x + \sum m\, xy\, \omega_z \omega_x$$

$$= A \frac{d\omega_x}{dt} - F \frac{d\omega_y}{dt} - E \frac{d\omega_z}{dt}$$

$$- D(\omega_y^2 - \omega_z^2) - (B - C) \omega_y \omega_z - E \omega_y \omega_x + F \omega_z \omega_x$$

251. Let $\bar{x}$, $\bar{y}$, $\bar{z}$ be the coördinates of the center of gravity of the rigid body, and let M be its total mass. Then

$$\sum mx = M\bar{x}, \text{ etc.}$$

As the value of the remaining part of the left-hand member of equation (127) we now have

$$\frac{d}{dt}\sum m\,[yw - zv] = \frac{d}{dt}(M\bar{y}w - M\bar{z}v)$$
$$= M\frac{d}{dt}(\bar{y}w - \bar{z}v).$$

Thus equation (127) now assumes the form

$$M\frac{d}{dt}(\bar{y}w - \bar{z}v) + A\frac{d\omega_x}{dt} - F\frac{d\omega_y}{dt} - E\frac{d\omega_z}{dt} - D(\omega_y^2 - \omega_z^2)$$
$$- (B - C)\omega_y\omega_z - E\omega_y\omega_x + F\omega_z\omega_x = L. \qquad (128)$$

If $\sum X$, $\sum Y$, $\sum Z$ denote the total components along the axes, we have, by § 180, the further equations

$$M\frac{d}{dt}\left(u + \frac{d\bar{x}}{dt}\right) = \sum X. \qquad (129)$$

Equations (128) and (129) and the two other pairs of equations corresponding to the two other axes are the equations of motion for a rigid body moving under any forces.

Euler's Equations

252. Let us now suppose that we have a second set of axes, which we shall denote by 1, 2, 3. Let these axes move so as always to retain the same position in the rigid body, the point O (which we have already supposed always to retain the same position in the rigid body) being the origin. Let the axes 1, 2, 3 coincide with the axes x, y, z at the instant under consideration. Then the values of the coefficients of inertia referred to axes 1, 2, 3 are the same as those referred to axes x, y, z, namely A, B, C, D, E, F.

EULER'S EQUATIONS

Moreover, all velocities referred to axes 1, 2, 3 have the same values as they would have if referred to axes x, y, z. Let us denote the rotations about the axes 1, 2, 3 by $\omega_1, \omega_2, \omega_3$, then at the instant under consideration we shall have

$$\omega_1 = \omega_x, \qquad \omega_2 = \omega_y, \qquad \omega_3 = \omega_z.$$

This is not necessarily true at any instant except the instant at which the axes coincide, so that it is not permissible to differentiate these equations with respect to the time and deduce that

$$\frac{d\omega_1}{dt} = \frac{d\omega_x}{dt}, \text{ etc.}$$

Nevertheless, it can be shown that this last result is true at the instant under consideration. Let OQ denote *any* line through O, let $\cos \alpha, \cos \beta, \cos \gamma$ be its direction cosines relative to axes 1, 2, 3, and let Ω_q be the component of angular velocity about OQ. If the resultant angular velocity is one of amount Ω about an axis OP of which the direction cosines referred to axes 1, 2, 3 are l, m, n, then we have

$$\begin{aligned}\Omega_q &= \Omega \cos POQ \\ &= \Omega(l \cos\alpha + m \cos\alpha + n \cos\beta) \\ &= \omega_1 \cos\alpha + \omega_2 \cos\beta + \omega_3 \cos\gamma.\end{aligned}$$

Whatever line OQ may be, this equation is always true; hence we may legitimately differentiate it with respect to the time, and so obtain

$$\frac{d\Omega_q}{dt} = \frac{d\omega_1}{dt}\cos\alpha + \frac{d\omega_2}{dt}\cos\beta + \frac{d\omega_3}{dt}\cos\gamma \\ - \omega_1 \sin\alpha \frac{d\alpha}{dt} - \omega_2 \sin\beta \frac{d\beta}{dt} - \omega_3 \sin\gamma \frac{d\gamma}{dt}. \quad (130)$$

Now let the line OQ be supposed to coincide with Ox, so that $\Omega_q = \omega_x$. At the instant under consideration, $\beta = \gamma = \frac{\pi}{2}, \alpha = 0$. Moreover, $\frac{d\beta}{dt}$ is the rate at which the angle between Ox and axis 1 increases, and clearly this is ω_3. Similarly, $\frac{d\gamma}{dt} = -\omega_2$ and $\frac{d\alpha}{dt} = 0$.

Making all these substitutions, we find that at the moment under consideration, at which the two sets of axes coincide, equation (130) assumes the form

$$\frac{d\omega_x}{dt} = \frac{d\omega_1}{dt} - \omega_2 \cdot \omega_3 + \omega_3 \cdot \omega_2$$

$$= \frac{d\omega_1}{dt}.$$

Thus at the instant at which the two sets of axes coincide, we have the relations

$$\omega_x = \omega_1, \text{ etc.,}$$

and also
$$\frac{d\omega_x}{dt} = \frac{d\omega_1}{dt}.$$

Let us introduce the further simplification of supposing that the origin is either a fixed point or the center of gravity of the body. In the former case we have

$$u = v = w = 0, \text{ always;}$$

in the latter case
$$\bar{x} = \bar{y} = \bar{z} = 0, \text{ always.}$$

Let us further suppose that the axes are chosen to be the principal axes of inertia through the origin, so that

$$D = E = F = 0.$$

Introducing all these simplifications into equation (128) and the two similar equations, we find that these assume the form

$$A\frac{d\omega_1}{dt} - (B-C)\omega_2\omega_3 = L, \qquad (131)$$

$$B\frac{d\omega_2}{dt} - (C-A)\omega_3\omega_1 = M, \qquad (132)$$

$$C\frac{d\omega_3}{dt} - (A-B)\omega_1\omega_2 = N. \qquad (133)$$

These equations are known as Euler's equations.

Rotation of a Planet

253. As a first example of the use of these equations, let us examine the motion of a rigid body, symmetrical about an axis, acted on by forces all of which pass through the center of gravity. These conditions approximately represent those which obtain when a planet moves in its orbit, or a star in space.

Let us take the center of gravity as origin and the axis of symmetry as axis 1. Let the moments of inertia be A, B, B. Then the equations of motion are

$$A\frac{d\omega_1}{dt} = 0, \tag{134}$$

$$B\frac{d\omega_2}{dt} = (B-A)\omega_3\omega_1, \tag{135}$$

$$B\frac{d\omega_3}{dt} = -(B-A)\omega_2\omega_1. \tag{136}$$

The first equation gives at once that ω_1 is constant, say equal to Ω. If we write

$$k = \frac{B-A}{B}\Omega,$$

equations (135) and (136) become

$$\frac{d\omega_2}{dt} = k\omega_3, \tag{137}$$

$$\frac{d\omega_3}{dt} = -k\omega_2. \tag{138}$$

Thus $$\frac{d^2\omega_2}{dt^2} = k\frac{d\omega_3}{dt} = -k^2\omega_2,$$

of which the solution is $\omega_2 = E\cos(kt + \epsilon)$;

and equation (137) now leads at once to

$$\omega_3 = -E\sin(kt + \epsilon).$$

Thus the components of angular velocity at the instant t are

$$\Omega, \quad E\cos(kt + \epsilon), \quad E\sin(kt + \epsilon),$$

and we see that the axis of rotation describes a cone in the solid, with period $\dfrac{2\pi}{k}$ or $\dfrac{2\pi}{\Omega}\dfrac{B}{B-A}$.

If B is very nearly equal to A, the period may be very great, and the motion consequently very slow. This happens in the case of the earth: the motion of the axis of rotation gives rise to the phenomenon known as the variation of latitude, of which the period is about 428 days. Since a period $\dfrac{2\pi}{\Omega}$ represents roughly one day, we conclude that for the earth $\dfrac{B-A}{B}$ is of the order of $\tfrac{1}{428}$.

The true value of this quantity is .00328, the discrepancy resulting from the imperfect rigidity of the earth.

Motion of a Top

254. As a second example of the methods of this chapter, let us consider the motion of a spinning top. This we shall suppose to be a solid of revolution spinning on a peg of which the end will be

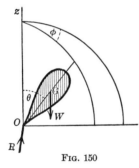

Fig. 150

treated as a point, the contact between the peg and the surface on which it rests being assumed rough enough to prevent slipping. The point of contact is now a fixed point O. Let us take axes Ox, Oy, Oz fixed in space, the axis of z being vertical, and also axes 1, 2, 3 fixed in the body, and coinciding with the principal axes of inertia through O. Let axis 1 be the axis of symmetry of the top, and let the moments of inertia about axes 1, 2, 3 be A, B, B.

The first of Euler's equations becomes

$$A\frac{d\omega_1}{dt} = 0,$$

since $B = C$ and $L = 0$. Thus ω_1 is a constant, say Ω.

Let the axis of the top cut a unit sphere about O at a point whose polar coördinates are 1, θ, ϕ, the axis of Oz being taken for pole, so that θ is the angle between the vertical and the axis of the top.

MOTION OF A TOP 311

The kinetic energy of the top is, by § 248,

$$\tfrac{1}{2}[A\Omega^2 + B(\omega_2^2 + \omega_3^2)],$$

while the potential energy is $Mgh \cos \theta$, where h is the distance of the center of gravity of the top from O. Thus the equation of energy is

$$A\Omega^2 + B(\omega_2^2 + \omega_3^2) + 2\,Mgh \cos \theta = E, \qquad (139)$$

where E is a constant. This may be put into a different form. For $\omega_2^2 + \omega_3^2$ is the square of the angular velocity of the axis of the top: it is therefore the square of the actual velocity of the point $1, \theta, \phi$ on the unit sphere, and hence we have

$$\omega_2^2 + \omega_3^2 = \left(\frac{d\theta}{dt}\right)^2 + \sin^2 \theta \left(\frac{d\phi}{dt}\right)^2.$$

The equation of energy now assumes the form

$$A\Omega^2 + B\left[\left(\frac{d\theta}{dt}\right)^2 + \sin^2 \theta \left(\frac{d\phi}{dt}\right)^2\right] + 2\,Mgh \cos \theta = E. \quad (140)$$

We can obtain a third equation from the fact that the angular momentum about Oz, the vertical, is constant. The angular momentum may be regarded as compounded of

(*a*) the momentum due to the rotation Ω about axis 3;
(*b*) the momentum due to the motion of the axis of the top.

The rotation Ω about axis 3 may be further decomposed into rotations $\Omega \cos \theta$, $\Omega \sin \theta$ about the horizontal and vertical, giving moments of momenta $A\Omega \cos \theta$, $A\Omega \sin \theta$ about the horizontal and vertical. Thus the moment of momentum contributed by part (*a*) is $A\Omega \cos \theta$.

The motion of the axis of the top may be resolved into a rotation of angular velocity $\sin \theta \dfrac{d\phi}{dt}$ about an axis making an angle $\dfrac{\pi}{2} - \theta$ with the vertical, and one of angular velocity $\dfrac{d\theta}{dt}$ about a

horizontal axis. The former may be replaced by $\sin^2\theta \dfrac{d\phi}{dt}$ about the vertical, and $\sin\theta\cos\theta \dfrac{d\phi}{dt}$ about a horizontal axis. Thus the moment of momentum about the vertical contributed by part (b) of the motion is
$$B\sin^2\theta \frac{d\phi}{dt};$$
and since the moment of momentum about the vertical has a constant value, say G, we have
$$A\Omega\cos\theta + B\sin^2\theta \frac{d\phi}{dt} = G. \tag{141}$$

If we eliminate $\dfrac{d\phi}{dt}$ from this equation and equation (140), we obtain
$$B\sin^2\theta\left[A\Omega^2 + B\left(\frac{d\theta}{dt}\right)^2 + 2Mgh\cos\theta - E\right] + (G - A\Omega\cos\theta)^2 = 0, \tag{142}$$
a differential equation giving the variations in the value of θ, and therefore allowing us to trace the changes in the inclination of the axis of the top to the vertical.

The maxima and minima of θ are given by putting $\dfrac{d\theta}{dt} = 0$, and are therefore the roots of
$$B(1 - \cos^2\theta)[A\Omega^2 + 2Mgh\cos\theta - E] + (G - A\Omega\cos\theta)^2 = 0. \tag{143}$$

Let us call the left hand of this equation $f(\cos\theta)$. Since f is a function of degree three, there will be three roots for $\cos\theta$. Let us suppose that the top is started at an angle $\theta = \theta_0$, and with the value of $\dfrac{d\theta}{dt}$ equal to $\left(\dfrac{d\theta}{dt}\right)_0$. Then, from equation (142),
$$B\sin^2\theta_0\left[A\Omega^2 + B\left(\frac{d\theta}{dt}\right)_0^2 + 2Mgh\cos\theta_0 - E\right] + (G - A\Omega\cos\theta_0)^2 = 0,$$
so that
$$f(\cos\theta_0) = B\sin^2\theta_0[A\Omega^2 + 2Mgh\cos\theta_0 - E] + (G - A\Omega\cos\theta_0)^2$$
$$= -B^2\sin^2\theta_0\left(\frac{d\theta}{dt}\right)_0^2,$$

so that $f(\cos \theta_0)$ is negative. We easily find, from equation (143), that
$$f(1) = (G - A\Omega)^2,$$
so that $f(1)$ is positive.

Again, $\qquad f(-1) = (G + A\Omega)^2,$

so that $f(-1)$ is positive, and
$$f(+\infty) = -2MghB(+\infty)^3,$$
which is negative. Thus we have seen that

when $\cos \theta = +\infty$, $\quad f(\cos \theta)$ is $-$;
when $\cos \theta = 1$, $\quad f(\cos \theta)$ is $+$;
when $\cos \theta = \cos \theta_0$, $\quad f(\cos \theta)$ is $-$;
when $\cos \theta = -1$, $\quad f(\cos \theta)$ is $+$.

Thus the three roots of the cubic $f(\cos \theta) = 0$ lie as follows:

a root $\theta = \theta_1$ between $\cos \theta = 1$ and $\cos \theta = \cos \theta_0$;

a root $\theta = \theta_2$ between $\cos \theta = \cos \theta_0$ and $\cos \theta = -1$;

a root for which $\cos \theta$ is numerically greater than unity, giving no real value for θ.

We see, therefore, that the only points at which $\dfrac{d\theta}{dt}$ can vanish are $\theta = \theta_1$, and $\theta = \theta_2$. Moreover, at these points $\dfrac{d\theta}{dt}$ vanishes, and as there are not coincident roots at either point $\dfrac{d\theta}{dt}$ changes sign on reaching these points, so that θ can range only between the values θ_1 and θ_2.

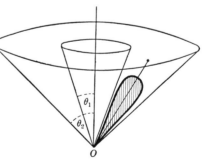

Fig. 151

Thus the axis of the top oscillates between the two cones $\theta = \theta_1$ and $\theta = \theta_2$.

255. Let us find what is the least angular momentum which the top must have so as to spin without falling over. To do this, we may assume that falling will occur if ever θ exceeds a certain limit θ_3, either through the peg slipping or through its side touching the ground. The condition that the top shall not fall over is that θ_2 must be less than θ_3, and hence that $f(\cos \theta_3)$ must be positive. Thus the values of E, G, and Ω must be such that

$$B \sin^2 \theta_3 (A\Omega^2 + 2 Mgh \cos \theta_3 - E) + (G - A\Omega \cos \theta_3)^2$$

is positive.

Suppose that the top is started at an inclination θ_0 to the vertical, having no motion except one of rotation Ω about its axis. We then have, from equations (140) and (141),

$$E = A\Omega^2 + 2 Mgh \cos \theta_0,$$
$$G = A\Omega \cos \theta_0.$$

Thus

$$\begin{aligned} f(\cos \theta_3) &= B \sin^2 \theta_3 (A\Omega^2 + 2 Mgh \cos \theta_3 - E) + (G - A\Omega \cos \theta_3)^2 \\ &= B \sin^2 \theta_3 \cdot 2 Mgh (\cos \theta_3 - \cos \theta_0) + A^2\Omega^2 (\cos \theta_3 - \cos \theta_0)^2 \\ &= (\cos \theta_3 - \cos \theta_0)[2 MghB \sin^2 \theta_3 + A^2\Omega^2 (\cos \theta_3 - \cos \theta_0)]. \end{aligned}$$
(144)

Since the top is necessarily started in a position in which it can spin, the value of $\cos \theta_3 - \cos \theta_0$ is necessarily negative. Thus in order that $f(\cos \theta_3)$ may be positive, we must have

$$A^2\Omega^2 (\cos \theta_0 - \cos \theta_3) - 2 MghB \sin^2 \theta_3 \qquad (145)$$

positive, or
$$\Omega^2 > \frac{2 MghB \sin^2 \theta_3}{A^2 (\cos \theta_0 - \cos \theta_3)}. \qquad (146)$$

We notice that if A is very small, the value of Ω required to keep the top from falling is very large. It is therefore very hard to spin a top of small cross section, such as a lead pencil or a pointed wire.

If we can choose the angle at which we start the top, $\cos \theta_0$ is at our disposal. We see that the necessary value for Ω is least

MOTION OF A TOP

when $\cos\theta_0$ is a maximum, i.e. when the top is started vertical. In this case the top will spin if

$$\Omega^2 > \frac{2\,Mgh B \sin^2\theta_3}{A^2(1-\cos\theta_3)},$$

or if

$$\Omega^2 > \frac{2\,Mgh B\,(1+\cos\theta_3)}{A^2}.$$

256. In general, for a top started vertically and with no velocity except one of pure rotation about its axis, we find, on putting $\cos\theta_0 = 1$ in equation (144),

$$f(\cos\theta) = (1-\cos\theta)^2[A^2\Omega^2 - 2\,Mgh B\,(1+\cos\theta)].$$

The roots of the equation $f(\cos\theta) = 0$ are

$$\cos\theta = +1,\ +1,\ \frac{A^2\Omega^2}{2\,Mgh B} - 1.$$

Let us write
$$\Omega_0^2 = \frac{4\,Mgh B}{A^2},$$

then when $\Omega^2 = \Omega_0^2$, the roots are

$$\cos\theta = +1,\ +1,\ +1.$$

When $\Omega^2 > \Omega_0^2$ the third root is greater than unity, and when $\Omega^2 < \Omega_0^2$ the third root is less than unity, say $\cos\theta = \cos\Theta$, where Θ is a real angle, given by

$$\cos\Theta = \frac{A^2\Omega^2}{2\,Mgh B} - 1 = 1 - \frac{\Omega_0^2 - \Omega^2}{2\,\Omega_0^2}. \tag{147}$$

Thus, as long as $\Omega^2 > \Omega_0^2$ the oscillations are confined within the coincident limits $\theta = 0$ and $\theta = 0$, so that the top remains vertical, but as soon as we have $\Omega^2 < \Omega_0^2$ the oscillations are between the limits $\theta = 0$ and $\theta = \Theta$.

Suppose we start a top with angular velocity Ω greater than Ω_0, so that at first its axis is vertical and the only motion of the top is one of rotation about its axis. Then the real roots for θ are 0, 0; there is therefore no range of oscillations, and the axis of the top remains strictly vertical, — in common language, the top is "asleep."

If the conditions were the ideal conditions supposed, this motion would continue forever, but in nature such ideal conditions cannot exist. The region of contact between the peg and the surface on which it spins is not strictly a point, but a small circle or ellipse, on account of the small compression which takes place at the point of contact. By making the peg of hard steel and spinning on a hard surface, this region is very small, but is still of finite dimensions. The consequence is that the reactions on the peg do not all meet the axis. There is a small frictional couple resisting the rotation of the top, and Ω gradually decreases.

When Ω has so far decreased as to be less than Ω_0, the ranges of oscillation are $\theta = 0$ and $\theta = \Theta$. The top is no longer asleep, but is now wobbling through an angle Θ. As Ω continues to decrease, Θ continually increases, as is clear from equation (147), and finally Θ reaches so large a value that the top rolls against the ground and so falls over.

257. The interest of these results will perhaps be enhanced by examining the form they assume when the top is of a very simple kind.

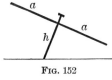

FIG. 152

Let us suppose that a top is formed by running a pin through the center of a uniform disk of mass M, radius a. Let the length of the pin which protrudes through the disk on its lower side be h, and let the mass of the pin be neglected in comparison with that of the disk. The h is the same as the h of our previous analysis. The values of A and B are

$$A = \tfrac{1}{2} Ma^2, \qquad B = \tfrac{1}{4} Ma^2,$$

so that

$$\Omega_0^2 = \frac{4\,Mgh\,B}{A^2} = \frac{4\,gh}{a^2}.$$

When spinning at the critical velocity Ω_0 at which wobbling sets in, the velocity of a point on the rim is $\Omega_0 a$ or $2\sqrt{gh}$. Thus wobbling begins when the velocity of a point on the rim is reduced to $2\sqrt{gh}$, a velocity which depends only on the height of the disk and not on its radius. We see that the lower the disk is, the slower it can spin without wobbling. If we take $h = 2$ inches, we find that wobbling begins when the rim velocity is about 4.7 feet a second.

The rim will touch the ground when the range of wobbling is given by $\tan\Theta = \dfrac{h}{a}$, and after this the top will roll on the ground. If we take $a = 6$ inches and $h = 2$ inches, as before, this gives $\tan\Theta = \tfrac{1}{3}$, so that $\cos\Theta = \dfrac{3}{\sqrt{10}}$ and $\dfrac{\Omega_0^2 - \Omega^2}{\Omega_0^2} = .106$, approximately. Thus $\Omega = \dfrac{19}{20}\Omega_0$, roughly. Thus such a top will "sleep" until its rim velocity is reduced to 4.7 feet a second. After this it will wobble, and as soon as its rim velocity is reduced to about 4.5 feet a second the top will begin to roll on the ground.

For an ordinary small, pear-shaped peg top we may take roughly $h = 1\tfrac{1}{2}$ inches, and the radii of gyration about axes through the point of contact as $\tfrac{3}{4}$ inch and 2 inches. Thus in inches

$$A = \tfrac{9}{16}M, \qquad B = 4M,$$
$$\Omega_0^2 = \frac{4\,Mgh\,B}{A^2} = \frac{2048}{27}g.$$

Taking $g = 386$ inches per second per second, this gives $\Omega_0 = 170$ revolutions per second. If thrown from a string of which the end is coiled round the top in circles of radius 1 inch, the string must be withdrawn with a velocity of about 60 miles an hour relative to the top to set up the required angular velocity.

GENERAL EXAMPLES

1. A fly wheel whose moment of inertia is I has a string wound round its axle of radius b. A tension equal to a weight w is applied to the string for 1 second. What is the angular velocity of the fly wheel at the end of 1 second?

2. A fleet of total displacement 200,000 tons steams from east to west along the equator, covering 20 minutes of longitude per hour. Regarding the earth as a homogeneous sphere of mass 6×10^{21} tons, find the change in the earth's angular velocity produced by the motion of the fleet. Show that the day is lengthened by, roughly, 16×10^{-14} seconds.

3. The earth's mass is 6×10^{21} tons, and icebergs and melted snow weighing 10^{10} tons move from the north pole to latitude 45 degrees. Find the change in the length of the day.

4. A train of mass m runs due north at 60 miles an hour. Show that there must be a pressure between the eastern rail and the flanges of the wheel in consequence of the earth's rotation, and find the amount of the pressure.

5. A thin layer of dust, thickness h feet, is formed on the earth by the fall of meteors, reaching the earth from all directions. Show that the

318   MOTION OF RIGID BODIES

change in the length of a day is about $\dfrac{5\,h\rho}{aD}$ of a day, where a is the radius of the earth in feet and D, ρ are the density of the earth and the meteoric dust respectively.

6. Two masses M and m suspended from a wheel and axle of radii a, b do not balance. Show that the acceleration of M is

$$\frac{Ma - mb}{Ma^2 + mb^2 + I}\, ag,$$

where I is the moment of inertia of the machine about its axis.

7. A uniform cylinder has coiled round its central section a light, perfectly flexible, inextensible string. One end of the string is attached to a fixed point, and the cylinder is allowed to fall. Show that it will fall with acceleration $\tfrac{2}{3} g$.

8. Two equal uniform rods of length $2\,a$, loosely joined at one extremity, are placed symmetrically upon a fixed sphere of radius $\dfrac{a\sqrt{2}}{3}$ and raised into a horizontal position so that the hinge is touching the sphere. They are then allowed to descend. Show that when they are first at rest they are inclined at an angle $\cos^{-1}\tfrac{1}{3}$ to the horizontal, that the pressure on the sphere at each point of contact is one quarter of the weight of a rod, and that there is no strain at the joint.

9. A rod rests with one extremity on a smooth horizontal plane and the other on a smooth vertical wall, the rod being inclined at an angle α to the horizon. If it is allowed to slip down, show that it will separate from the wall when its inclination to the horizontal is $\sin^{-1}(\tfrac{2}{3}\sin\alpha)$.

10. If the sun gradually contracts in such a way as always to remain similar to itself in constitution and form, show that when every radius has contracted an nth part of its length, where n is large, the angular velocity will have increased to $\left(1 + \dfrac{2}{n}\right)$ times its former value. Examine the change in the kinetic energy of rotation.

11. An elastic band of natural length $2\,\pi a$, mass m, modulus λ, rests against a rough wheel of radius a in a horizontal plane. The string is held against the circumference of the wheel, which is made to rotate with angular velocity Ω. If the string is left to itself, show that it will expand, and that when its radius is r its angular velocity will be $\dfrac{a^2\Omega}{r^2}$, and that its radial velocity will be

$$\left[\frac{a^2\Omega^2}{r^2}(r^2 - a^2) - \frac{2\,\pi\lambda\,(r - a)^2}{ma}\right]^{\frac{1}{2}}.$$

12. A uniform triangular disk ABC is so supported that it can oscillate in its own plane, which is vertical, about A. Show that the length of the simple equivalent pendulum is

$$\frac{1}{4}\frac{3(b^2+c^2)-a^2}{\sqrt{2(b^2+c^2)-a^2}}.$$

13. A spherical hollow of radius a is made in a cube of glass of mass M, and a particle of mass m is placed inside. The cube is then projected with velocity V on a smooth horizontal plane. If the particle just gets round the sphere, remaining in contact with it all the way, show that

$$V^2 = 5\,ag + 4\,ag\frac{m}{M}.$$

14. Three equal particles are attached to the ends and middle point of a rod of negligible mass, and one of the end particles is struck by a blow at right angles to the rod. Show that the velocities of the particles at starting are in the ratio
$$5:2:1.$$

15. A rough horizontal cylinder of mass M and radius a is free to turn about its axis. Round it is coiled a string, to the free extremity of which is attached a chain of mass m and length l. The chain is gathered close up and then let go. Show that if θ is the angle through which the cylinder has turned after a time t before the chain is fully stretched, then

$$Ma\theta = \frac{m}{l}\left(\frac{1}{2}gt^2 - a\theta\right)^2.$$

16. A uniform flat circular disk is projected on a rough horizontal table, the friction on any element moving with velocity V being $cV^3 \times$ (mass of element), in a direction opposite to that of V. Find the path of the center of the disk.

CHAPTER XII

GENERALIZED COÖRDINATES

258. So far we have dealt with the mechanics (dynamics and statics) of material bodies on the supposition that these bodies consist of innumerable small particles which, in the case of a rigid body, are held firmly in position and serve the purpose of transmitting force from one part of the body to another.

259. Even when dealing with rigid bodies this conception of the structure of matter has not led to entirely consistent results. For instance, we have found that after an impact between two imperfectly elastic bodies, or after sliding between two imperfectly smooth bodies, a certain amount of energy disappears from view, and we have had to suppose that this energy is used in starting small motions, relative to one another, of the ultimate particles of which the bodies are composed. In other words, after an impact or sliding has taken place, a rigid body can no longer be supposed to satisfy the conditions postulated for a rigid body.

When dealing with bodies which are obviously not rigid the case is worse. Here the conceptions which we have introduced into the study of rigid bodies do not help at all, and very little progress is possible without introducing some other conceptions to replace these.

260. There are two ways of proceeding at this stage. We may introduce new conceptions which seem plausible, and in this way try to form a picture of the structure of the matter with which we are dealing. We cannot be certain that the results obtained in this way will be true, for we can never be sure that our conceptions of the nature of the ultimate structure of matter are accurate. But it may be worth trying what results are obtained by introducing a set of provisional conceptions as to the structure of matter. If these results are in agreement with the phenomena observed in nature, the probability that our provisional conceptions are near to the truth is strengthened. If, on the contrary, the results obtained are not found to agree with what is observed in nature, the provisional conceptions from which these results have been deduced must be either modified or withdrawn.

Different sets of conceptions as to the structure of the matter dealt with will lead to different branches of mathematical physics. As instances of such branches of mathematical physics may be mentioned the theory of elastic solids which is based upon certain provisional conceptions as to the behavior of the particles of which solid bodies are composed, and the kinetic theory of gases which is based upon certain provisional conceptions as to the behavior of the particles of a gas. The tracing out of the consequences of different sets of provisional conceptions as to the structure of matter cannot, however, be regarded as coming within the scope of a book such as the present one.

261. There is, however, an alternative way of proceeding. We have taken Newton's laws of motion as the material supplied by experimental science for theoretical science to work upon. The truth of these laws as applied to the ultimate particles of the material universe is by no means certain, because we cannot obtain the ultimate particles to experiment upon. Suppose, however, that we examine whether any further progress can be made in the study of mechanics without introducing any hypothesis beyond the single one (admittedly uncertain) that Newton's laws apply to the ultimate particles. If we can make progress in this direction, the results obtained will of course apply to all further extensions of mechanics, whether or not additional hypotheses are introduced as to the nature and arrangement of the ultimate particles.

262. The standpoint from which we are regarding the matter can, perhaps, be explained by an analogy, first suggested by Professor Clerk Maxwell. Suppose that we have a complicated machine in a closed room, and that the only connection between this machine and the outer world is by means of a number of ropes which hang through holes in the floor into the room beneath.

A man introduced into the room beneath will have no opportunity of inspecting the machinery above, but he can manipulate it to a certain extent by pulling the different ropes. If on pulling one rope he finds that the others are set into motion, he will understand that the different ropes must be connected above by some kind of mechanism, but will not be able to discover the exact nature of the mechanism.

This concealed mechanism may be supposed to represent those parts of the mechanism of the universe which are hidden from our view, while the ropes represent those parts which we can manipulate. In nature, there are certain acts which we can perform, corresponding to the pulling of the ropes in our analogy, and we see that these are followed by certain consequences, analogous to the motion of the other ropes; but the ultimate mechanism by which the cause produces the effect remains entirely unknown to us. For instance, if we press the key of an electric circuit, we may find that the needle of a distant galvanometer is moved, but the mechanical processes which transmit the action through the wires of the circuit and through the ether surrounding the galvanometer needle remain unknown.

263. Now suppose that the imaginary man is at liberty to handle the ropes and that he wishes to study the connection between them. He may begin by conjecturing that the connecting mechanism in the room above consists of arrangements of, say, levers, pulleys, and cogwheels, and he may work out for himself the manner in which the ropes ought to move if his conjectures are correct. This procedure would be analogous to that we have described in § 260; it is not the procedure we are going to follow here.

On the other hand, without any conjecture at all as to the nature of the mechanism above, the man will know that certain laws will govern any manipulation of the ropes, if the ropes are connected by mechanism of any kind whatever, such that each particle obeys Newton's laws of motion.

To explain this, let us take the simplest case, and suppose that there are two ropes only and that when A is pulled down half an inch, then B invariably rises through two inches. The mechanism may be a lever, an arrangement of pulleys, or clockwork. But whether it is any one of these, or something entirely different from any of them, it will be known that the motion of rope A downwards can be restrained by exerting on rope B a force equal to half of that applied to A. This fact follows from the principle of virtual work, quite apart from any conjecture as to the nature of the hidden mechanism. Now the question before us is as follows: Can we, without any knowledge of the hidden mechanism, discover what motion of the ropes will ensue, if they are started in any given way. And the answer is that we can, provided we know the amount of energy involved in a motion of any kind, — i.e. provided we know the kinetic energy of every motion, and also the potential energy of every configuration.

So also, to pass from analogies to realities, we can, without any knowledge of the ultimate mechanism of the universe, discover what motion will ensue from any initial conditions, provided that we know the kinetic and potential energies of all configurations of the portion of the universe with which we are dealing.

Hamilton's Principle

264. Let us suppose that any single particle of a material system has at any instant coördinates x_1, y_1, z_1, its mass being m_1, and that it is acted upon by forces of which the resultant has components X_1, Y_1, Z_1. Let the velocity of this particle have components u_1, v_1, w_1, so that
$$u_1 = \frac{dx_1}{dt}, \text{ etc.}$$

Then, if the motion of this particle is governed by Newton's laws, we shall have
$$m_1 \frac{du_1}{dt} = X_1, \qquad (148)$$

$$m_1 \frac{dv_1}{dt} = Y_1, \qquad (149)$$

$$m_1 \frac{dw_1}{dt} = Z_1. \qquad (150)$$

Let us compare this motion with a slightly different motion in which Newton's laws are not obeyed. In this second motion let the coördinates of m_1, at the instant at which they are x_1, y_1, z_1 in the actual motion, be supposed to be x_1', y_1', z_1', and let the components of velocity at this instant be u_1', v_1', w_1', so that

$$u_1' = \frac{dx_1'}{dt}, \text{ etc.}$$

Let us agree that the modified motion is to differ so slightly from the actual, that any quantity such as $x_1' - x_1$, $u_1' - u_1$, which measures part of this difference, may be treated as a small quantity. Let us denote $x_1' - x_1$ by δx_1, and use a similar notation for the other differences.

Multiply equations (148), (149), (150), which are true at every instant, by δx_1, δy_1, δz_1, and add. We obtain

$$m_1 \frac{du_1}{dt} \delta x_1 + m_1 \frac{dv_1}{dt} \delta y_1 + m_1 \frac{dw_1}{dt} \delta z_1$$
$$= X_1 \delta x_1 + Y_1 \delta y_1 + Z_1 \delta z_1. \tag{151}$$

Now
$$\frac{du_1}{dt} \delta x_1 = \frac{d}{dt}(u_1 \delta x_1) - u_1 \frac{d}{dt}(\delta x_1)$$
$$= \frac{d}{dt}(u_1 \delta x_1) - u_1 \frac{d}{dt}(x_1' - x_1)$$
$$= \frac{d}{dt}(u_1 \delta x_1) - u_1(u_1' - u_1)$$
$$= \frac{d}{dt}(u_1 \delta x_1) - u_1 \delta u_1.$$

Hence

$$m_1 \frac{du_1}{dt} \delta x_1 + m_1 \frac{dv_1}{dt} \delta y_1 + m_1 \frac{dw_1}{dt} \delta z_1$$
$$= m_1 \left[\frac{d}{dt}(u_1 \delta x_1 + v_1 \delta y_1 + w_1 \delta z_1) - (u_1 \delta u_1 + v_1 \delta v_1 + w_1 \delta w_1) \right]$$
$$= X_1 \delta x_1 + Y_1 \delta y_1 + Z_1 \delta z_1, \tag{152}$$

by equation (151).

HAMILTON'S PRINCIPLE

An equation of this kind is true for each particle of the system and at every instant of the motion. It is moreover true whatever the displaced motion may be. On summing this equation for all particles we obtain

$$\sum m_1 \left[\frac{d}{dt}(u_1 \delta x_1 + v_1 \delta y_1 + w_1 \delta z_1) - (u_1 \delta u_1 + v_1 \delta v_1 + w_1 \delta w_1) \right]$$
$$= \sum (X_1 \delta x_1 + Y_1 \delta y_1 + Z_1 \delta z_1). \qquad (153)$$

Now let T denote the kinetic energy of the motion, so that

$$T = \tfrac{1}{2} \sum m_1 (u_1^2 + v_1^2 + w_1^2).$$

Then
$$\delta T = \tfrac{1}{2} \sum m_1 (u_1'^2 - u_1^2 + v_1'^2 - v_1^2 + w_1'^2 - w_1^2).$$

Now $\quad u_1'^2 - u_1^2 = (u_1 + \delta u_1)^2 - u_1^2 = 2u_1 \delta u_1,$

if we neglect the small quantity of the second order $(\delta u_1)^2$, so that we have
$$\delta T = \sum m_1 (u_1 \delta u_1 + v_1 \delta v_1 + w_1 \delta w_1).$$

265. Assuming for the moment that the system of forces is conservative, let W denote the potential energy of the system at the instant under consideration, and W' that of the imaginary system in the slightly displaced configuration. Then, by § 118, we have

$$\delta W = W' - W$$
$$= \text{(work done in moving system from actual to displaced configuration)}$$
$$= - \sum (X_1 \delta x_1 + Y_1 \delta y_1 + Z_1 \delta z_1). \qquad (154)$$

Substituting into equation (153) for the expressions which have been found to be equal to δT and δW, we find that this equation reduces to the simpler form

$$\sum m_1 \frac{d}{dt}(u_1 \delta x_1 + v_1 \delta y_1 + w_1 \delta z_1) - \delta T = -\delta W,$$

or again,
$$\frac{d}{dt} \sum m_1 (u_1 \delta x_1 + v_1 \delta y_1 + w_1 \delta z_1) = \delta (T - W).$$

This equation is true at every instant of the motion. Let us integrate it between any two instants of the motion, say from $t = t_1$ to $t = t_2$. We obtain

$$\left[\sum m_1 (u_1 \delta x_1 + v_1 \delta y_1 + w_1 \delta z_1)\right]_{t_1}^{t_2} = \int_{t_1}^{t_2} \delta(T - W)\, dt. \quad (155)$$

The displaced motion has so far been subject to no restrictions except that the difference between it and the actual motion must always remain small. Let us now introduce the further restriction that at times t_1 and t_2 the configurations in the displaced motion are to be identical with those in the actual motion. The displaced motion is now one in which the imaginary system starts in the same configuration as the actual system at time $t = t_1$, swerves from the course of the actual system from time t_1 to time t_2 (because the actual system obeys Newton's laws, while the imaginary system does not), and ultimately ends in the same position as the actual system at time t_2.

In consequence of this restriction on the motion of the imaginary system, we have at times t_1 and t_2,

$$\delta x_1 = \delta y_1 = \delta z_1 = 0,$$

and similar relations for the other particles. Thus

$$\left[\sum m_1 (u_1 \delta x_1 + v_1 \delta y_1 + w_1 \delta z_1)\right]_{t_1}^{t_2} = 0,$$

and equation (155) reduces to

$$\int_{t_1}^{t_2} \delta(T - W)\, dt = 0. \quad (156)$$

Here we have an equation which depends only on the amounts of the kinetic and potential energies of the system, and not on the mechanism of the system. We shall find that from this single equation we can determine the motion of all the known parts of the system as soon as T and W are known, without any knowledge of the mechanism of the unknown parts.

PRINCIPLE OF LEAST ACTION 327

266. Before proving this, however, we may attempt to interpret equation (156). Let us denote $T-W$ by L. Then

$$\int_{t_1}^{t_2} \delta(T-W)\,dt = \int_{t_1}^{t_2} \delta L\, dt$$
$$= \int_{t_1}^{t_2} (L'-L)\,dt$$
$$= \int_{t_1}^{t_2} L'\,dt - \int_{t_1}^{t_2} L\,dt$$
$$= \delta\left(\int_{t_1}^{t_2} L\,dt\right).$$

If we denote $\int_{t_1}^{t_2} L\,dt$ by S, the equation becomes $\delta S = 0$, or

$$S' = S.$$

Thus the value of the function S for the actual motion is the same, except for small quantities of the second and higher orders, as the corresponding function S' for any slightly different motion, which begins and ends with the same configuration at the same instants. In other words, the function S is either a maximum or a minimum when the series of configurations is that which actually occurs in nature.

Principle of Least Action

267. The total energy will, by the theorem of § 143, remain constant during the actual motion, say equal to E, so that at every instant we shall have

$$T+W = E, \quad T-W = L.$$

In the slightly varied series of configurations it is not true that the total energy remains constant throughout the motion, but out of the infinite number of slightly varied series of configurations there will still be an infinite number for which the conditions already postulated are satisfied, together with the condition that the total energy at every instant shall have the value E. For such

a series we have
$$T' + W' = E, \quad T' - W' = L'.$$
Thus we have $L = 2T - E, \quad L' = 2T' - E,$

so that
$$S = \int_{t_1}^{t_2} L\, dt$$
$$= \int_{t_1}^{t_2} (2T - E)\, dt$$
$$= \int_{t_1}^{t_2} 2T\, dt - (t_2 - t_1)E.$$

Thus if S is a maximum or a minimum, it follows that
$$\int_{t_1}^{t_2} 2T\, dt$$
is a maximum or a minimum. This integral is called the *action* of the motion. We now see that of all possible series of configurations which bring the system from one configuration to another in a given time, and in such a way that the total energy has always a specified constant value, that one which can be described by a natural system is the one on which the action is a maximum or a minimum. Since the action is in general a minimum, this principle is known as the *principle of least action*.

The statement of this principle was first given by Maupertius (1690–1759), who did not deduce it by mathematical reasoning, but believed it could be proved by theological arguments that all changes in the universe must take place so as to make the action a minimum (*Essai de Cosmologie*, 1751).

Non-Conservative Forces

268. If the forces are non-conservative, we may no longer, as in equation (154), replace
$$\sum (X_1 \delta x_1 + Y_1 \delta y_1 + Z_1 \delta z_1)$$
by $-\delta W$, and consequently, instead of equation (156), we shall have
$$\int_{t_1}^{t_2} \left[\delta T + \sum (X \delta x + Y \delta y + Z \delta z) \right] dt = 0. \qquad (157)$$

LAGRANGE'S EQUATIONS

269. If the coördinates x_1, y_1, z_1, etc., of every particle of the system are known, we know not only the configuration of the system but also the mechanism by which the different parts of the system are connected. It may, however, be that we can determine the configuration of the system by knowing a smaller number of quantities which do not give us a knowledge of the mechanism.

For instance, in our former illustration we imagined two ropes to hang from an unknown machine, the ropes being connected in such a way that a motion of one inch in the one invariably produced a motion of two inches in the other. In this case the configuration is fully determined when we know the single coördinate which measures the position of the end of the first rope, but a knowledge of this coördinate does not imply a knowledge of the mechanism connecting the ropes.

Again, the position of a rigid body is, as we have seen (§ 65), determined by the values of sufficient quantities (six) to fix the positions in space of three non-collinear particles of the body, but a knowledge of these quantities does not give us information as to the arrangement of the particles of which the body is formed.

Let θ_1, θ_2, $\cdots$, θ_n be a set of quantities such that when their value is known, the configuration of a system of bodies is fully determined. Then the quantities θ_1, θ_2, $\cdots$, θ_n are called *generalized coördinates* of the system.

270. Let x, y, z be the coördinates of any particle of the system. Then x is fully determined by the values of θ_1, θ_2, $\cdots$, θ_n, so that it is a function of these quantities, say

$$x = f(\theta_1, \theta_2, \cdots, \theta_n). \tag{158}$$

If the system is in motion, all the quantities which enter in equation (158) are functions of the time. We have, on differentiation with respect to the time,

$$\frac{dx}{dt} = \frac{\partial f}{\partial \theta_1}\frac{d\theta_1}{dt} + \frac{\partial f}{\partial \theta_2}\frac{d\theta_2}{dt} + \cdots + \frac{\partial f}{\partial \theta_n}\frac{d\theta_n}{dt}.$$

330 GENERALIZED COÖRDINATES

To abbreviate, let us denote $\dfrac{dx}{dt}, \dfrac{d\theta_1}{dt}, \cdots$ by $\dot{x}, \dot{\theta}_1, \cdots$. Then the equation just obtained may be written

$$\dot{x} = \frac{\partial f}{\partial \theta_1} \dot{\theta}_1 + \frac{\partial f}{\partial \theta_2} \dot{\theta}_2 + \cdots + \frac{\partial f}{\partial \theta_n} \dot{\theta}_n, \tag{159}$$

so that $\dot{x}$ is a linear function of $\dot{\theta}_1, \dot{\theta}_2, \cdots, \dot{\theta}_n$, the coefficients being functions of $\theta_1, \theta_2, \cdots, \theta_n$.

The kinetic energy,

$$T = \tfrac{1}{2} \sum m (\dot{x}^2 + \dot{y}^2 + \dot{z}^2),$$

is now seen to be a quadratic function of $\dot{\theta}_1, \dot{\theta}_2, \cdots, \dot{\theta}_n$, the coefficients being functions of $\theta_1, \theta_2, \cdots, \theta_n$.

The potential energy W depends only on the configuration of the system, so that W is a function of $\theta_1, \theta_2, \cdots, \theta_n$ only.

Thus the function L, or $T - W$, is a function of

$$\theta_1, \theta_2, \cdots, \theta_n, \dot{\theta}_1, \dot{\theta}_2, \cdots, \dot{\theta}_n,$$

say $\qquad L = \phi(\theta_1, \theta_2, \cdots, \theta_n, \dot{\theta}_1, \dot{\theta}_2, \cdots, \dot{\theta}_n). \tag{160}$

The corresponding function L' in the displaced motion is the same function of

$$\theta_1 + \delta\theta_1, \ \theta_2 + \delta\theta_2, \ \cdots, \text{ etc.,}$$

so that

$$L' = \phi(\theta_1 + \delta\theta_1, \ \theta_2 + \delta\theta_2, \ \cdots, \ \theta_n + \delta\theta_n, \ \dot{\theta}_1 + \delta\dot{\theta}_1, \ \cdots).$$

By Taylor's theorem, we may expand L' in the form

$$L' = \phi(\theta_1, \theta_2, \cdots, \theta_n, \dot{\theta}_1, \cdots)$$
$$+ \delta\theta_1 \frac{\partial \phi}{\partial \theta_1} + \delta\theta_2 \frac{\partial \phi}{\partial \theta_2} + \cdots + \delta\theta_n \frac{\partial \phi}{\partial \theta_n} + \delta\dot{\theta}_1 \frac{\partial \phi}{\partial \dot{\theta}_1} + \cdots,$$

or, from equation (160),

$$L' = L + \sum_1^n \delta\theta_1 \frac{\partial L}{\partial \theta_1} + \sum_1^n \delta\dot{\theta}_1 \frac{\partial L}{\partial \dot{\theta}_1}. \tag{161}$$

Equation (156), namely

$$\int_{t_1}^{t_2} \delta(T-W)\,dt = 0,$$

may be written in the form

$$\int_{t_1}^{t_2} (L' - L)\,dt = 0,$$

and this, we now see, may be replaced by

$$\int_{t_1}^{t_2} \left(\sum_1^n \delta\theta_1 \frac{\partial L}{\partial \theta_1} + \sum_1^n \delta\dot\theta_1 \frac{\partial L}{\partial \dot\theta_1} \right) dt = 0. \tag{162}$$

Now we have

$$\delta\dot\theta_1 = \dot\theta_1' - \dot\theta_1 = \frac{d}{dt}(\theta_1 + \delta\theta_1) - \frac{d}{dt}\theta_1 = \frac{d}{dt}(\delta\theta_1),$$

so that

$$\int_{t_1}^{t_2} \frac{\partial L}{\partial \dot\theta_1} \delta\dot\theta_1\,dt = \int_{t_1}^{t_2} \frac{\partial L}{\partial \dot\theta_1} \frac{d}{dt}(\delta\theta_1)\,dt,$$

and, integrated by parts, this becomes

$$\left[\frac{\partial L}{\partial \dot\theta_1} \delta\theta_1 \right]_{t_1}^{t_2} - \int_{t_1}^{t_2} \frac{d}{dt}\left(\frac{\partial L}{\partial \dot\theta_1} \right) \delta\theta_1\,dt. \tag{163}$$

Since the disturbed configuration, by hypothesis, coincides with the actual configuration, we have $\delta\theta_1 = 0$ at times t_1 and t_2. Thus the first term in expression (163) vanishes, and leaves

$$\int_{t_1}^{t_2} \frac{\partial L}{\partial \dot\theta_1} \delta\dot\theta_1\,dt = -\int_{t_1}^{t_2} \frac{d}{dt}\left(\frac{\partial L}{\partial \dot\theta_1} \right) \delta\theta_1\,dt.$$

Equation (162) now assumes the form

$$\int_{t_1}^{t_2} \left\{ \sum_1^n \delta\theta_1 \left[\frac{\partial L}{\partial \theta_1} - \frac{d}{dt}\left(\frac{\partial L}{\partial \dot\theta_1} \right) \right] \right\} dt = 0. \tag{164}$$

The limits t_1 and t_2 are entirely at our disposal; the equation is true whatever values we assign to them. In other words, the sum of a number of small differentials vanishes, no matter how many

332 GENERALIZED COÖRDINATES

of them are included in the sum. It follows that each term of the sum must vanish. Thus we must have

$$\sum_1^n \delta\theta_1 \left[\frac{\partial L}{\partial \theta_1} - \frac{d}{dt}\left(\frac{\partial L}{\partial \dot\theta_1}\right) \right] = 0 \tag{165}$$

at every instant.

271. At this point we have to consider two alternatives. It may be that whatever values are assigned to $\delta\theta_1, \delta\theta_2, \cdots, \delta\theta_n$, the new configuration, specified by coördinates

$$\theta_1 + \delta\theta_1, \ \theta_2 + \delta\theta_2, \ \cdots, \ \theta_n + \delta\theta_n,$$

will be a possible configuration; that is to say, will be one which the system can assume without violating the constraints imposed by the mechanism of the system. In this case the system is said to have *n degrees of freedom*.

If the system has n degrees of freedom, equation (165) is true for all values of $\delta\theta_1, \delta\theta_2, \cdots, \delta\theta_n$. For instance, it is true if we take

$$\delta\theta_1 = \epsilon, \ \delta\theta_2 = \delta\theta_3 = \cdots = \delta\theta_n = 0,$$

where ϵ is any small quantity. In this case we must have

$$\epsilon \left[\frac{\partial L}{\partial \theta_1} - \frac{d}{dt}\left(\frac{\partial L}{\partial \dot\theta_1}\right) \right] = 0,$$

and therefore
$$\frac{d}{dt}\left(\frac{\partial L}{\partial \dot\theta_1}\right) - \frac{\partial L}{\partial \theta_1} = 0.$$

A similar equation will, of course, hold for each of the coördinates $\theta_1, \theta_2, \cdots, \theta_n$. These equations are known as *Lagrange's equations*. There are n equations between the n unknown quantities $\theta_1, \theta_2, \cdots, \theta_n$ and their differential coefficients with respect to the time. Thus they enable us to find the way in which $\theta_1, \theta_2, \cdots, \theta_n$ change with the time. To use the equations we require a knowledge only of the function L, and therefore only of the kinetic and potential energies of the system; we do not need a knowledge of the internal mechanism of the system. Thus the problem proposed in § 263 is solved, if we can solve Lagrange's equations.

LAGRANGE'S EQUATIONS

ILLUSTRATIVE EXAMPLE

Common pendulum. As a simple example of the use of Lagrange's equations, let us consider the problem of the motion of the common pendulum. A rigid body is constrained to move so that one point O remains fixed, while the line OG joining O to the center of gravity moves in a vertical plane. Let θ be the inclination of OG to the vertical; then the position of the system is entirely fixed as soon as the value of θ is known. The kinetic and potential energies are, in the notation of § 245,

$$T = \tfrac{1}{2} M k^2 \dot{\theta}^2, \quad W = Mgh(1 - \cos\theta),$$

so that $\quad L = \tfrac{1}{2} M k^2 \dot{\theta}^2 - Mgh(1 - \cos\theta).$

Thus $\dfrac{\partial L}{\partial \dot{\theta}} = M k^2 \dot{\theta}$, and Lagrange's equation,

$$\frac{d}{dt}\left(\frac{\partial L}{\partial \dot{\theta}}\right) = \frac{\partial L}{\partial \theta},$$

becomes $\quad M k^2 \dfrac{d^2 \theta}{dt^2} = -Mgh \sin\theta,$

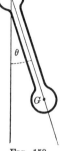

FIG. 153

the same equation as was obtained in § 245, and from this the motion can be deduced.

We notice, however, that Lagrange's method shows that the motion is independent of the method of suspension of the pendulum, provided only that it is constrained to move in the way described. For instance, the result is true if there is no pivot at all at O, the constraints being imposed by a suspension of strings.

272. Let us now consider the second alternative to that examined in § 271. It may be that if we assign arbitrary values to $\delta\theta_1, \delta\theta_2, \cdots, \delta\theta_n$, the new configuration obtained is not in every case a possible one. It may be that there are certain relations which must be satisfied, in order that the constraints imposed by the mechanism may not be violated.

For instance, in the illustration already employed, let there be two ropes hanging from a ceiling of a room, such that on pulling one down one inch the mechanism compels the second to rise two inches. Let θ_1, θ_2 denote the lengths of ropes below the ceiling. Then a displacement in which $\delta\theta_1 = \tfrac{1}{10}$ inch, $\delta\theta_2 = \tfrac{1}{50}$ inch, is not a possible displacement; such a displacement is not permitted by the mechanism above. We must always have $\delta\theta_1$, $\delta\theta_2$ connected by the relation

$$\delta\theta_1 + \tfrac{1}{2} \delta\theta_2 = 0.$$

In general let us suppose that we have certain relations imposed by the mechanism, these being of the form

$$a_1\delta\theta_1 + a_2\delta\theta_2 + \cdots + a_n\delta\theta_n = 0, \qquad (166)$$

$$b_1\delta\theta_1 + b_2\delta\theta_2 + \cdots + b_n\delta\theta_n = 0, \cdots. \qquad (167)\cdots$$

Then equation (165), namely

$$\sum_1^n \left[\frac{d}{dt}\left(\frac{\partial L}{\partial \dot\theta_1}\right) - \frac{\partial L}{\partial \theta_1}\right]\delta\theta_1 = 0, \qquad (168)$$

is true only if $\delta\theta_1, \delta\theta_2, \cdots, \delta\theta_n$ satisfy relations (166), (167), $\cdots$.

For a *possible* displacement, however, $\delta\theta_1, \delta\theta_2, \cdots, \delta\theta_n$ will be such that equations (166), (167) $\cdots$, and (168) are all true. Let us multiply by $\lambda, \mu, \cdots$ and unity, and add, $\lambda, \mu, \cdots$ being quantities as yet undetermined — *undetermined multipliers*, we may call them. Then we have the equation

$$\left[\frac{d}{dt}\left(\frac{\partial L}{\partial \dot\theta_1}\right) - \frac{\partial L}{\partial \theta_1} + \lambda a_1 + \mu b_1 + \cdots\right]\delta\theta_1$$
$$+ \left[\frac{d}{dt}\left(\frac{\partial L}{\partial \dot\theta_2}\right) - \frac{\partial L}{\partial \theta_2} + \lambda a_2 + \mu b_2 + \cdots\right]\delta\theta_2$$
$$+ \cdots + \left[\frac{d}{dt}\left(\frac{\partial L}{\partial \dot\theta_n}\right) - \frac{\partial L}{\partial \theta_n} + \lambda a_n + \mu b_n + \cdots\right]\delta\theta_n = 0. \quad (169)$$

The quantities $\delta\theta_1, \delta\theta_2, \cdots, \delta\theta_n$ are not at our disposal. If, however, the relations of the type (166) are m in number, we may say that of the quantities $\delta\theta_1, \delta\theta_2, \cdots, \delta\theta_n$ all except m are at our disposal, and after arbitrary values have been assigned to $n - m$ of these quantities, the remaining m quantities must be obtained by solving equations (166), (167) $\cdots$. The configuration obtained in this way must necessarily be a possible one.

Let us assign arbitrary values to

$$\delta\theta_{m+1}, \delta\theta_{m+2}, \cdots, \delta\theta_n,$$

LAGRANGE'S EQUATIONS 335

and then find the values of $\delta\theta_1, \delta\theta_2, \cdots, \delta\theta_m$ from equations (166), (167) $\cdots$. Let us, moreover, choose the m undetermined multipliers $\lambda, \mu, \cdots$ so that they satisfy the m equations

$$\frac{d}{dt}\left(\frac{\partial L}{\partial \dot{\theta}_1}\right) - \frac{\partial L}{\partial \theta_1} + \lambda a_1 + \mu b_1 + \cdots = 0, \qquad (170)$$

$$\cdots \cdots \cdots \cdots \cdots \cdots \cdots$$

$$\frac{d}{dt}\left(\frac{\partial L}{\partial \dot{\theta}_m}\right) - \frac{\partial L}{\partial \theta_m} + \lambda a_m + \mu b_m + \cdots = 0, \qquad (171)$$

the suffixes ranging from 1 to m. Then equation (169) reduces to

$$\sum_{m+1}^{n} \left[\frac{d}{dt}\left(\frac{\partial L}{\partial \dot{\theta}_{m+1}}\right) - \frac{\partial L}{\partial \theta_{m+1}} + \lambda a_{m+1} + \mu b_{m+1} + \cdots \right]\delta\theta_{m+1} = 0.$$

Inasmuch as $\delta\theta_{m+1}, \delta\theta_{m+2}, \cdots, \delta\theta_n$ are all arbitrary, we may take

$$\delta\theta_{m+1} = \epsilon, \ \delta\theta_{m+2} = \delta\theta_{m+3} = \cdots = \delta\theta_n = 0,$$

and obtain

$$\frac{d}{dt}\left(\frac{\partial L}{\partial \dot{\theta}_{m+1}}\right) - \frac{\partial L}{\partial \theta_{m+1}} + \lambda a_{m+1} + \mu b_{m+1} + \cdots = 0;$$

and similarly we may obtain the same equation for all suffixes from $m+1$ to n. The equation has, however, already been supposed true for suffixes 1 to m [cf. equations (170) $\cdots$, (171)].

Thus we have the complete system of equations

$$\frac{d}{dt}\left(\frac{\partial L}{\partial \dot{\theta}_1}\right) - \frac{\partial L}{\partial \theta_1} + \lambda a_1 + \mu b_1 + \cdots = 0,$$

$$\cdots \cdots \cdots \cdots \cdots \cdots \cdots$$

$$\frac{d}{dt}\left(\frac{\partial L}{\partial \dot{\theta}_n}\right) - \frac{\partial L}{\partial \theta_n} + \lambda a_n + \mu b_n + \cdots = 0,$$

in which the suffixes range from 1 to n. On eliminating the m multipliers $\lambda, \mu, \cdots$ from these n equations, we are left with $n-m$ equations, which enable us to determine the changes in the coördinates.

ILLUSTRATIVE EXAMPLES

1. *A homogeneous sphere of radius a rolls down the outer surface of a fixed sphere of radius b without sliding. Find the motion.*

At any instant let the line of centers make an angle ϕ with the vertical, and let the angular velocity of the rolling sphere be $\dot{\theta}$. The velocity of the center of the rolling sphere is $(a+b)\dot{\phi}$, so that

$$T = \tfrac{1}{2} m [(a+b)^2 \dot{\phi}^2 + \tfrac{2}{5} a^2 \dot{\theta}^2].$$

The potential energy is

$$W = mg(a+b)\cos\phi,$$

so that $L = T - W$

$$= \tfrac{1}{2} m (a+b)^2 \dot{\phi}^2 + \tfrac{1}{5} m a^2 \dot{\theta}^2 - mg(a+b)\cos\phi. \quad (a)$$

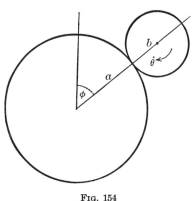

FIG. 154

The variations in θ and ϕ are not capable of having any values we please, for the velocity of the center of the moving sphere is $(a+b)\dot{\phi}$, and also must be $a\dot{\theta}$, since the sphere is rolling with angular velocity $\dot{\theta}$ without sliding. Thus

$$a\dot{\theta} = (a+b)\dot{\phi}. \quad (b)$$

This is true at every instant of every possible motion, so that we must have, on integrating with respect to the time,

$$a\theta = (a+b)\phi + \text{a constant},$$

and hence we must suppose changes in the coördinates θ, ϕ to be connected by the relation

$$a\,\delta\theta = (a+b)\,\delta\phi.$$

Thus Lagrange's equations are

$$\frac{d}{dt}\left(\frac{\partial L}{\partial \dot{\theta}}\right) - \frac{\partial L}{\partial \theta} + \lambda a = 0,$$

$$\frac{d}{dt}\left(\frac{\partial L}{\partial \dot{\phi}}\right) - \frac{\partial L}{\partial \phi} - \lambda(a+b) = 0.$$

Eliminating λ, we obtain

$$(a+b)\left[\frac{d}{dt}\left(\frac{\partial L}{\partial \dot{\theta}}\right) - \frac{\partial L}{\partial \theta}\right] + a\left[\frac{d}{dt}\left(\frac{\partial L}{\partial \dot{\phi}}\right) - \frac{\partial L}{\partial \phi}\right] = 0.$$

Substituting from equation (a), this becomes

$$(a+b)\left[\frac{d}{dt}\left(\tfrac{2}{5} m a^2 \dot{\theta}\right)\right] + a\left[\frac{d}{dt}(m(a+b)^2 \dot{\phi}) - mg(a+b)\sin\phi\right] = 0.$$

ILLUSTRATIVE EXAMPLES

After replacing $a\dot{\theta}$ by $(a + b)\dot{\phi}$ from equation (b), we have

$$\frac{7}{5} ma(a+b)^2 \frac{d^2\phi}{dt^2} = mga(a+b)\sin\phi,$$

or
$$(a+b)\frac{d^2\phi}{dt^2} = \frac{5}{7}g\sin\phi,$$

showing that the center of the moving sphere moves with five sevenths of the acceleration of a smooth particle sliding down a sphere of radius $a + b$.

The same result could have been obtained by eliminating $\dot{\theta}$ from equations (a) and (b), and then regarding ϕ as a single Lagrangian coördinate.

2. *A flywheel is connected by a crank and rod to a piston moving in a horizontal cylinder. When there is no steam in the engine the flywheel rests in its position of equilibrium. Find its motion if displaced.*

Let a, b denote the length of the crank and rod, and let θ, ϕ be the angles they make with the horizontal in any position of the flywheel. Then the position of the engine is known fully when θ and ϕ are known. Not only do the

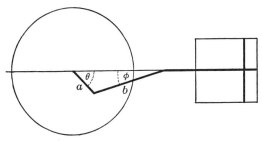

Fig. 155

values of θ and ϕ suffice to determine the position of the engine, but if we assign arbitrary values to θ and ϕ, we do not necessarily obtain a possible position for the engine.

The velocity of rotation of flywheel, axle, and crank is $\dot{\theta}$, so that the kinetic energy of this motion is $\frac{1}{2}I\dot{\theta}^2$, where I is the moment of inertia of this part of the engine about the axis of the flywheel. The coördinates of the center of gravity of the rod, which we shall assume to be its middle point, measured from the axis of the flywheel, are:

horizontal: $a\cos\theta + \frac{1}{2}b\cos\phi$,
vertical: $\frac{1}{2}b\sin\phi$.

Thus the velocity of its center of gravity has components

$$-(a\sin\theta \cdot \dot{\theta} + \tfrac{1}{2}b\sin\phi \cdot \dot{\phi})$$

horizontally, and $\frac{1}{2} b \cos \phi \cdot \dot\phi$ vertically. The whole velocity v of the center of gravity of the rod is therefore given by

$$v^2 = (a \sin\theta \cdot \dot\theta + \tfrac{1}{2} b \sin\phi \cdot \dot\phi)^2 + (\tfrac{1}{2} b \cos\phi \cdot \dot\phi)^2$$
$$= a^2 \sin^2\theta \cdot \dot\theta^2 + ab \sin\theta \sin\phi \cdot \dot\phi\dot\theta + \tfrac{1}{4} b^2 \dot\phi^2.$$

The angular velocity of the rod is $\dot\phi$, and its radius of gyration k is given by

$$k^2 = \frac{1}{3}\left(\frac{b}{2}\right)^2 = \frac{1}{12} b^2.$$

Thus, if m is the mass of the rod, the kinetic energy of the rod is

$$\tfrac{1}{2} m (v^2 + k^2 \dot\phi^2)$$
$$= \tfrac{1}{2} m (a^2 \sin^2\theta \cdot \dot\theta^2 + ab \sin\theta \sin\phi \cdot \dot\phi\dot\theta + \tfrac{1}{3} b^2 \dot\phi^2).$$

Lastly, the horizontal distance of the end of the piston rod from the center of the flywheel is $a \cos\theta + b \cos\phi$, so that the velocity of the piston and piston rod is
$$- a \sin\theta \cdot \dot\theta - b \sin\phi \cdot \dot\phi.$$

If M is the mass of the piston and piston rod, the kinetic energy of this part of the engine is
$$\tfrac{1}{2} M (a \sin\theta \cdot \dot\theta + b \sin\phi \cdot \dot\phi)^2.$$

We now have, for the whole kinetic energy T,

$$2T = I\dot\theta^2 + m(a^2 \sin^2\theta \cdot \dot\theta^2 + ab \sin\theta \sin\phi \cdot \dot\phi\dot\theta + \tfrac{1}{3} b^2 \dot\phi^2)$$
$$+ M(a \sin\theta \cdot \dot\theta + b \sin\phi \cdot \dot\phi)^2. \qquad (a)$$

The potential energy W, measured from a standard configuration in which $\theta = \phi = 0$, is
$$W = - \mathcal{M}gh \sin(\theta + \epsilon) - \tfrac{1}{2} mgb \sin\phi. \qquad (b)$$

Here $\mathcal{M}$ is the total mass of flywheel and crank, and h, ϵ are the polar coördinates of its center of gravity when $\theta = 0$.

The changes in θ and ϕ are not independent. A glance at the figure shows that we must always have
$$a \sin\theta = b \sin\phi, \qquad (c)$$

and on differentiating this, we can see that if θ and ϕ are taken as generalized coördinates, we must suppose them connected by

$$a \cos\theta\, \delta\theta - b \cos\phi\, \delta\phi = 0.$$

Thus Lagrange's equations will be

$$\frac{d}{dt}\left(\frac{\partial L}{\partial \dot\theta}\right) - \frac{\partial L}{\partial \theta} + \lambda a \cos\theta = 0, \qquad (d)$$

$$\frac{d}{dt}\left(\frac{\partial L}{\partial \dot\phi}\right) - \frac{\partial L}{\partial \phi} - \lambda b \cos\phi = 0. \qquad (e)$$

LAGRANGE'S EQUATIONS

The elimination of λ from these equations gives

$$b \cos \phi \left[\frac{d}{dt}\left(\frac{\partial L}{\partial \dot\theta}\right) - \frac{\partial L}{\partial \theta} \right] + a \cos \theta \left[\frac{d}{dt}\left(\frac{\partial L}{\partial \dot\phi}\right) - \frac{\partial L}{\partial \phi} \right] = 0,$$

and on substituting for L from equations (a) and (b), this equation becomes an equation between θ, ϕ, and their differential coefficients with respect to the time. From this and the geometrical relation (c),

$$a \sin \theta = b \sin \phi, \qquad (f)$$

we can proceed to determine θ and ϕ in terms of the time.

Using equation (f), we can transform equation (b) into

$$W = -\mathcal{M}gh \sin(\theta + \epsilon) - \tfrac{1}{2} mga \sin \theta.$$

It will be possible to arrange counterpoises on the flywheel in such a way as to make
$$\epsilon = 0, \quad \mathcal{M}h + \tfrac{1}{2} ma = 0,$$

and if this is done, the center of gravity will always be at the same height. This is called *balancing* the engine.

If we suppose the engine balanced in this way, we have $W = 0$ and therefore $L = T$. We can, however, determine the motion much more simply than by using Lagrange's equations, for we know that T must remain constant throughout the motion; and by differentiation of equation (f) we have

$$a \cos \theta \, \dot\theta = b \cos \phi \, \dot\phi,$$

so that we can replace equation (a) by

$$2T = I\dot\theta^2 + m(a^2 \sin^2\theta \, \dot\theta^2 + a^2 \sin\theta \cos\theta \tan\phi \, \dot\theta^2 + \tfrac{1}{3} a^2 \cos^2\theta \sec^2\phi \, \dot\theta^2)$$
$$+ M(a \sin\theta \cdot \dot\theta + a \cos\theta \tan\phi \cdot \dot\theta)^2$$
$$= \dot\theta^2 [I + ma^2 \sin\theta \sin(\theta + \phi) \sec\phi + \tfrac{1}{3} ma^2 \cos^2\theta \sec^2\phi$$
$$+ Ma^2 \sin^2(\theta + \phi) \sec^2\phi].$$

This is constant throughout the motion, but we see that it does not follow that $\dot\theta$ is constant. Thus, although the engine is balanced so as to remain at rest in any position, it will not necessarily run evenly if started into motion.

Lagrange's Equations for Non-Conservative Systems

273. For non-conservative systems it has been shown (§ 268) that equation (156), namely

$$\int_{t_1}^{t_2} \delta L \, dt = 0, \qquad (172)$$

must be replaced by

$$\int_{t_1}^{t_2} [\delta T + \sum (X \delta x + Y \delta y + Z \delta z)] \, dt = 0. \qquad (173)$$

340 GENERALIZED COÖRDINATES

Now since, as in equation (158),
$$x = f(\theta_1, \theta_2, \cdots, \theta_n),$$
we must have
$$\begin{aligned}\delta x &= x' - x \\ &= f(\theta_1 + \delta\theta_1, \theta_2 + \delta\theta_2, \cdots) - f(\theta_1, \theta_2, \cdots) \\ &= \frac{\partial f}{\partial \theta_1}\delta\theta_1 + \frac{\partial f}{\partial \theta_2}\delta\theta_2 + \cdots,\end{aligned}$$
neglecting small quantities of the second order.

Thus
$$\sum (X\,\delta x + Y\,\delta y + Z\,\delta z) = \Theta_1\,\delta\theta_1 + \Theta_2\,\delta\theta_2 + \cdots + \Theta_n\,\delta\theta_n,$$
where $\Theta_1, \Theta_2, \cdots, \Theta_n$ depend on the configuration of the system, and therefore are functions of $\theta_1, \theta_2, \cdots, \theta_n$ only.

Equation (173) now becomes
$$\int_{t_1}^{t_2} \delta T\,dt + \int_{t_1}^{t_2} (\Theta_1\,\delta\theta_1 + \Theta_2\,\delta\theta_2 + \cdots + \Theta_n\,\delta\theta_n)\,dt = 0. \quad (174)$$

Just as in § 270 we found that
$$\int_{t_1}^{t_2} \delta L\,dt$$
could be transformed into
$$\int_{t_1}^{t_2} \left\{ \sum_1^n \delta\theta_1 \left[\frac{\partial L}{\partial \theta_1} - \frac{d}{dt}\left(\frac{\partial L}{\partial \dot\theta_1}\right)\right] \right\} dt,$$
so the first term of equation (174) can now be transformed into
$$\int_{t_1}^{t_2} \left\{ \sum_1^n \delta\theta_1 \left[\frac{\partial T}{\partial \theta_1} - \frac{d}{dt}\left(\frac{\partial T}{\partial \dot\theta_1}\right)\right] \right\} dt.$$

Substituting this, the equation becomes
$$\int_{t_1}^{t_2} \left\{ \sum_1^n \delta\theta_1 \left[\frac{\partial T}{\partial \theta_1} - \frac{d}{dt}\left(\frac{\partial T}{\partial \dot\theta_1}\right) + \Theta_1 \right] \right\} dt = 0.$$

LAGRANGE'S EQUATIONS

Since this is true for all possible ranges of time, we must have

$$\sum_1^n \delta\theta_1 \left[\frac{\partial T}{\partial \theta_1} - \frac{d}{dt}\left(\frac{\partial T}{\partial \dot\theta_1}\right) + \Theta_1 \right] = 0 \tag{175}$$

at every instant.

If the θ's can vary independently, each coefficient must vanish, and the system of equations will be

$$\frac{d}{dt}\left(\frac{\partial T}{\partial \dot\theta_1}\right) - \frac{\partial T}{\partial \theta_1} - \Theta_1 = 0, \tag{176}$$

etc., while if $\delta\theta_1, \delta\theta_2, \cdots$ are connected by the constraints implied in equations (166), (167), $\cdots$, we find, as in § 272, that the system of equations must be replaced by

$$\frac{d}{dt}\left(\frac{\partial T}{\partial \dot\theta_1}\right) - \frac{\partial T}{\partial \theta_1} - \Theta_1 + \lambda a_1 + \mu b_1 + \cdots = 0. \tag{177}$$

274. These systems of equations reduce to those previously obtained in the special case in which the forces are conservative. For in this case consider the work done in a slight displacement in which θ_1 alone varies, θ_1 being increased by $\delta\theta_1$. It is $\Theta_1 \delta\theta_1$, and is also $-\frac{\partial W}{\partial \theta_1}\delta\theta_1$, so that $\Theta_1 = -\frac{\partial W}{\partial \theta_1}$.

Thus
$$\frac{\partial T}{\partial \theta_1} + \Theta_1 = \frac{\partial T}{\partial \theta_1} - \frac{\partial W}{\partial \theta_1} = \frac{\partial L}{\partial \theta_1},$$

and
$$\frac{\partial T}{\partial \dot\theta_1} = \frac{\partial (L+W)}{\partial \dot\theta_1} = \frac{\partial L}{\partial \dot\theta_1},$$

since W does not contain $\dot\theta_1$. Thus equation (175) reduces to

$$\frac{d}{dt}\left(\frac{\partial L}{\partial \dot\theta_1}\right) - \frac{\partial L}{\partial \theta_1} = 0,$$

as before, and in the same way, equations (177), $\cdots$ can be transformed into the equations obtained in § 272.

Lagrange's Equations by Direct Transformation

275. Instead of deducing Lagrange's equations from equation (156), they may be obtained directly by transformation of the equations of motion.

We have, as before,
$$x = f(\theta_1, \theta_2, \ldots, \theta_n),$$
so that, on differentiation,
$$\frac{dx}{dt} = \frac{\partial f}{\partial \theta_1}\frac{d\theta_1}{dt} + \frac{\partial f}{\partial \theta_2}\frac{d\theta_2}{dt} + \cdots,$$
or
$$\dot{x} = \frac{\partial x}{\partial \theta_1}\dot{\theta}_1 + \frac{\partial x}{\partial \theta_2}\dot{\theta}_2 + \cdots. \qquad (178)$$

Thus $\dot{x}$ is a linear function of $\dot{\theta}_1, \dot{\theta}_2, \ldots$, and
$$\frac{\partial \dot{x}}{\partial \dot{\theta}_1} = \frac{\partial x}{\partial \theta_1}. \qquad (179)$$

We have $\quad T = \tfrac{1}{2}\sum m(\dot{x}^2 + \dot{y}^2 + \dot{z}^2),$

so that T, as before, is a quadratic function of $\dot{\theta}_1, \dot{\theta}_2, \ldots$, which also involves $\theta_1, \theta_2, \ldots$. By differentiation,
$$\frac{\partial T}{\partial \dot{\theta}_1} = \sum m\left(\dot{x}\frac{\partial \dot{x}}{\partial \dot{\theta}_1} + \dot{y}\frac{\partial \dot{y}}{\partial \dot{\theta}_1} + \dot{z}\frac{\partial \dot{z}}{\partial \dot{\theta}_1}\right),$$
or, by equation (179),
$$\frac{\partial T}{\partial \dot{\theta}_1} = \sum m\left(\dot{x}\frac{\partial x}{\partial \theta_1} + \dot{y}\frac{\partial y}{\partial \theta_1} + \dot{z}\frac{\partial z}{\partial \theta_1}\right).$$

Thus
$$\frac{d}{dt}\left(\frac{\partial T}{\partial \dot{\theta}_1}\right) = \sum m\left(\frac{d^2x}{dt^2}\frac{\partial x}{\partial \theta_1} + \frac{d^2y}{dt^2}\frac{\partial y}{\partial \theta_1} + \frac{d^2z}{dt^2}\frac{\partial z}{\partial \theta_1}\right)$$
$$+ \sum m\left[\dot{x}\frac{d}{dt}\left(\frac{\partial x}{\partial \theta_1}\right) + \dot{y}\frac{d}{dt}\left(\frac{\partial y}{\partial \theta_1}\right) + \dot{z}\frac{d}{dt}\left(\frac{\partial z}{\partial \theta_1}\right)\right]. \quad (180)$$

Since $\dfrac{\partial x}{\partial \theta_1}$ is a function of $\theta_1, \theta_2, \cdots$, we have

$$\frac{d}{dt}\left(\frac{\partial x}{\partial \theta_1}\right) = \frac{\partial^2 x}{\partial \theta_1^2}\frac{d\theta_1}{dt} + \frac{\partial^2 x}{\partial \theta_1 \partial \theta_2}\frac{d\theta_2}{dt} + \cdots, \qquad (181)$$

while, by differentiation of equation (178),

$$\frac{\partial \dot{x}}{\partial \theta_1} = \frac{\partial^2 x}{\partial \theta_1^2}\dot{\theta}_1 + \frac{\partial^2 x}{\partial \theta_1 \partial \theta_2}\dot{\theta}_2 + \cdots. \qquad (182)$$

The right-hand members of equations (181) and (182) are seen to be identical, so that

$$\frac{d}{dt}\left(\frac{\partial x}{\partial \theta_1}\right) = \frac{\partial \dot{x}}{\partial \theta_1},$$

and the last line of equation (180) transforms into

$$\sum m\left(\dot{x}\frac{\partial \dot{x}}{\partial \theta_1} + \dot{y}\frac{\partial \dot{y}}{\partial \theta_1} + \dot{z}\frac{\partial \dot{z}}{\partial \theta_1}\right),$$

of which the value is

$$\frac{\partial}{\partial \theta_1}\sum m\left[\tfrac{1}{2}(\dot{x}^2 + \dot{y}^2 + \dot{z}^2)\right],$$

or $\qquad\qquad\qquad\qquad \dfrac{\partial T}{\partial \theta_1}.$

Equation (180) now becomes

$$\frac{d}{dt}\left(\frac{\partial T}{\partial \dot{\theta}_1}\right) - \frac{\partial T}{\partial \theta_1} = \sum m\left(\frac{d^2 x}{dt^2}\frac{\partial x}{\partial \theta_1} + \frac{d^2 y}{dt^2}\frac{\partial y}{\partial \theta_1} + \frac{d^2 z}{dt^2}\frac{\partial z}{\partial \theta_1}\right).$$

From the equations of motion,

$$X = m\frac{d^2 x}{dt^2}, \text{ etc.,}$$

so that this again becomes

$$\frac{d}{dt}\left(\frac{\partial T}{\partial \dot{\theta}_1}\right) - \frac{\partial T}{\partial \theta_1} = \sum\left(X\frac{\partial x}{\partial \theta_1} + Y\frac{\partial y}{\partial \theta_1} + Z\frac{\partial z}{\partial \theta_1}\right).$$

344 GENERALIZED COÖRDINATES

If we give the system a small displacement, in which θ_1 is increased to $\theta_1 + \delta\theta_1$, θ_2 to $\theta_2 + \delta\theta_2$, etc., we have, on equating two different expressions for the work done:

$$\Theta_1 \delta\theta_1 + \Theta_2 \delta\theta_2 + \cdots$$
$$= \sum (X\,\delta x + Y\,\delta y + Z\,\delta z).$$

On substituting the value of δx obtained on page 340, this

$$= \sum \left[X\left(\frac{\partial x}{\partial \theta_1}\delta\theta_1 + \frac{\partial x}{\partial \theta_2}\delta\theta_2 + \cdots\right) + Y(\cdots) + Z(\cdots) \right]$$

$$= \delta\theta_1 \sum \left(X\frac{\partial x}{\partial \theta_1} + Y\frac{\partial y}{\partial \theta_1} + Z\frac{\partial z}{\partial \theta_1} \right) + \delta\theta_2 \sum (\cdots) + \cdots$$

$$= \delta\theta_1 \left[\frac{d}{dt}\left(\frac{\partial T}{\partial \dot{\theta}_1}\right) - \frac{\partial T}{\partial \theta_1} \right] + \delta\theta_2 [\cdots] + \cdots,$$

and the equation reduces to

$$\sum \delta\theta_1 \left[\frac{d}{dt}\left(\frac{\partial T}{\partial \dot{\theta}_1}\right) - \frac{\partial T}{\partial \theta_1} - \Theta_1 \right] = 0.$$

This is the same equation as equation (175), and the different forms of Lagrange's equations can be deduced as before.

Lagrange's Equations for Impulsive Forces

276. Let a system of impulses act during the short interval from $t = t_1$ to $t = t_2$. Let $\theta_1, \theta_2, \cdots, \theta_n$ now be supposed to be independent coördinates, so that Lagrange's equations are

$$\frac{d}{dt}\left(\frac{\partial T}{\partial \dot{\theta}_1}\right) - \frac{\partial T}{\partial \theta_1} = \Theta_1, \text{ etc.}$$

If we multiply by dt, and integrate from $t = t_1$ to $t = t_2$, we have

$$\int_{t_1}^{t_2} \frac{d}{dt}\left(\frac{\partial T}{\partial \dot{\theta}_1}\right) dt - \int_{t_1}^{t_2} \frac{\partial T}{\partial \theta_1} dt = \int_{t_1}^{t_2} \Theta_1\, dt.$$

The value of the first term is

$$\left(\frac{\partial T}{\partial \dot{\theta}_1}\right)_{t=t_2} - \left(\frac{\partial T}{\partial \dot{\theta}_1}\right)_{t=t_1},$$

LAGRANGE'S EQUATIONS

and when the interval from t_1 to t_2 is made vanishingly small, this measures simply the change in $\dfrac{\partial T}{\partial \dot{\theta}_1}$ produced by the impulse.

In the second term $\displaystyle\int_{t_1}^{t_2} \dfrac{\partial T}{\partial \theta_1} dt$, the integrand $\dfrac{\partial T}{\partial \theta_1}$ is finite, so that when the interval of time is supposed to vanish, this term will vanish with it. Thus the equation becomes

$$\text{change in } \frac{\partial T}{\partial \dot{\theta}_1} = \int_{t_1}^{t_2} \Theta_1 \, dt. \tag{183}$$

277. If F is an ordinary force acting impulsively through the interval t_1 to t_2, we call $\displaystyle\int_{t_1}^{t_2} F dt$ the impulse. By analogy we call

$$\int_{t_1}^{t_2} \Theta_1 \, dt$$

the *generalized impulse*, corresponding to the generalized coördinate θ_1. Thus we have equation (183) in the form

$$\text{change in } \frac{\partial T}{\partial \dot{\theta}_1} = \text{generalized impulse.}$$

From analogy with the relation,

change in momentum of a particle = impulse on particle,

we call $\dfrac{\partial T}{\partial \dot{\theta}_1}$ the *generalized momentum* corresponding to the coördinate θ_1. Thus with these meanings attached to the terms "impulse" and "momentum," the relation

change of momentum = impulse

is true in generalized coördinates.

When our coördinates are x, y, z, the coördinates in space of a moving particle, the generalized momenta will of course become identical with the ordinary components of momentum. We have

$$T = \tfrac{1}{2} m (\dot{x}^2 + \dot{y}^2 + \dot{z}^2),$$

so that

$$\frac{\partial T}{\partial \dot{x}} = m\dot{x}, \text{ etc.}$$

EULER'S EQUATIONS FOR A RIGID BODY

278. Euler's equations (§ 252) can be derived from those of Lagrange.

Let the moments of a rigid body about its principal axes of inertia at a point O, which is fixed in the body and is also either fixed in space or is the center of gravity of the body, be A, B, C. Then if ω_1, ω_2, ω_3 are the components of rotation about these axes, we have, as in § 248,

$$T = \tfrac{1}{2}(A\omega_1^2 + B\omega_2^2 + C\omega_3^2). \tag{184}$$

As Lagrangian coördinates, let us take θ, ϕ the spherical polar coördinates of the third axis OC of the body, and ψ a third coördinate which measures the angle between the first axis OA of the rigid body and the plane through OC and the axis $\theta = 0$, say the plane COz.

We have first to find ω_1, ω_2 and ω_3 in terms of θ, ϕ and ψ, so as to express $2T$ as a function of these coördinates. The motion of the body is compounded of the motion relative to the plane COz, together with the motion of the plane COz relative to fixed axes. The former motion consists of a rotation $\dot\psi$ about OC, and this, resolved along the axes OA, OB, OC, has components

$$0, \quad 0, \quad \dot\psi.$$

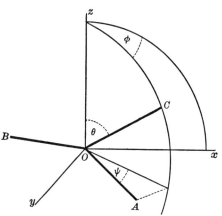

Fig. 156

The motion of the plane COz is compounded of

(a) a rotation $\dot\theta$ about an axis at right angles to its plane;
(b) a rotation $\dot\phi$ about the axis $\theta = 0$.

EULER'S EQUATIONS FOR A RIGID BODY 347

Resolved along the axes OA, OB, OC, the first part has components
$$\dot\theta\sin\psi,\qquad \dot\theta\cos\psi,\qquad 0,$$
while the second has components
$$-\dot\phi\sin\theta\cos\psi,\qquad \dot\phi\sin\theta\sin\psi,\qquad \dot\phi\cos\theta.$$

Compounding these motions, we obtain
$$\left.\begin{array}{l}\omega_1 = \dot\theta\sin\psi - \dot\phi\sin\theta\cos\psi\\ \omega_2 = \dot\theta\cos\psi + \dot\phi\sin\theta\sin\psi\\ \omega_3 = \dot\psi\qquad\ \ + \dot\phi\cos\theta\end{array}\right\}. \qquad(185)$$

Let the work done in a small displacement be
$$\Theta\,\delta\theta + \Phi\,\delta\phi + \Psi\,\delta\psi\,;$$
then Lagrange's equation for the coördinate Ψ is
$$\frac{\partial}{\partial t}\left(\frac{\partial T}{\partial\dot\psi}\right) - \frac{\partial T}{\partial\psi} = \Psi. \qquad(186)$$

We have, by differentiation of equation (184),
$$\frac{\partial T}{\partial\dot\psi} = A\omega_1\frac{\partial\omega_1}{\partial\dot\psi} + B\omega_2\frac{\partial\omega_2}{\partial\dot\psi} + C\omega_3\frac{\partial\omega_3}{\partial\dot\psi}$$
$$= C\omega_3,$$
on substituting from equations (185). Also
$$\frac{\partial T}{\partial\psi} = A\omega_1\frac{\partial\omega_1}{\partial\psi} + B\omega_2\frac{\partial\omega_2}{\partial\psi} + C\omega_3\frac{\partial\omega_3}{\partial\psi}$$
$$= A\omega_1(\dot\theta\cos\psi + \dot\phi\sin\theta\sin\psi)$$
$$\quad + B\omega_2(-\dot\theta\sin\psi + \dot\phi\sin\theta\cos\psi)$$
$$= (A - B)\,\omega_1\omega_2.$$

Finally $\Psi\,\delta\psi$ is the work done by external forces in a small rotation $\delta\psi$, and therefore, by § 121, Ψ is equal to N, the sum of the moments of these forces about the axis OC.

348 GENERALIZED COÖRDINATES

Making all these substitutions, equation (186) becomes

$$C\frac{d\omega_3}{dt} - (A - B)\omega_1\omega_2 = N,$$

which is Euler's third equation, and the other two equations follow from symmetry.

SMALL OSCILLATIONS

279. Let $\theta_1, \theta_2, \cdots, \theta_n$ be generalized coördinates of any system, and let it be supposed that these coördinates are all independent, so that any set of values of $\theta_1, \theta_2, \cdots, \theta_n$ gives a possible configuration of the system.

Suppose that the configuration

$$\theta_1 = \bar{\theta}_1, \qquad \theta_2 = \bar{\theta}_2, \qquad \cdots, \qquad \theta_n = \bar{\theta}_n \qquad (187)$$

is known to be a configuration of equilibrium. Then, if

$$\phi_1 = \theta_1 - \bar{\theta}_1, \qquad \phi_2 = \theta_2 - \bar{\theta}_2, \qquad \cdots, \qquad \phi_n = \theta_n - \bar{\theta}_n,$$

the quantities $\phi_1, \phi_2, \cdots, \phi_n$ may be taken to be generalized coördinates of the system, and will possess the property of all vanishing in the position of equilibrium.

Let W_0 denote the value of the potential energy in the configuration of equilibrium. The potential energy in any other configuration may, by Taylor's theorem, be expanded in the form

$$W = W_0 + \phi_1\frac{\partial W}{\partial \theta_1} + \phi_2\frac{\partial W}{\partial \theta_2} + \cdots + \phi_n\frac{\partial W}{\partial \theta_n}$$
$$+ \tfrac{1}{2}\left(\phi_1^2\frac{\partial^2 W}{\partial \theta_1^2} + 2\phi_1\phi_2\frac{\partial^2 W}{\partial \theta_1 \partial \theta_2} + \cdots\right) + \cdots,$$

where all the differential coefficients are evaluated in the position of equilibrium. In this position of equilibrium, however, we have, by the theorem of § 135,

$$\frac{\partial W}{\partial \theta_1} = \frac{\partial W}{\partial \theta_2} = \cdots = \frac{\partial W}{\partial \theta_n} = 0,$$

SMALL OSCILLATIONS 349

so that we can write the value of W in the form

$$W = W_0 + a_{11}\phi_1^2 + 2\,a_{12}\phi_1\phi_2 + \cdots + a_{nn}\phi_n^2, \qquad (188)$$

in which powers of $\phi_1, \phi_2, \cdots$ higher than the second are left out of account, because we are going to confine our attention to motions in which $\phi_1, \phi_2, \cdots$ are all small quantities.

The kinetic energy, as before (§ 270), is a quadratic function of $\dot{\phi}_1, \dot{\phi}_2, \cdots, \dot{\phi}_n$. Let us say

$$T = b_{11}\dot{\phi}_1^2 + 2\,b_{12}\dot{\phi}_1\dot{\phi}_2 + \cdots + b_{nn}\dot{\phi}_n^2. \qquad (189)$$

The coefficients $b_{11}, b_{12}, \cdots, b_{nn}$ are, strictly speaking, functions of $\theta_1, \theta_2, \cdots, \theta_n$, but we may regard their values as being equal to the values in the configuration of equilibrium, and so may treat them as constants.

280. Now consider two quadratic functions of n variables $x_1, x_2, \cdots, x_n$, defined by

$$f(x_1, x_2, \cdots, x_n) = a_{11}x_1^2 + 2\,a_{12}x_1x_2 + \cdots + a_{nn}x_n^2,$$
$$F(x_1, x_2, \cdots, x_n) = b_{11}x_1^2 + 2\,b_{12}x_1x_2 + \cdots + b_{nn}x_n^2.$$

Since the function T defined by equation (189) is necessarily positive, it follows that $F(x_1, x_2, \cdots, x_n)$ is positive for all values of $x_1, x_2, \cdots, x_n$. Hence, by a known theorem in algebra, we can find a transformation of the type

$$\left.\begin{aligned}x_1 &= \kappa_{11}\xi_1 + \kappa_{12}\xi_2 + \cdots + \kappa_{1n}\xi_n \\ x_2 &= \kappa_{21}\xi_1 + \kappa_{22}\xi_2 + \cdots + \kappa_{2n}\xi_n\end{aligned}\right\}, \qquad (190)$$

in which the coefficients κ_{11}, etc., are real, which is such that f and F transform into expressions of the type

$$f(x_1, x_2, \cdots, x_n) = \alpha_1\xi_1^2 + \alpha_2\xi_2^2 + \cdots + \alpha_n\xi_n^2,$$
$$F(x_1, x_2, \cdots, x_n) = \beta_1\xi_1^2 + \beta_2\xi_2^2 + \cdots + \beta_n\xi_n^2,$$

and all the coefficients $\beta_1, \beta_2, \cdots, \beta_n$ will be positive.

Algebraic proofs of this theorem will be found in treatises on analysis, or in Salmon's *Higher Algebra*, Lesson VI. The theorem will be readily understood on considering a geometrical interpretation in the case in which the number of variables is three. Calling the variables x, y, and z, the equations

$$f(x, y, z) = 1, \quad F(x, y, z) = 1 \tag{191}$$

will be the equations of concentric quadrics; and since F is positive for all values of x, y, z, the second quadric will be an ellipsoid. It is known that two concentric quadrics, of which one is an ellipsoid, always have one real set of mutually conjugate diameters in common. A transformation of the type expressed by equation (194) enables us to transform to these axes as axes of coördinates, and the equations of the quadrics are then of the required forms

$$\alpha_1 \xi_1^2 + \alpha_2 \xi_2^2 + \alpha_3 \xi_3^2 = 1, \quad \beta_1 \xi_1^2 + \beta_2 \xi_2^2 + \beta_3 \xi_3^2 = 1. \tag{192}$$

[Simple reasoning will show the truth of the geometrical theorem that an ellipsoid and a second quadric always have one common set of real mutually conjugate diameters. For a real linear transformation will transform the ellipsoid into a sphere, and the second quadric into a new, but still real, quadric. The principal axes of this real quadric are now real mutually conjugate diameters for the sphere and the quadric, and on transforming back, real mutually conjugate diameters remain real mutually conjugate diameters.]

The algebraic proof that equations (191) could be transformed into equations (192) would, however, clearly not be limited to the case of three variables, so that the theorem must be true for any number of variables.

281. This theorem proves that we can find new coördinates $\psi_1, \psi_2, \cdots, \psi_n$ connected with $\phi_1, \phi_2, \cdots, \phi_n$ by relations of the type

$$\phi_1 = \kappa_{11} \psi_1 + \kappa_{12} \psi_2 + \cdots + \kappa_{1n} \psi_n, \tag{193}$$

$$\dot{\phi}_1 = \kappa_{11} \dot{\psi}_1 + \kappa_{12} \dot{\psi}_2 + \cdots + \kappa_{1n} \dot{\psi}_n, \tag{194}$$

such that, expressed in terms of these coördinates, the potential and kinetic energies assume the forms

$$W = W_0 + \alpha_1 \psi_1^2 + \alpha_2 \psi_2^2 + \cdots + \alpha_n \psi_n^2, \tag{195}$$

$$T = \beta_1 \dot{\psi}_1^2 + \beta_2 \dot{\psi}_2^2 + \cdots + \beta_n \dot{\psi}_n^2. \tag{196}$$

The coördinates $\psi_1, \psi_2, \cdots, \psi_n$ are called the *principal coördinates* of the system, or, by some writers, the *normal coördinates*.

SMALL OSCILLATIONS

Lagrange's equations, in terms of these coördinates, are

$$\frac{d}{dt}\left(\frac{\partial T}{\partial \dot{\psi}_1}\right) - \frac{\partial T}{\partial \psi_1} = -\frac{\partial W}{\partial \psi_1}, \text{ etc.,}$$

which become $\quad \beta_1 \dfrac{d^2\psi_1}{dt^2} = -\alpha_1\psi_1$, etc. $\qquad(197)$

Stable Equilibrium

282. If α_1 is positive, let us put $\dfrac{\alpha_1}{\beta_1} = k_1^2$, so that k_1 will be real. The equation is now

$$\frac{d^2\psi_1}{dt^2} = -k_1^2\psi_1,$$

of which the solution is

$$\psi_1 = A_1\cos(k_1 t - \epsilon_1), \qquad (198)$$

as in § 208. Thus the motion is a simple harmonic motion of frequency k_1. If all the coefficients $\alpha_1, \alpha_2, \cdots, \alpha_n$ are positive, the complete solution of the equations will be of the form

$$\psi_1 = A_1\cos(k_1 t - \epsilon_1),$$
$$\psi_2 = A_2\cos(k_2 t - \epsilon_2), \text{ etc.,}$$

and the coördinate x of any particle, of which the value in the equilibrium position is x_0, will be

$$x = x_0 + (\theta_1 - \bar{\theta}_1)\frac{\partial x}{\partial \theta_1} + (\theta_2 - \bar{\theta}_2)\frac{\partial x}{\partial \theta_2} + \cdots$$
$$= x_0 + \phi_1\frac{\partial x}{\partial \theta_1} + \phi_2\frac{\partial x}{\partial \theta_2} + \cdots$$
$$= x_0 + B_1\cos(k_1 t - \epsilon_1) + B_2\cos(k_2 t - \epsilon_2) + \cdots,$$

where $B_1, B_2, \cdots$ are new constants.

Thus the motion of any single particle will be a motion compounded of a number of simple harmonic motions.

283. The potential energy corresponding to any principal coördinate ψ_1 is $\alpha_1 \psi_1^2$, or, if we suppose ψ_1 given by equation (198), is

$$\alpha_1 A_1^2 \cos^2(k_1 t - \epsilon_1).$$

Similarly, the kinetic energy corresponding to this principal vibration is $\beta_1 \dot\psi_1^2$, or

$$\beta_1 A_1^2 k_1^2 \sin^2(k_1 t - \epsilon_1).$$

Averaged over a very long time, the average values of $\cos^2(k_1 t - \epsilon_1)$ and of $\sin^2(k_1 t - \epsilon_1)$ are each $\tfrac{1}{2}$, so that the average potential and kinetic energies are respectively

$$\tfrac{1}{2}\alpha_1 A_1^2, \qquad \tfrac{1}{2}\beta_1 A_1^2 k_1^2,$$

and these are equal since $k_1^2 = \dfrac{\alpha_1}{\beta_1}$. Thus *in any vibration the average kinetic and potential energies are equal.*

Unstable Equilibrium

284. Suppose now that any one of the coefficients in equation (195) is negative, say α_1. Let us put $\dfrac{\alpha_1}{\beta_1} = -k_1^2$, so that k_1 will be real. Equation (197) now assumes the form

$$\frac{d^2\psi_1}{dt^2} = k_1^2 \psi_1,$$

and this has as solution

$$\psi_1 = A_1 e^{k_1 t} + B_1 e^{-k_1 t},$$

showing that ψ_1 increases indefinitely with the time, and does not oscillate about the value $\psi_1 = 0$. Thus the motion is unstable, and we now see that the motion can only be stable provided *all* the coefficients $\alpha_1, \alpha_2, \cdots, \alpha_n$ are positive. In other words,

For stable equilibrium the potential energy in the configuration of equilibrium must be an absolute minimum.

This is the result which has already been stated without proof in § 153.

Forced Oscillations

285. The oscillations which have so far been considered are of the type known as *free vibrations*, — that is to say, the forces acting arise entirely from the potential energy of the system itself.

A second type of oscillation occurs when the system is acted on by forces from outside, in addition to those arising from its own potential energy. These oscillations are known as *forced oscillations*.

Let us suppose that the potential and kinetic energies of the system are given by equations (195) and (196), and that the system of external forces acting at any instant is such that the work done in a small displacement is

$$\Psi_1 \delta\psi_1 + \Psi_2 \delta\psi_2 + \cdots.$$

Then Lagrange's equations for this system are

$$\frac{d}{dt}\left(\frac{\partial T}{\partial \dot\psi_1}\right) - \frac{\partial T}{\partial \psi_1} = -\frac{\partial W}{\partial \psi_1} + \Psi_1,$$

which becomes

$$2\beta_1 \frac{d^2\psi_1}{dt^2} = -2\alpha_1\psi_1 + \Psi_1, \qquad (199)$$

in which Ψ_1, it must be remembered, is now a function of the time. This equation can be solved according to the rules given in any treatise on differential equations. If, as before, we take $k_1^2 = \dfrac{\alpha_1}{\beta_1}$, the general solution is found to be

$$\psi_1 = A_1 \cos(k_1 t - \epsilon_1) + \frac{1}{\sqrt{2\alpha_1\beta_1}} \int_{t'=-\infty}^{t'=t} (\Psi_1)_{t=t'} \sin k_1(t-t')\, dt',$$

the lower limit of integration being either $t' = -\infty$, or the instant of which the external forces first came into operation.

286. A case of extreme importance occurs when Ψ_1 is simply periodic with respect to the time, say

$$\Psi_1 = E \cos(p_1 t - \gamma_1).$$

The solution is then found to be
$$\psi_1 = A_1 \cos(k_1 t - \epsilon_1) + \frac{E}{2\alpha_1 - 2\beta_1 p_1^2} \cos(p_1 t - \gamma_1),$$
or, since $\alpha_1 = \beta_1 k_1^2$,
$$\psi_1 = A_1 \cos(k_1 t - \epsilon_1) + \frac{E}{2\alpha_1\left(1 - \dfrac{p_1^2}{k_1^2}\right)} \cos(p_1 t - \gamma_1).$$

Thus the variation in ψ_1 is now compounded of a simple harmonic motion of frequency k_1, and also one of frequency p_1, the frequency of the impressed force.

We notice that if p_1 is very nearly equal to k_1, then the second vibration is of very large amplitude. In the limiting case in which $p_1 = k_1$, the amplitude of the second vibration becomes infinite, but now the two vibrations are of the same period, so that they may be compounded, and we cannot say that the resultant vibration is one of infinite amplitude, because we do not know the values of A_1 and ϵ_1, and these may just be such as to destroy the infinite amplitude of the second term. The result we have obtained may be enunciated in the following form:

When a system is acted on by a periodic force, of frequency very nearly equal to that of one of the principal vibrations of the system, then the forced oscillations will be of very great amplitude.

This is known as the *principle of resonance*.

The principle is one of which many applications appear in nature. For instance, a bridge, not being absolutely rigid, may be regarded as a system having a number of free vibrations. A body of men marching over the bridge in regular step will apply a periodic force, and if the period of their step happens to nearly coincide with one of the free periods of the bridge, the amplitude of the vibrations forced in the bridge may be so large as to endanger the bridge. For this reason troops are ordered to "break step" when crossing a bridge.

Again, a ship is not perfectly rigid, and so will possess a number of free vibrations. The motion of its engines will apply a periodic force of period equal to that of its revolution, and if this coincides with that of one of the vibrations of the ship, large pulsations will be set up. This can be remedied by altering the speed of the engine until it no longer is in resonance with the free vibrations of the ship.

THE CANONICAL EQUATIONS 355

As a last example, it may be noticed that a ship will have a free period of rolling about its vertical position. If it is in a rolling sea, the waves which meet it will apply external forces which may be regarded as approximately periodic. If the period of the waves happens to coincide with that of the ship, the ship will roll heavily even though the waves may be comparatively small. This danger can be remedied by altering the course of the ship and so causing it to meet the waves at a different interval. Another way is to give the ship a list by spreading canvas, and so causing it to oscillate about a different position of equilibrium, about which the periods of free vibrations are different.

The Canonical Equations

287. If $\theta_1, \theta_2, \cdots$ are Lagrangian coördinates of any system, the kinetic energy T is a quadratic function of $\dot{\theta}_1, \dot{\theta}_2, \dot{\theta}_3, \cdots$. Let the corresponding momenta be $u_1, u_2, \cdots, u_n$, these being given by

$$u_1 = \frac{\partial T}{\partial \dot{\theta}_1}, \text{ etc.} \qquad (200)$$

Now let us introduce a function T', defined by

$$T' = u_1 \dot{\theta}_1 + u_2 \dot{\theta}_2 + \cdots - T,$$

so that T' is a function of $u_1, u_2, \cdots, \dot{\theta}_1, \dot{\theta}_2, \cdots, \theta_1, \theta_2, \cdots$; and $u_1, u_2, \cdots$ are of course functions of $\dot{\theta}_1, \dot{\theta}_2, \cdots, \theta_1, \theta_2, \cdots$.

On differentiation of T' we have

$$\begin{aligned}
dT' = & \; u_1 d\dot{\theta}_1 + u_2 d\dot{\theta}_2 + \cdots \\
& + \dot{\theta}_1 du_1 + \dot{\theta}_2 du_2 + \cdots \\
& - \frac{\partial T}{\partial \theta_1} d\theta_1 - \frac{\partial T}{\partial \theta_2} d\theta_2 - \cdots \\
& - \frac{\partial T}{\partial \dot{\theta}_1} d\dot{\theta}_1 - \frac{\partial T}{\partial \dot{\theta}_2} d\dot{\theta}_2 - \cdots,
\end{aligned}$$

and this, by equation (200), reduces to

$$dT' = \dot{\theta}_1 du_1 + \dot{\theta}_2 du_2 + \cdots - \frac{\partial T}{\partial \theta_1} d\theta_1 - \frac{\partial T}{\partial \theta_2} d\theta_2 - \cdots. \qquad (201)$$

356 GENERALIZED COÖRDINATES

Since the differentials $d\dot\theta_1$, $d\dot\theta_2$, $\cdots$ do not occur, it appears that T' can be expressed as a function of $u_1, u_2, \cdots, \theta_1, \theta_2, \cdots$ only. We can easily find its value; we have

$$u_1\dot\theta_1 + u_2\dot\theta_2 + \cdots + u_n\dot\theta_n$$
$$= \dot\theta_1 \frac{\partial T}{\partial \dot\theta_1} + \dot\theta_2 \frac{\partial T}{\partial \dot\theta_2} + \cdots + \dot\theta_n \frac{\partial T}{\partial \dot\theta_n}$$
$$= 2\,T, \text{ since } T \text{ is a homogeneous quadratic function of } \dot\theta_1, \dot\theta_2, \cdots + \dot\theta_n.$$

Thus $\quad T' = 2\,T - T = T,$

showing that T' is equal to T, but is expressed as a function of $u_1, u_2, \cdots, \theta_1, \theta_2, \cdots$. Thus

$$T' = T = \tfrac{1}{2}(u_1\dot\theta_1 + u_2\dot\theta_2 + \cdots + u_n\dot\theta_n).$$

To illustrate, let

$$T = a\dot\theta_1^2 + 2h\dot\theta_1\dot\theta_2 + b\dot\theta_2^2,$$

so that $\quad u_1 = 2(a\dot\theta_1 + h\dot\theta_2), \quad u_2 = 2(h\dot\theta_1 + b\dot\theta_2).$

Then, by definition,

$$T' = u_1\dot\theta_1 + u_2\dot\theta_2 + \cdots - T$$
$$= 2\,\dot\theta_1(a\dot\theta_1 + h\dot\theta_2) + 2\,\dot\theta_2(h\dot\theta_1 + b\dot\theta_2) - T$$
$$= T = \tfrac{1}{2}u_1\dot\theta_1 + \tfrac{1}{2}u_2\dot\theta_2.$$

From equation (201), we have

$$\frac{\partial T'}{\partial \theta_1} = -\frac{\partial T}{\partial \theta_1},$$
$$\frac{\partial T'}{\partial u_1} = \dot\theta_1. \tag{202}$$

In Lagrange's equations

$$\frac{d}{dt}\left(\frac{\partial L}{\partial \dot\theta_1}\right) - \frac{\partial L}{\partial \theta_1} = 0,$$

we have $\quad \dfrac{\partial L}{\partial \dot\theta_1} = \dfrac{\partial(T - W)}{\partial \dot\theta_1} = \dfrac{\partial T}{\partial \dot\theta_1} = u_1.$

THE CANONICAL EQUATIONS 357

Thus Lagrange's equations may be written as

$$\frac{du_1}{dt} = \frac{\partial L}{\partial \theta_1} = \frac{\partial (T-W)}{\partial \theta_1} = -\frac{\partial (T'+W)}{\partial \theta_1},$$

while, by equation (202), $\quad \dfrac{d\theta_1}{dt} = \dfrac{\partial T'}{\partial u_1}.$

If we write $H = T' + W$, these equations assume the symmetrical form

$$\frac{d\theta_1}{dt} = \frac{\partial H}{\partial u_1},$$

$$\frac{du_1}{dt} = -\frac{\partial H}{\partial \theta_1}, \text{ etc.}$$

288. This is known as the canonical form of the dynamical equations. The function H is called the Hamiltonian function, and since $H = T' + W$, we notice that H is the total energy expressed as a function of the coördinates $\theta_1, \theta_2, \cdots, \theta_n$ and of the momenta $u_1, u_2, \cdots, u_n$.

The canonical form is the simplest and most perfect form in which the generalized dynamical equations can be expressed. For this reason the canonical system of equations forms the starting point of a great many investigations in higher dynamics, mathematical physics, and mathematical astronomy.

289. We may appropriately terminate the present book by giving illustrations of the use of generalized coördinates from two branches of mathematical physics.

Illustration from hydrodynamics. Let a solid of any shape be in a stream of water flowing with uniform velocity V. If the solid is at a sufficient depth from the surface, its presence will not disturb the flow at the surface, and the only disturbance in the flow of the water will be in the neighborhood of the solid. It can be proved, from elementary hydrodynamical principles, that there is only one way in which the water can flow past the solid. Hence it follows that the kinetic energy of the flow of the water is given by

$$T = T_0 + \alpha V^2,$$

where T_0 is the value which the kinetic energy would have if the solid were removed. Suppose that the solid is acted on by external forces, besides the pressure of the water. Let the sum of the moments of these forces about any axis be Θ, and let θ be a coördinate which measures the angle turned through about this axis. Then Lagrange's equation corresponding to the coördinate θ is

$$\frac{d}{dt}\left(\frac{\partial T}{\partial \dot\theta}\right) - \frac{\partial T}{\partial \theta} = \Theta.$$

If the external forces just suffice to hold the body at rest in the liquid, we have $\frac{d}{dt}\left(\frac{\partial T}{\partial \dot\theta}\right) = 0$, so that

$$\Theta = -\frac{\partial T}{\partial \theta} = -\frac{\partial \alpha}{\partial \theta} V^2.$$

Hence the sum of the moments of the liquid pressure must be $-\Theta$, or

$$\frac{\partial \alpha}{\partial \theta} V^2.$$

We can calculate α from the shape of the solid, and so can obtain a knowledge of the couples acting on the solid.

Illustration from electromagnetism. The energy required to establish the flow of two steady currents of electricity of strengths i, i' in two given closed circuits is known to be of the form

$$E = \tfrac{1}{2} L i^2 + M i i' + \tfrac{1}{2} N i'^2,$$

where L and N depend on the shape of the first and second circuits respectively, while M depends on the shape of both circuits, and also on their positions relative to one another.

Suppose that the second circuit is free to move along any line towards the first circuit. Let x be a coördinate measured along this line, and let the force required to hold the second circuit at rest be X in the direction in which x is measured.

Let L denote the usual function $T - W$, and let the second circuit be acted on by an externally applied force X. Then Lagrange's equation for the coördinate x is

$$\frac{d}{dt}\left(\frac{\partial L}{\partial \dot x}\right) - \frac{\partial L}{\partial x} = X,$$

so that, since there is no acceleration,

$$X = -\frac{\partial L}{\partial x}.$$

As a matter of experiment, it is found that

$$X = -ii'\frac{\partial M}{\partial x} = -\frac{\partial E}{\partial x}.$$

EXAMPLES 359

If the energy of the two currents were potential energy, we should have

$$-\frac{\partial L}{\partial x} = +\frac{\partial W}{\partial x}, \quad -\frac{\partial E}{\partial x} = -\frac{\partial W}{\partial x},$$

so that the force X would be exactly opposite to that observed.

On the other hand, if the energy is kinetic energy, we have

$$-\frac{\partial L}{\partial x} = -\frac{\partial T}{\partial x} = -\frac{\partial E}{\partial x},$$

so that the value of X agrees with that observed.

Hence we conclude that the energy of an electric current is wholly kinetic.

GENERAL EXAMPLES

1. The friction of an engine is such that one horse power can run it at 250 revolutions per second when it is doing no external work. The inertia of its moving parts is such that when running at 125 revolutions per second, and acted on by one horse power, its speed is accelerated at the rate of 10 revolutions per second. If the engine is left to itself when running at its full speed of 250 revolutions per second, find how many revolutions it will make before coming to rest.

2. A square is moving freely about a diagonal with angular velocity ω, when suddenly one of the angular points not in that diagonal becomes fixed. Determine the impulsive pressure on the fixed point, and show that the new angular velocity will be $\frac{1}{7}\omega$.

3. Four equal rods, each of length $2a$ and mass m, are freely jointed so as to form a rhombus. The system falls from rest with one diagonal vertical, and strikes a fixed horizontal inelastic plane. Find the impulse and the subsequent motion.

4. Two particles connected by a rigid rod move on a smooth vertical circle. Find the time of a small oscillation.

5. A uniform rod of length l has the two points at distance c from its middle point connected by equal strings of length L to two fixed points at distances $2c$ apart in the same horizontal line.

Find the principal coördinates and the corresponding periods of vibration.

6. If the rod of the last question receives a horizontal blow of impulse I at one extremity and at right angles to its length, find the subsequent motion.

7. A rough uniform cylinder of radius a has an inextensible string coiled round its central section. One end of the string is fastened to a fixed point P, and the cylinder is rolled up the string until it is touching P, with the tangent to the cylinder at P vertical. The cylinder is then let go. Find the motion.

8. In the last question, find the motion if the tangent at P is perpendicular to the axis of the cylinder, but is not quite vertical.

9. In spherical polar coördinates prove that the kinetic energy of a moving particle of unit mass is given by

$$T = \tfrac{1}{2}(\dot{r}^2 + r^2\dot{\theta}^2 + r^2\sin^2\theta\,\dot{\phi}^2).$$

Hence, prove that the acceleration of the particle has components in the direction of r, θ, ϕ increasing, of amounts

$$\frac{d}{dt}\left(\frac{\partial T}{\partial \dot{r}}\right) - \frac{\partial T}{\partial r}, \quad \frac{1}{r}\left[\frac{d}{dt}\left(\frac{\partial T}{\partial \dot{\theta}}\right) - \frac{\partial T}{\partial \theta}\right], \quad \frac{1}{r\sin\theta}\frac{d}{dt}\left(\frac{\partial T}{\partial \dot{\phi}}\right).$$

Show that the actual values of these accelerations are

$$\frac{d^2r}{dt^2} - r\dot{\theta}^2 - r\sin^2\theta\,\dot{\phi}^2, \quad \frac{1}{r}\frac{d}{dt}(r^2\dot{\theta}) - r\sin\theta\cos\theta\,\dot{\phi}^2, \quad \frac{1}{r\sin\theta}\frac{d}{dt}(r^2\sin^2\theta\,\dot{\phi}).$$

10. The velocity of a particle in its orbit is found to vary in the inverse square of its distance from a fixed point. Apply the principle of least action to find the orbit, and thence the law of attraction.

Deduce the same results from the law of conservation of energy.

11. Suppose that all forces are annihilated in the universe, and that there is a concealed mechanism capable of possessing kinetic energy. Suppose that the amount of this kinetic energy depends only on the positions of the material bodies in the universe, being equal in magnitude except for a constant, and opposite in sign, to the potential energy which the system would have if the forces had not been annihilated.

Show that the dynamical phenomena of a universe of this kind will be identical with those of a universe in which both forces and kinetic energy exist, the changes in the latter being determined by Newton's laws of motion.

12. A number of spheres without mass, of radii a, b, c, $\cdots$, move in a straight line through an infinite ocean of density ρ_0, the distances apart of their centers being r_{ab}, r_{bc}, etc., and their velocities v_a, v_b, v_c, $\cdots$. When a, b, c, $\cdots$ are small compared to r_{ab}, etc., the kinetic energy of the motion of the ocean is given by

$$2T = \tfrac{2}{3}\,\pi\rho u^3 v_a^2 + \cdots + 2\pi\rho\frac{a^3b^3}{r_{ab}^3}v_a v_b + \cdots.$$

Show that to an observer who is unconscious of the presence of the ocean, the spheres will appear to move as though having masses $\tfrac{2}{3}\pi\rho a^3$, $\tfrac{2}{3}\pi\rho b^3$, etc., and as though forces of attraction acted between every pair of spheres, proportional to the product of the masses, to the product of their velocities, and to the inverse fourth power of the distance between them.

INDEX

(The numbers refer to pages.)

Absolute units, of force, 30; of work, 146.
Acceleration, 12; parallelogram of, 13; in circular motion, 14, 18.
Action, 328; principle of least, 328.
Amplitude, of a pendulum, 261; of simple harmonic motion, 265.
Angle of friction, 47.
Angular momentum, 297; conservation of, 297.
Angular velocity, 286; composition of, 287.
Arc, center of gravity of circular, 125.
Atwood's machine, 195.
Average velocity, 6.
Axes of inertia, 303.
Axis of rotation, 92.

Balancing, of an engine, 337.
Belt, center of gravity of spherical, 130.

Canonical equations, 355.
Catenary, 80.
Center of force, motion of point about, 269.
Center of gravity, 117; of a lamina, 121, 135; of a solid, 132, 135; of a triangle, 121; of a pyramid, 132; of a circular arc, 126; of a segment and sector of a circle, 128, 129; of a spherical belt and cap, 130; of a sector of a sphere, 130; motion of, of a system, 224.
Central axis, of a system of forces, 107.
Centroid, 20.

Circular arc, center of gravity of, 125.
Coefficient, of friction, 47; of elasticity, 240.
Coefficients of inertia, 301.
Composition, of motions, 4; of velocities, 7; of accelerations, 13; of forces acting on a particle, 37; of forces acting in a plane, 95; of parallel forces, 99; of couples, 105; of rotations, 286.
Compression, moment of greatest, 238.
Conical pendulum, 271.
Conservation, of energy, 171; of linear momentum, 223; of angular momentum, 297.
Conservative system, of forces, 163.
Coördinates, generalized, 320, 329; normal or principal, 350.
Couples, 101; in parallel planes, 104; composition of, 105; work performed against, 154.
Cycloidal pendulum, 265.

Degrees of freedom, number of, 184, 332.
Descent, line of quickest, 193.
Diagram, indicator, 151.
Differential equations, of orbits, 275.
Double stars, 280.

Earth's rotation, 198, 310.
Elasticity, of a string, 45; modulus of, 45; of a solid, 238; coefficient of, 240.
Ellipsoid of inertia, 302.

362　INDEX

Energy, potential, 163; kinetic, 168, 228; total, 171; conservation of, 171; of motion of a system, 230; of a rigid body, 290.
Envelope of paths, of projectiles, 211.
Equation, of energy, 171, 256; of motion of a particle, 254; of orbit of a particle, 275; of a rigid body, 304.
Equations, Euler's, 306, 346; Lagrange's, 329, 342; canonical (Hamilton's), 355.
Equilibrium, of a particle, 38, 41; of a system of particles, 63; of a rigid body, 93; stability and instability of, 174, 351.
Euler's equations, 306, 346.
Extensibility, of strings, 44.

Flexibility, of strings, 43.
Force, 26; measurement of, 30; transmissibility of, 94.
Forced oscillations, 353.
Forces, composition and resolution of, 37–39; in one plane, 66, 95; parallel, 96, 99; in space, 106; impulsive, 233.
Frame of reference, 3, 33; motion referred to moving, 197; kinetic energy referred to moving, 228.
Frequency, of a vibration, 263.
Friction, 46; coefficient of, 47. Reaction between moving rough bodies, 200.

Generalized coördinates, 320, 329; impulse, 345; momentum, 345.
Gravitation, law of, 279.
Gravity, work performed against, 153; motion of body falling under, 189; variation with latitude, 200.
Gyration, radii of, 290.

Hamilton's Principle, 323.
Harmonic motion, simple, 261.
Hooke's law, 44.
Horse power, 146.

Impact, 238; of particle on fixed surface, 241; of any two moving bodies, 244; of two smooth spheres, 246.
Impulse, 233; of compression, 239; of restitution, 240; generalized, 345.
Impulsive forces, 233, 345.
Inclined plane, motion of particle on, 192.
Indicator diagram, 151.
Inertia, moment of, 290; coefficients and products of, 301, 302; ellipsoid of, 302; principal axes of, 303.
Inverse square, law of, 276.

Kepler's laws, 279.
Kinetic energy, 168; of system of particles, 228; of rotation, 289; of a rigid body, 290.

Lagrange's equations, 329; for impulsive forces, 344; for non-conservative systems, 339.
Lamina, center of gravity of, 121, 135.
Latitude, variation of gravity with, 200; variation of terrestrial, 310.
Laws, of nature, 1; of motion, 26.
Least action, 327.
Line of action, of a force, 60.

Mass, and measurement of mass, 29.
Measurement, of velocity, 6; of acceleration, 12; of mass, 29; of force, 30; of work, 145; of acceleration due to gravity, 195; of an impulse, 235.
Modulus of elasticity, of a string, 45.
Moment, of a force, 60; of a velocity, 274; of inertia, 290; of momentum, 295.
Moment of greatest compression, 238.
Moments, principal, of inertia, 301.
Momentum, 29; conservation of linear, 223; moment of, 295; conservation of angular, 297; generalized, 345.
Motion, referred to frame of reference, 3; of a rigid body, 91, 286; referred to moving frame of reference, 197;

INDEX 363

of system of particles, 220; of center of gravity of any system, 224; simple harmonic, 261; of particle about a center of force, 269; of particle under law of inverse square, 276.

Neutral equilibrium, 182.
Newton's law, of elasticity, 245.
Newton's laws, of motion, 26.
Normal coördinates, 350.

Orbit, general theory, 273; differential equation of, 275.
Orbit of a particle, law of direct distance, 269; law of inverse square of distance, 276.
Oscillations, of a pendulum, 298; small, of a general dynamical system, 348; forced, 353.

Parallel forces, 96, 99.
Parallelogram law, velocities, 9; accelerations, 13; forces, 38; couples, 105; angular velocity, 286.
Pendulum, simple, 259; seconds, 261; cycloidal, 265; general motion of, 298.
Period, of vibration, 261; of simple harmonic motion, 265.
Plane, composition of forces in one, 95; orbit about a center of force confined to one, 273.
Planet, rotation of a, 309.
Point of application, of a force, 60, 95.
Potential energy, 163.
Principal axes of inertia, 303.
Principal coördinates, 350.
Principle, of least action, 327; Hamilton's, 323.
Products of inertia, 302.
Projectiles, 205; range on a horizontal plane, 209; range on an inclined plane, 209; envelope of paths with given initial velocity, 211.

Pulleys, systems of, 157.
Pyramid, center of gravity of, 132.

Quickest descent, line of, 193.

Radius of gyration, 290.
Range, of a projectile, 209.
Reaction, 31; frictional, between bodies at rest, 46; frictional, between bodies in motion, 200.
Reference, frame of, 3, 33; motion of frame of, 197.
Relative motion, 4.
Resonance, principle of, 354.
Rest, 3.
Restitution, impulse of, 240.
Retardation, 12.
Rigidity, 90.
Rotation, axis of, 92; of earth, 198; of a rigid body, kinetic energy of, 289; of a planet, 309.

Sag, of a string, 85.
Sector, of a circle, center of gravity of, 129; of a sphere, center of gravity of, 134.
Segment, of a circle, center of gravity of, 128.
Simple harmonic motion, 261.
Spherical cap, center of gravity of, 130.
Spinning top, motion of, 310.
Stability and instability of equilibrium, 174, 351.
Stars, double, orbits of, 280.
Strings, tension of, 42; flexibility of, 43; extensibility of, 44; on surface, 74; sag of a stretched, 85; work of stretching, 150.
Suspension bridge, 78.
System, of pulleys, 157; conservative, of forces, 163.
System of particles, statics of, 59; motion of, 220; kinetic energy of, 230.

Tension, of a string, 42, 74.
Top, motion of, 310.

Transmissibility of force, 94.
Triangle, of velocities, 10.
Triangular lamina, center of gravity of, 121.

Uniformity, of nature, 1.
Unit, of velocity, 6; of force, 30; of work, 145.

Variation, of value of g, 200; of terrestrial latitude, 310.
Vectors, 16; in one plane, 16; in space, 19.
Velocity, uniform and variable, 6; average, 6; composition of, 7; moment of, 274; angular, 286.

Vibrations, 348, 353.
Virtual work, principle of, 155.

Weight, of a particle, 42; of a system of particles, 118.
Wheel and axle, 65.
Work, measurement of, 145; against a variable force, 148; of stretching a string, 148; represented by an area, 150; against an oblique force, 152; performed against gravity, 153; performed by a couple, 154; principle of virtual, 155; performed by an impulse, 235.
Wrench, 107.

DOVER PHOENIX EDITIONS

A series of hardcover reprints of major works in mathematics, science and engineering.
All editions are 5⅝ × 8½ unless otherwise noted.

Mathematics

Theory of Approximation, N. I Achieser. Unabridged republication of the 1956 edition. 320pp. 49543-4
The Origins of the Infinitesimal Calculus, Margaret E. Baron. Unabridged republication of the 1969 edition. 320pp. 49544-2
A Treatise on the Calculus of Finite Differences, George Boole. Unabridged republication of the 2nd and last revised edition. 352pp. 49523-X
Space and Time, Emile Borel. Unabridged republication of the 1926 edition. 15 figures. 256pp. 49545-0
An Elementary Treatise on Fourier's Series, William Elwood Byerly. Unabridged republication of the 1893 edition. 304pp. 49546-9
Substance and Function & Einstein's Theory of Relativity, Ernst Cassirer. Unabridged republication of the 1923 double volume. 480pp. 49547-7
A History of Geometrical Methods, Julian Lowell Coolidge. Unabridged republication of the 1940 first edition. 13 figures. 480pp. 49524-8
Linear Groups with an Exposition of Galois Field Theory, Leonard Eugene Dickson. Unabridged republication of the 1901 edition. 336pp. 49548-5
Continuous Groups of Transformations, Luther Pfahler Eisenhart. Unabridged republication of the 1933 first edition. 320pp. 49525-6
Transcendental and Algebraic Numbers, A. O. Gelfond. Unabridged republication of the 1960 edition. 208pp. 49526-4
Lectures on Cauchy's Problem in Linear Partial Differential Equations, Jacques Hadamard. Unabridged reprint of the 1923 edition. 320pp. 49549-3
The Theory of Branching Processes, Theodore E. Harris. Unabridged, corrected republication of the 1963 edition. xiv+230pp. 49508-6
The Continuum, Edward V. Huntington. Unabridged republication of the 1917 edition. 4 figures. 96pp. 49550-7
Lectures on Ordinary Differential Equations, Witold Hurewicz. Unabridged republication of the 1958 edition. xvii+122pp. 49510-8
Mathematical Methods and Theory in Games, Programming, and Economics: Two Volumes Bound as One, Samuel Karlin. Unabridged republication of the 1959 edition. 848pp. 49527-2
Famous Problems of Elementary Geometry, Felix Klein. Unabridged reprint of the 1930 second edition, revised and enlarged. 112pp. 49551-5
Lectures on the Icosahedron, Felix Klein. Unabridged republication of the 2nd revised edition, 1913. 304pp. 49528-0
On Riemann's Theory of Algebraic Functions, Felix Klein. Unabridged republication of the 1893 edition. 43 figures. 96pp. 49552-3
A Treatise on the Theory of Determinants, Thomas Muir. Unabridged republication of the revised 1933 edition. 784pp. 49553-1
A Survey of Minimal Surfaces, Robert Osserman. Corrected and enlarged republication of the work first published in 1969. 224pp. 49514-0
The Variational Theory of Geodesics, M. M. Postnikov. Unabridged republication of the 1967 edition. 208pp. 49529-9

DOVER PHOENIX EDITIONS

An Introduction to the Approximation of Functions, Theodore J. Rivlin. Unabridged republication of the 1969 edition. 160pp. 49554-X

An Essay on the Foundations of Geometry, Bertrand Russell. Unabridged republication of the 1897 edition. 224pp. 49555-8

Elements of Number Theory, I. M. Vinogradov. Unabridged republication of the first edition, 1954. 240pp. 49530-2

Asymptotic Expansions for Ordinary Differential Equations, Wolfgang Wasow. Unabridged republication of the 1976 corrected, slightly enlarged reprint of the original 1965 edition. 384pp. 49518-3

Physics

Semiconductor Statistics, J. S. Blakemore. Unabridged, corrected, and slightly enlarged republication of the 1962 edition. 141 illustrations. xviii+318pp. 49502-7

Wave Propagation in Periodic Structures, L. Brillouin. Unabridged republication of the 1946 edition. 131 illustrations. 272pp. 49556-6

The Conceptual Foundations of the Statistical Approach in Mechanics, Paul and Tatiana Ehrenfest. Unabridged republication of the 1959 edition. 128pp. 49504-3

The Analytical Theory of Heat, Joseph Fourier. Unabridged republication of the 1878 edition. 20 figures. 496pp. 49531-0

States of Matter, David L. Goodstein. Unabridged republication of the 1975 edition. 154 figures. 4 tables. 512pp. 49506-X

The Principles of Mechanics, Heinrich Hertz. Unabridged republication of the 1900 edition. 320pp. 49557-4

Thermodynamics of Small Systems, Terrell L. Hill. Unabridged and corrected republication in one volume of the two-volume edition published in 1963–1964. 32 illustrations. 408pp. 6½ x 9¼. 49509-4

Theoretical Physics, A. S. Kompaneyets. Unabridged republication of the 1961 edition. 56 figures. 592pp. 49532-9

Quantum Mechanics, H. A. Kramers. Unabridged republication of the 1957 edition. 14 figures. 512pp. 49533-7

The Theory of Electrons, H. A. Lorentz. Unabridged reproduction of the 1915 edition. 9 figures. 352pp. 49558-2

The Principles of Physical Optics, Ernst Mach. Unabridged republication of the 1926 edition. 279 figures. 10 portraits. 336pp. 49559-0

The Scientific Papers of James Clerk Maxwell, James Clerk Maxwell. Unabridged republication of the 1890 edition. 197 figures. 39 tables. Total of 1,456pp.
Volume I (640pp.) 49560-4; *Volume II* (816pp.) 49561-2

Vectors and Tensors in Crystallography, Donald E. Sands. Unabridged and corrected republication of the 1982 edition. xviii+228pp. 49516-7

Principles of Mechanics and Dynamics, Sir William (Lord Kelvin) Thompson and Peter Guthrie Tait. Unabridged republication of the 1912 edition. 168 diagrams. Total of 1,088pp. *Volume I* (528pp.) 49562-0; *Volume II* (560pp.) 49563-9

Treatise on Irreversible and Statistical Thermophysics: An Introduction to Nonclassical Thermodynamics, Wolfgang Yourgrau, Alwyn van der Merwe, and Gough Raw. Unabridged, corrected republication of the 1966 edition. xx+268pp. 49519-1

Engineering

Principles of Aeroelasticity, Raymond L. Bisplinghoff and Holt Ashley. Unabridged, corrected republication of the original 1962 edition. xi+527pp. 49500-0

Statics of Deformable Solids, Raymond L. Bisplinghoff, James W. Mar, and Theodore H. H. Pian. Unabridged and corrected Dover republication of the edition published in 1965. 376 illustrations. xii+322pp. 6½ x 9¼. 49501-9